AF531495

Nanoelectronics

Nanoelectronics

Er. Nishit Mathur

SHREE PUBLISHERS & DISTRIBUTORS
NEW DELHI-110 002

Edition : 2013

ISBN : 978–81–8329–499–7

Published by :
SHREE PUBLISHERS & DISTRIBUTORS
22/4735 Prakash Deep Building,
Ansari Road, Darya Ganj,
New Delhi–110 002

Printed in India.

Preface

Nanoelectronics is to method, transmit and store in sequence by taking improvement of properties of matter that are definitely different from macroscopic properties. Nanoelectronics is predictable to be cheaper to fabricate than silicon/gallium/arsenic based electronics. It also strength be small enough to not require a power source as it is possible to abstract a small amount of energy from the surrounding heat by a molecular level energy scavenging system. At current, growth in electronic devices means a race for a constant decrease in the arrange of dimension. The general public is well aware of the fact that we live in the age of microelectronics, an expression which is derived from the size of a device's active zone, *e.g.*, the channel length of a field effect transistor or the thickness of a gate dielectric. However, there are convincing indications that we are entering another era, namely the age of nanotechnology. The expression "nanotechnology" is again derived from the typical geometrical dimension of an electronic device, which is the nanometer and which is one billionth (10^9) of ameter. 30,000 nm are approximately equal to the thickness of a human hair. It is worthwhile comparing electrical machines, such as a motor or a telephone with their typical dimensions of 10 cm.

Although a broad meaning, we categorise nanomaterials as those which have structured components with at least one dimension less than 100nm. Materials that have one dimension in the nanoscale are layers, such as a thin films or surface coatings. Some of the features on computer chips come in this category. Materials that are nanoscale in two dimensions include nanowires and nanotubes. Materials that are nanoscale in three dimensions are particles, for example precipitates, colloids and quantum dots. Nanocrystalline materials, made up of nanometre-sized grains, also fall into this category. Some of these materials have been available for some time; others are genuinely new. The aim of this book is to give an overview of the properties, and the significant foreseeable applications of some key nanomaterials.

Nanoelectronics research is currently looking not only for the successor to CMOS processing but also for a replacement for the transistor itself. On the scale of 10-nm dimensions, components have a wavelength comparable to that of an electron at the Fermi energy. The confinement and coherence of the electron gives rise to gross deviations from the classical charge transport found in conventional devices. Quantum-mechanical laws become increasingly dominant on the nanoscale, and it is probable that nanoelectronics will operate on quantum principles. One popular approach has been to use small conducting islands in which electrons are confined and quantized in a definite state. These islands are typically connected to electrodes by thin tunnelling barriers. Quantum dots, resonant tunnel devices, and single-electron transistors are examples of devices that use this basic concept, albeit in different ways. Single-electron transistors are an example of a three-terminal device in which the charge of a single electron is sufficient to switch the source-to-drain current. The tiny energy required to drive single-electron transistor devices makes this approach very appealing. Nevertheless, a variety of drawbacks and obstacles limit the application of such devices in solid-state nanoelectronics, for example, their sensitivity to small fluctuations in voltage and background charges, which tend to accumulate in semiconductors. Such problems suggest that single-electron devices may be used for storage rather than logic functions. Currently, such devices operate at cryogenic temperatures, although there are a few examples of room temperature operation.

Nanotubes are members of the fullerene structural family, which also includes the spherical buckyballs, and the ends of a nanotube may be capped with a hemisphere of the buckyball structure. Their name is derived from their long, hollow structure with the walls formed by one-atom-thick sheets of carbon, called graphene. These sheets are rolled at specific and discrete ("chiral") angles, and the combination of the rolling angle and radius decides the nanotube properties; for example, whether the individual nanotube shell is a metal or semiconductor. Nanotubes are categorized as single-walled nanotubes (SWNTs) and multi-walled nanotubes (MWNTs). Individual nanotubes naturally align themselves into "ropes" held together by van der Waals forces, more specifically, pi-stacking. Applied quantum chemistry, specifically, orbital hybridization best describes chemical bonding in nanotubes. The chemical bonding of nanotubes is composed entirely of sp^2 bonds, similar to those of graphite. These bonds, which are stronger than the sp^3 bonds found in alkanes and diamond, provide nanotubes with their unique strength.

This book is improved understand drivers and command patterns for nanoelectronics and nano-optical products, such as the tendency in computing, communications, optical satellite telecommunications or sensor technology and IT.

The present work is purely a research compilation of based on authoritative publications/readings/excerpts/notes/reviews/researches authored by eminent scientists.

Assitance has been sought from Websites through Internet.

The author, humbly, acknowledges the contributions of all those eminent scientists alongwith their respective publishers and Websites in preparation of this book.

—Er. Nishit Mathur

Contents

1 Introduction

Nanoelectronics is predictable to be cheaper to fabricate than silicon/gallium/arsenic based electronics. It also strength be small enough to not require a power source as it is possible to abstract a small amount of energy from the surrounding heat by a molecular level energy scavenging system. At current, growth in electronic devices means a race for a constant decrease in the arrange of dimension. The general public is well aware of the fact that we live in the age of microelectronics, an expression which is derived from the size (1 μm) of a device's active zone, *e.g.*, the channel length of a field effect transistor or the thickness of a gate dielectric. However, there are convincing indications that we are entering another era, namely the age of nanotechnology. The expression "nanotechnology" is again derived from the typical geometrical dimension of an electronic device, which is the nanometer and which is one billionth (10^9) of ameter. 30,000 nm are approximately equal to the thickness of a human hair. It is worthwhile comparing electrical machines, such as a motor or a telephone with their typical dimensions of 10 cm.

WHAT IS NANOELECTRONICS?

Semiconductor electronics have seen a sustained exponential reduce in size and cost and a similar augment in performance and level of integration over the last thirty years. The Silicon Roadmap is laid out for the next ten years. After that, either economical or physical barriers will pose a huge challenge. The former is connected to the difficulty of making a profit in view of the exorbitant costs of building the necessary manufacturing capabilities if present day technologies are extrapolated. The latter is a direct consequence of the shrinking device size, leading to physical phenomena impeding the operation of current devices.

Quantum and coherence effects, high electric fields creating avalanche dielectric breakdowns, heat dissipation problems in closely packed structures as well as the non-uniformity of dopant atoms and the relevance of single atom defects are all roadblocks along the current road of miniaturization. These phenomena are characteristic for structures a few nanometers in size and, instead of being viewed as an obstacle to future progress might form the basis of post-silicon information processing technologies. It is even far from clear that electrons will be the technique of choice for signal processing or computation in the long term - quantum computing, spin electronics, optics or even computing based on (nano-) mechanics are actively being discussed. *Nanoelectronics* thus needs to be understood as a general field of research aimed at developing an understanding of the phenomena characteristic of nanometer sized objects with the aim of exploiting them for information processing purposes.

Specifically, by electronics we mean the handling of complicated electrical wave forms for *communicating information* (as in cellular phones), *probing* (as in radar) and *data processing* (as in computers). Concepts at the fundamental research level are being persued worldwide to find nano-solutions to these three characteristic applications of electronics. One can group these concepts into three main categories:

- Molecular electronics: Electronic effects (*e.g.* electrical conductance of C60). Synthesis (DNA computing as a buzz word)
- *Quantum Electronics, Spintronics* (*e.g.* quantum dots, magnetic effects)
- Quantum computing.

At present the most active field of investigate is the fabrication and characterization of individual components that could replace the macroscopic silicon components with nanoscale systems. Examples are molecular diodes, single atom switches or the increasingly better control and understanding of the transport of electrons in quantum dot structures. A second field with substantial activity is the investigation of potential interconnects. Here, mostly carbon nanotubes and self-assembled metallic or organic structures are being investigated. Very little work is being performed on architecture. Furthermore, modelling with predictive power is in a very juvenile stage of development. This understanding is necessary to develop engineering rules of thumb to design complex systems. One needs to appreciate that currently the best calculations of the conductance of a simple molecule such as C60 are off by a factor of more than 30. This has to do with the difficult to model, but non-trivial influence of the electronic contact leads. The situation in quantum computing is somewhat different. The main activities are on theoretical development of core concepts and algorithms. Experimental implementations are only starting. An exception is the field of cryptography (information transportation), where entangled photon states propagating in a conventional optical fibre have been demonstrated experimentally.

What Needs to be Done ?

First, nanoelctronics is a wide release field with vast potential for breakthroughs coming from fundamental research. Some of the major issues that need to be addressed are the following:

- Understand nanoscale transport! (closed loop between theory and experiment necessary). Most experiments and modelling concentrate on DC properties, AC properties at THz frequencies are however expected to be relevant.
- Develop/understand self-assembly techniques to do conventional things cheaper. This has the future potential to displace a large fraction of conventional semiconductor applications. One needs to solve the interconnect problem and find a replacement of the transistor. If this can be done by self-assembly, a major cost advantage compared to conventional silicon technology would result.
- Find new ways of responsibility electronics and find ways of implementing them (*e.g.* quantum computing; electronics modelled after living systems; hybrid Si-biological systems; cellular automata). Do not try and duplicate a transistor, but instead investigate new electronics paradigms! Do research as a graduate student in this field and lay the foundation for the Intel of the New Millennium.

How Nanoelectronics Works?

Nanoelectronics is one of the most important technologies of nanotechnology. It plays vital role in the field of engineering and electronics. Nanoelectronics make use of scientific techniques at atomic scale for developing the nano machines. The main target is to reduce the size, risk factor and surface areas of the materials and molecules. Machines under nanoelectronic process under goes the long range of manufacturing steps each with accurate molecular treatment.

Nanoelectronics develop its products and circuits at super miniature level. Individual part of the machine is handled atomically. Much kind of integrated circuits are present in the nanoelectronic devices. These integrated circuits are known as microelectronics.

HISTORICAL CONDITIONS

Before the innovation of nanoelectronics two famous electrical engineers named Ari and Mark started dealing with the single molecular electronics. They first gave the theory that single molecule of an atom can be processed to advent new structure. Later on concept of nanoelectronics was officially presented at the University of California in early 1990's.

NANOELECTRONICS APPLICATIONS

Nanoelectronics have wide range of applications which includes the field of electronics, artificial intelligence, fiction, technology, telecommunications, medical sciences and space exploration. With the help of nanoelectronics more complex and integrated devices have been created such as nanoelectronic robots with self replication, bacteria and material purifiers which automatically treat the agent or element according to the boiling points. Advanced sensors, trackers and detection apparatus are also the inventions of nanoelectronics

ADVANTAGES OF NANO ELECTRONICS

- One of the obvious benefit is that nanoelectronics reduces size and scale of the machine with the help of complex integration on the circuit silicon chips.
- Advanced propertied of semiconductors can be determined with the help of nanoelectronics.
- Molecular scale nanoelectronics is also known as "the next step" in the miniaturization of electronic devices, with latest electronics theory and research in the field of nanoelectronics it is possible to explore the diverse properties of molecules.
- Extreme fabrication also supported the multiple use of single machine. Parallel processing is also empowered by nanoelectronics.

Disturbed Issues of Nanoelectronics

- It is difficult to understand self assembling techniques which are used in nanoelectronics.
- Nanoelectronics are not cost effective as it involves complex integrations of circuits in it machines.
- Hug research and technical knowledge may lead criminal brains to develop destructive weapons which could be the threat to the living world.
- Nanoelectronics as we know is providing world with efficient and reliable machines which is the major cause of unemployment in the leading countries.

NANOELECTROMECHANICAL SYSTEMS

Intended for instance single electron transistors which include transistor operation based on a single electron based on a single electron. Nanoelectromechanical systems also come under this classification. Nanofabrication can be used to build ultra-dense parallel arrays of nanowires, as an alternative to synthesizing nanowires individually.

And besides being small and permitting more transistors to be packed into a single chip, the uniform and symmetrical structure of nanotubes permits higher electron mobility, a higher dielectric continuous and a symmetrical electron feature.

Positive Reconfigurable Computing

Separate molecule devices are another possibility. These schemes would make heavy use of molecular self-assembly, designing the device elements to build a larger structure or even a complete system on their own. This can be very constructive for reconfigurable computing and may even completely substitute present FPGA technology. Molecular electronics is a new technology which is still in its infancy, but also brings hope for truly atomic scale electronic systems in the future.

Molecular Electronics

One of the additional promising applications of molecular electronics was proposed by the IBM researcher Ari Aviram and the theoretical chemist Mark Ratner in their year 1974 and 1988 papers molecules for memory, logic and amplification. This is one of many possible ways in which a molecular level diode might be synthesized by organic chemistry. A model system was proposed with a spiro carbon structure giving a molecular diode about half a nanometer across which could be connected by polythiophene molecular wires.

APPROACHES TO NANOELECTRONICS

Nanofabrication

For instance, single electron transistors, which involve transistor operation based on a single electron. Nanoelectromechanical systems also fall under this category. Nanofabrication can be used to construct ultradense parallel arrays of nanowires, as an alternative to synthesizing nanowires individually.

Nanomaterials Electronics

Besides life form small and allowing more transistors to be packed into a single chip, the uniform and symmetrical structure of nanotubes allows a higher electron mobility (faster electron movement in the material), a higher dielectric constant (faster frequency), and a symmetrical electron/hole characteristic.

Also, nanoparticles can be used as quantum dots.

Molecular Electronics

Single molecule devices are another possibility. These schemes would make heavy use of molecular self-assembly, designing the device components to construct a larger structure or even a complete system on their own. This can be very useful for reconfigurable computing, and may even completely replace present FPGA technology. Molecular electronics is a new technology which is still in its infancy, but also brings hope for truly atomic scale electronic systems in the future. One of the more promising applications of molecular electronics was proposed by the IBM researcher Ari Aviram and the theoretical chemist Mark Ratner in their 1974 and 1988 papers *Molecules for Memory, Logic and Amplification*.

This is one of several possible ways in which a molecular level diode/ transistor might be synthesized by organic chemistry. A model system was proposed with a spiro carbon structure giving a molecular diode about half a nanometre across which could be connected by polythiophene molecular wires. Theoretical calculations showed the design to be sound in principle and there is still hope that such a system can be made to work.

Other Approaches

Nanoionics studies the transport of ions quite than electrons in nanoscale systems. Nanophotonics studies the behaviour of light on the nanoscale, and has the goal of developing devices that take advantage of this behaviour.

NANOELECTRONIC DEVICES

Radios

Nanoradios have been developed structured around carbon nanotubes.

Computers

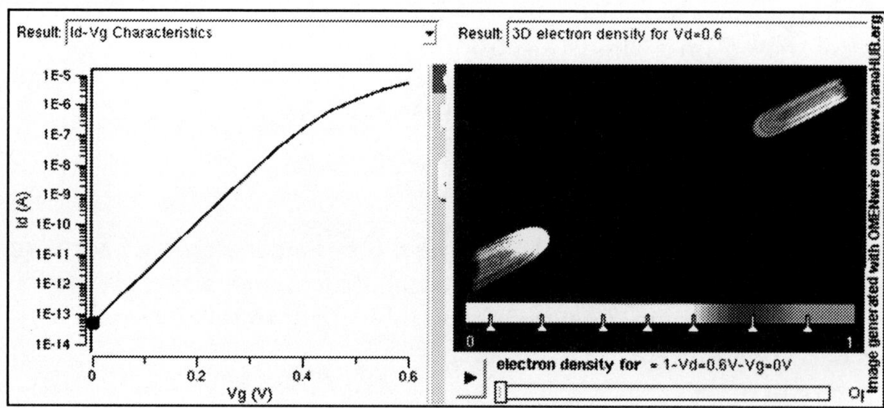

Figure: Simulation result for formation of inversion channel (electron density) and attainment of threshold voltage (IV) in a nanowire MOSFET. Note that the threshold voltage for this device lies around 0.45V.

Nanoelectronics holds the assure of making computer processors more powerful than are possible with conventional semiconductor fabrication techniques. A number of approaches are currently being researched, including new forms of nanolithography, as well as the use of nanomaterials such as nanowires or small molecules in place of traditional CMOS components. Field effect transistors have been made using both semiconducting carbon nanotubes and with heterostructured semiconductor nanowires.

Energy Production

Investigate is ongoing to use nanowires and other nanostructured materials with the hope to create cheaper and more efficient solar cells than are possible with conventional planar silicon solar cells. It is believed that the invention of more efficient solar energy would have a great effect on satisfying global energy needs.

There is also research into energy production for devices that would operate *in vivo,* called bio-nano generators. A bio-nano generator is a nanoscale electrochemical device, like a fuel cell or galvanic cell, but drawing power from blood glucose in a living body, much the same as how the body generates energy from food. To achieve the effect, an enzyme is used that is capable of stripping glucose of its electrons, freeing them for use in electrical devices. The average person's body could, theoretically, generate 100 watts of electricity (about 2000 food calories per day) using a bio-nano generator. However, this estimate is only true if all food was converted to electricity, and the human body needs some energy consistently, so possible power generated is likely much lower. The electricity generated by such a device could power devices embedded in the body (such as pacemakers), or sugar-fed nanorobots. Much of the research done on bio-nano generators is still experimental, with Panasonic's Nanotechnology Research Laboratory among those at the forefront.

Medical Diagnostics

There is enormous interest in constructing nanoelectronic devices that could detect the concentrations of biomolecules in real time for use as medical diagnostics, thus falling into the category of nanomedicine. A parallel line of research seeks to create nanoelectronic devices which could interact with single cells for use in basic biological research. These devices are called nanosensors. Such miniaturization on nanoelectronics towards in vivo proteomic sensing should enable new approaches for health monitoring, surveillance, and defence technology.

NANOELECTRONICS: TRANSISTORS ARRIVE AT THE ATOMIC LIMIT

One cannot make a example of material that is smaller than an atom: this sets a lower limit on the size of any part of a solid-state device. Even before this atomic limit is reached, however, the properties of devices containing tens or hundreds of atoms differ significantly from their large-scale counterparts, which presents an increasingly difficult challenge for the semiconductor industry. Nevertheless, silicon transistors have been relentlessly scaled down in size over recent decades, resulting in devices with gate oxides that are only several atomic layers thick and channels that are just tens of nanometres long. One of the main challenges now is caused by natural variations in the number and position of the dopant atoms in the channel, which leads to device-to-device fluctuations that are detrimental to operation.

But as top-down approaches to machine fabrication run into these problems, opportunities open for bottom-up approaches that build devices atom by atom, including single-atom transistors and quantum-information-processing devices. However, creating single-atom devices in silicon — and controlling the position of every atom — is a daunting experimental task. Now, writing in *Nature Nanotechnology*, Michelle Simmons and co-workers report that they have reached an important milestone in atomic-scale device fabrication by building the first silicon-based single-atom transistor from the bottom-up.

The device is fabricated using a combination of scanning tunnelling microscopy (STM) and hydrogen-resist lithography. This involves covering the surface of a clean silicon wafer with a layer of hydrogen, and then using the tip of an STM (with a suitable voltage applied) to remove the hydrogen from specific regions of the surface. When the wafer is subsequently exposed to phosphine (PH_3), the remaining hydrogen acts as a mask, so that the phosphorus can only bind to the surface in those regions from where the hydrogen has been removed. As such, Simmons and co-workers — who are based at the University of New South Wales (UNSW), the Korea Institute of Science and Technology Information, Purdue University and the universities of Sydney and Melbourne — are able to 'write' a nanostructure made of phosphorus atoms onto the silicon wafer. Finally, the nanostructure is encapsulated by growing a fresh layer of silicon on top of it, which keeps the phosphorus atoms fixed at their positions and also protects them from the environment.

Earlier work at UNSW has demonstrated that the combination of STM and hydrogen-resist lithography can be used to make a wide range of device elements. For example, back in 2004, the UNSW team showed that (metallic) structures with a variety of different shapes could be fabricated with this approach. As is well known, phosphorus is a group V atom that acts as a donor atom in silicon — it adds a single electronic charge to the host lattice. Because the structures in their initial experiments were many phosphorus atoms wide, the overlap of so many donor-electron wave-functions created a metallic system. And in subsequent work, Simmons and co-workers at UNSW and the University of Wisconsin-Madison showed that this approach could also be used to fabricate quantum-dot devices. This involved defining two metallic leads (and two gate electrodes) with a region of phosphorus atoms between them that was small enough for the electron charging energy to be higher than the thermal energy: the electric current through this device displayed Coulomb blockade effect, which is a classic hallmark of a quantum dot. And recently, the UNSW–Purdue–Melbourne collaboration reported that Ohm's law remains valid in phosphorus wires just four atoms wide.

The central achievement of the most recent work is to use the STM–hydrogen-resist lithography approach to position a single phosphorus atom between source and drain contacts and two (more distant) gate electrodes. The electrical characteristics of the resulting single-atom transistor are, in a way, similar to those of a conventional transistor in that the current between the source and drain can be controlled by applying a voltage to the gate electrodes. However, when the gate voltage is just below the threshold for transistor operation, the single-atom transistor behaves as a small quantum dot. Schematic of the single-atom transistor made by Simmons and co-workers.

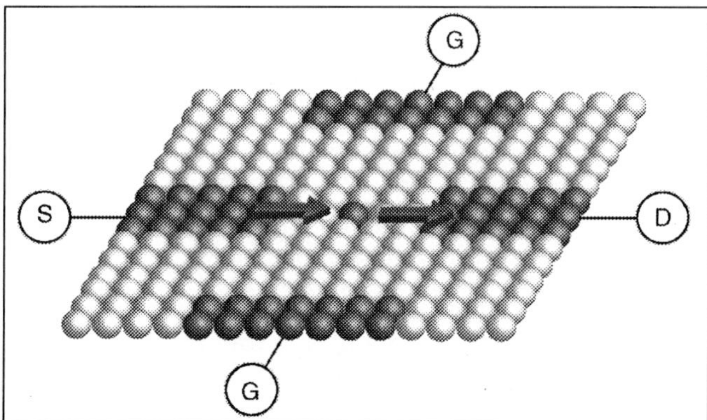

Figure : A single phosphorus atom (blue sphere) is placed with atomic precision on the surface of a silicon crystal (yellow spheres) between the metallic source (S) and drain (D) electrodes, which are formed by phosphorus wires that are multiple atoms wide. Electric charge flows (red arrows) from the source to the drain through the phosphorus atom when an appropriate voltage is applied across the gate electrodes (G). This schematic is not to scale: there are several tens of rows of silicon atoms between the phosphorus atom and the source and drain electrodes, and more than 100 rows of silicon atoms between the phosphorus atom and the gate electrodes.

In this quantum-dot regime the source–drain current shows Coulomb blockade effects that reflect the three possible charge states of the phosphorus atom: the zero-electron state (D^+), the one-electron state (D^0) and the two-electron state (D^-). Simmons and co-workers carefully measured the current characteristics and the energies associated with these different charge states, and compared these values with the results of large-scale tight-binding calculations of their entire device structure. The strong agreement between experiment and simulation, without the use of any fitting parameters, unequivocally demonstrates that a single-atom transistor was fabricated. Furthermore, they found that the energy levels associated with the three charge states were close to their 'bulk' values, which is a strong sign that the nearby electrodes did not influence the donor atom significantly.

Single-atom transistors represent the ultimate limit in solid-state device miniaturization, but they are also interesting for another reason. Deterministically positioned single-dopant atoms in silicon, electrically addressable by metallic leads, are at the heart of a number of promising proposals for quantum-information-processing devices. The long coherence and relaxation times associated with single dopants make them very attractive candidates for quantum-device architectures.

The atom-by-atom fabrication technique developed by Simmons and co-workers therefore fulfils a long-standing need for a technique that is capable of atomic-scale device fabrication in silicon. And although the technique is not directly applicable on an industrial scale, it does bring the development of truly atomistic electronics — and the possibilities they offer — into the experimental realm.

FUTURE OF NANOELECTRONICS

Nanoelectronics based on conventional concepts of microelectronics. Due to the extreme changes of nanoelectronic developments are needed in the future, however, new theories required in order to maintain the ongoing trend continues. Besides the development of nanoelectronics itself, it is expected that they will encourage technological progress in all sectors. Thus, at lower cost higher functionality in almost every area of electronics could be achieved. Due to the progressive approximation of the size of the nanoelectronic components of biological molecules, it could also lead to a future cooperation between nanoelectronics and biotechnology. Here, it is expected that the development of novel biosensors research is a first area of cooperation would be. Even in the organic sector could be in future nanoelectronics helpful.

The carbon nanotubes and semiconductor nanowires that became available to scientists in the 1990s captured my imagination and attracted me to the field now called nanoelectronics. For an inorganic materials chemist like me, these newly discovered tiny building blocks were like Tinkertoys that could potentially be used to make all kinds of gadgets and widgets. The desire to build something, to invent new structures out of them, spoke to me as a chemist, and I've been fascinated by the possibilities of these new nanomaterials ever since. Thanks to Moore's Law, though, the electronic devices produced by the conventional semiconductor process are now also "nanoelectronics." So what will be the role of new nanoelectronics based on chemically synthesized nanostructures, like carbon nanotubes and nanowires? Thermodynamics dictates that chemically synthesized nanostructures will probably never achieve the uniformity and perfection of electronic devices carved out of silicon crystals using conventional lithography. Rather, the strength of the new nanomaterials is in their chemical diversity and flexibility.

Those of us working on novel nanoelectronics should be trying to do things conventional technology simply cannot do, rather then trying to compete directly with the well-established technology. There are opportunities using physical mechanisms that have not been utilized in solid-state electronics. My research group, for example, is working on nanoscale magnetic semiconducting materials that could enable the demonstration of spintronics, which seeks to exploit the spin properties of electrons in computing devices.

The silicon microelectronics industry has been remarkably conservative, using only a very small fraction of the elements in the periodic table. For a long time, after all, reducing the size of transistors, which was made possible because of advances in the lithographic process, was sufficient to sustain the performance of devices as they were scaled down. But size reduction itself simply cannot deliver the necessary increases in performance anymore. New chemically synthesized nanostructures might be suitable for various advanced functions that are difficult to achieve with more conventional -silicon-based materials.

In such a scenario, we would use exotic materials for new functions that are needed only at certain places on a very large-scale circuit that is generated by a conventional semiconductor process. Newly synthesized nanostructures can also compete in applications where perfection is not of paramount importance but cheap processing, chemical flexibility, and function are. Electronic biosensors in which carbon nanotubes or nanowires are used to detect specific molecules are certainly one of those promising areas, but there may well be others that we have not imagined yet.

How can we organize these new nanostructures and integrate them into useful systems? This is an enormously challenging problem, and though I have worked on it before, I do not pretend that we have fully addressed it. But now that researchers have the Tinkertoys in hand, we can begin to come up with more creative ideas and strategies to manipulate these building blocks.

Employment Opportunities for Nanoelectronics

Nanoelectronics is now second-hand in many areas. It occurs among others in the consumer electronics, medical technology and automation engineering. Today's communication technology would be almost unthinkable without microelectronic developments, both the classical and the mobile phone is dependent on them. The nanoelectronic production is further necessary basis for computers. By the nanoelectronic improvement of computer chips, for example, the speed and computing capacity is improved. This results in considerable cost savings and increased reliability of electronic items. In addition, find nanoelectronic work place, especially for the development of improved techniques and associated equipment for electrical production. The building makes use of the advantage now nanoelectronics, such as industrial manufacturing, nanoelectronic components used for example in the machine control.

Benefits of Nanoelectronics

In addition to cost investments and performance optimization, offers the additional advantages nanoelectronics. By scaled transistors, higher switching frequencies can be generated which in turn lead to reduction of the signal propagation times. This is possible due to the smaller structures more efficient and faster circuits. The increased system integration that is obtained by nano-electronics, requires fewer individual components on a printed circuit board, which may go to less damage and higher reliability is obtained. Without the possibility of size reduction technology equipment such as mobile phones, notebooks, tablet PCs or smartphones today would be unthinkable.

REFERENCES

- Bhatia, A.S. : *Nano Electronics*, Deep and Deep, Delhi, 2011.
- Branda Paz: *A Handbook on Nanoelectronics*, Dominant, Delhi, 2008.
- Drexler, K. Eric: *Nanosystems: Molecular Machinery, Manufacturing, and Computation*, New York: Wiley, 1992.
- Gregory Kovacs: *Micromachined Transducers Sourcebook*, McGraw-Hill, London, 1998.
- Rao, M. Balakrishna : *Nanotechnology in Electronics*, Campus Books International, Delhi, 2007.
- Sergey Edward Lyshevski: *Nano- and Microelectromechanical Systems*, CRC Press, Delhi, 2000.
- Sharma, P.K. : *NanoMEMS Physics : Applications and Principles*, Vista International Pub, Delhi, 2011.

2

Nanotechnology and Nanomachines

INTRODUCTION TO NANOTECHNOLOGY

Truthfully *revolutionary* nanotechnology products, materials and applications, such as nanorobotics, are years in the future (some say only a few years; some say many years). What qualifies as "nanotechnology" today is basic research and development that is happening in laboratories all over the world. "Nanotechnology" products that are on the market today are mostly gradually improved products (using *evolutionary* nanotechnology) where some form of nanotechnology enabled material (such as carbon nanotubes, nanocomposite structures or nanoparticles of a particular substance) or nanotechnology process (*e.g.* nanopatterning or quantum dots for medical imaging) is used in the manufacturing process. In their ongoing quest to improve existing products by creating smaller components and better performance materials, all at a lower cost, the number of companies that will manufacture "nanoproducts" (by this definition) will grow very fast and soon make up the majority of all companies across many industries. Evolutionary nanotechnology should therefore be viewed as a process that gradually will affect most companies and industries.

DEFINITION OF NANOTECHNOLOGY

So what precisely is *nanotechnology*? One of the problems facing nanotechnology is the confusion about its definition. Most definitions revolve around the study and control of phenomena and materials at length scales below 100 nm and quite often they make a comparison with a human hair, which is about 80,000 nm wide. Some definitions include a reference to molecular systems and devices and nanotechnology 'purists' argue that any definition of nanotechnology needs to include a reference to "functional systems". The inaugural issue of *Nature Nanotechnology* asked 13 researchers from different areas what nanotechnology means to them and the responses, from enthusiastic to sceptical, reflect a variety of perspectives.

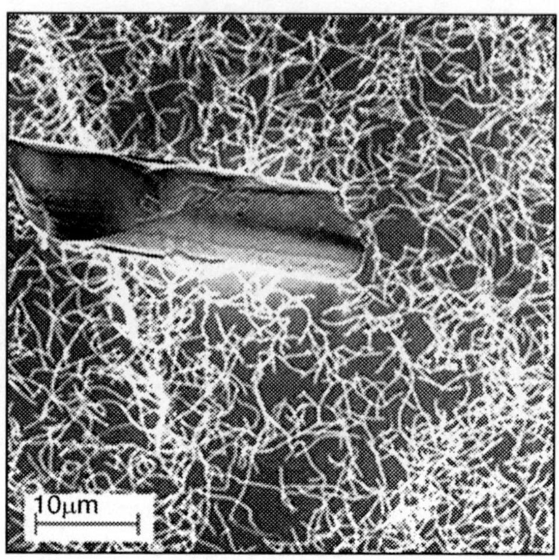

Figure: Human hair fragment and a network of single-walled carbon nanotubes

It seems that a size restriction of nanotechnology to the 1-100 nm range, the area where size-dependant quantum effects come to bear, would exclude numerous materials and devices, especially in the pharmaceutical area, and some experts caution against a rigid definition based on a sub-100 nm size. Another important criteria for the definition is the requirement that the nanostructure is man-made. Otherwise you would have to include every naturally formed biomolecule and material particle, in effect redefining much of chemistry and molecular biology as 'nanotechnology.'

The most important requirement for the nanotechnology definition is that the nanostructure has special properties that are exclusively due to its nanoscale proportions.

THE SIGNIFICANCE OF THE NANOSCALE

A nanometer (nm) is one thousand millionth of a meter. For judgment, a red blood cell is approximately 7,000 nm wide and a water molecule is almost 0.3nm across. People are interested in the nanoscale (which we define to be from 100nm down to the size of atoms (approximately 0.2nm)) because it is at this scale that the properties of materials can be very different from those at a larger scale. We define nanoscience as the study of phenomena and manipulation of materials at atomic, molecular and macromolecular scales, where properties differ significantly from those at a larger scale; and nanotechnologies as the design, characterisation, production and application of structures, devices and systems by controlling shape and size at the nanometer scale. In some senses, nanoscience and nanotechnologies are not new. Chemists have been making polymers, which are large molecules made up of nanoscale subunits, for many decades and nanotechnologies have been used to create the tiny features on computer chips for the past 20 years. However, advances in the tools that now allow atoms and molecules to be examined and probed with great precision have enabled the expansion and development of nanoscience and nanotechnologies.

The bulk properties of materials frequently change dramatically with nano ingredients. Composites made from particles of nano-size ceramics or metals smaller than 100 nanometers can suddenly become much stronger than predicted by existing materials-science models. For example, metals with a so-called grain size of around 10 nanometers are as much as seven times harder and tougher than their ordinary counterparts with grain sizes in the hundreds of nanometers. The causes of these drastic changes stem from the weird world of quantum physics. The bulk properties of any material are merely the average of all the quantum forces affecting all the atoms. As you make things smaller and smaller, you eventually reach a point where the averaging no longer works.

The properties of materials can be dissimilar at the nanoscale for two main reasons:

First, nanomaterials have a relatively larger surface area when compared to the same mass of material produced in a larger form. This can make materials more chemically reactive (in some cases materials that are inert in their larger form are reactive when produced in their nanoscale form), and affect their strength or electrical properties.

Second, quantum effects can begin to dominate the behaviour of matter at the nanoscale - particularly at the lower end - affecting the optical, electrical and magnetic behaviour of materials. Materials can be produced that are nanoscale in one dimension (for example, very thin surface coatings), in two dimensions (for example, nanowires and nanotubes) or in all three dimensions (for example, nanoparticles).

NANOMATERIALS

Much of nanoscience and numerous nanotechnologies are concerned with producing new or enhanced materials. Nanomaterials can be constructed by 'top down' techniques, producing very small structures from larger pieces of material, for example by etching to create circuits on the surface of a silicon microchip. They may also be constructed by 'bottom up' techniques, atom by atom or molecule by molecule. One way of doing this is self-assembly, in which the atoms or molecules arrange themselves into a structure due to their natural properties. Crystals grown for the semiconductor industry provide an example of self assembly, as does chemical synthesis of large molecules. A second way is to use tools to move each atom or molecule individually. Although this 'positional assembly' offers greater control over construction, it is currently very laborious and not suitable for industrial applications. It has been 25 years since the scanning tunnelling microscope (STM) was invented, followed four years later by the atomic force microscope, and that's when nanoscience and nanotechnology really started to take off. Various forms of scanning probe microscopes based on these discoveries are essential for many areas of today's research. Scanning probe techniques have become the workhorse of nanoscience and nanotechnology research. Here is a Scanning Electron Microscope (SEM) image of a gold tip for Near-field Scanning Optical Microscopy (SNOM) obtained by Focused Ion Beam (FIB) milling. The small tip at the centre of the structure measures some tens of nanometers.

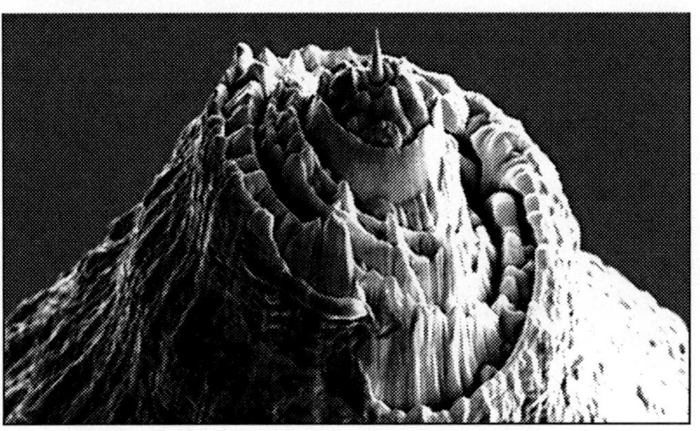

Figure: Gold Tip for SNOM.

Present applications of nanoscale materials include very thin coatings used, for example, in electronics and active surfaces (for example, self-cleaning windows). In most applications the nanoscale components will be fixed or embedded but in some, such as those used in cosmetics and in some pilot environmental remediation applications, free nanoparticles are used. The ability to machine materials to very high precision and accuracy (better than 100nm) is leading to considerable benefits in a wide range of industrial sectors, for example in the production of components for the information and communication technology, automotive and aerospace industries.

DEFINITION OF NANOMATERIALS

Although a broad meaning, we categorise nanomaterials as those which have structured components with at least one dimension less than 100nm. Materials that have one dimension in the nanoscale (and are extended in the other two dimensions) are layers, such as a thin films or surface coatings. Some of the features on computer chips come in this category. Materials that are nanoscale in two dimensions (and extended in one dimension) include nanowires and nanotubes. Materials that are nanoscale in three dimensions are particles, for example precipitates, colloids and quantum dots (tiny particles of semiconductor materials). Nanocrystalline materials, made up of nanometre-sized grains, also fall into this category. Some of these materials have been available for some time; others are genuinely new. The aim of this chapter is to give an overview of the properties, and the significant foreseeable applications of some key nanomaterials.

Two principal factors cause the properties of nanomaterials to differ significantly from other materials: increased relative surface area, and quantum effects. These factors can change or enhance properties such as reactivity, strength and electrical characteristics. As a particle decreases in size, a greater proportion of atoms are found at the surface compared to those inside. For example, a particle of size 30 nm has 5% of its atoms on its surface, at 10 nm 20% of its atoms, and at 3 nm 50% of its atoms. Thus nanoparticles have a much greater surface area per unit mass compared with larger particles. As growth and catalytic chemical reactions occur at surfaces, this means that a given mass of material in nanoparticulate form will be much more reactive than the same mass of material made up of larger particles.

PROPERTIES OF NANOMATERIALS

In tandem with surface-area effects, quantum effects can commence to dominate the properties of matter as size is reduced to the nanoscale. These can affect the optical, electrical and magnetic behaviour of materials, particularly as the structure or particle size approaches the smaller end of the nanoscale. Materials that exploit these effects include quantum dots, and quantum well lasers for optoelectronics.

For other materials such as crystalline solids, as the size of their structural components decreases, there is much greater interface area within the material; this can greatly affect both mechanical and electrical properties. For example, most metals are made up of small crystalline grains; the boundaries between the grain slow down or arrest the propagation of defects when the material is stressed, thus giving it strength. If these grains can be made very small, or even nanoscale in size, the interface area within the material greatly increases, which enhances its strength. For example, nanocrystalline nickel is as strong as hardened steel. Understanding surfaces and interfaces is a key challenge for those working on nanomaterials, and one where new imaging and analysis instruments are vital.

NANOMATERIAL SCIENCE

Nanomaterials are not basically one more step in the miniaturization of materials. They often require very different production approaches. There are several processes to create nanomaterials, classified as 'top-down' and 'bottom-up'. Although many nanomaterials are currently at the laboratory stage of manufacture, a few of them are being commercialised.

NANOSCALE IN ONE DIMENSION

Thin Films, Layers and Surfaces

One-dimensional nanomaterials, such as thin films and engineered surfaces, have been developed and used for decades in fields such as electronic device manufacture, chemistry and engineering. In the silicon integrated-circuit industry, for example, many devices rely on thin films for their operation, and control of film thicknesses approaching the atomic level is routine. Monolayers (layers that are one atom or molecule deep) are also routinely made and used in chemistry. The formation and properties of these layers are reasonably well understood from the atomic level upwards, even in quite complex layers (such as lubricants). Advances are being made in the control of the composition and smoothness of surfaces, and the growth of films.

Engineered surfaces with tailored properties such as large surface area or specific reactivity are used routinely in a range of applications such as in fuel cells and catalysts. The large surface area provided by nanoparticles, together with their ability to self assemble on a support surface, could be of use in all of these applications.

Although they characterize incremental developments, surfaces with enhanced properties should find applications throughout the chemicals and energy sectors. The benefits could surpass the obvious economic and resource savings achieved by higher activity and greater selectivity in reactors and separation processes, to enabling small-scale distributed processing (making chemicals as close as possible to the point of use). There is already a move in the chemical industry towards this. Another use could be the small-scale, on-site production of high value chemicals such as pharmaceuticals.

NANOSCALE IN TWO DIMENSIONS

Two dimensional nanomaterials such as tubes and wires have generated considerable interest among the scientific community in recent years. In particular, their novel electrical and mechanical properties are the subject of intense research.

Carbon Nanotubes

Carbon nanotubes (CNTs) were first observed by Sumio Iijima in 1991. CNTs are extended tubes of rolled graphene sheets. There are two types of CNT: single-walled (one tube) or multi-walled (several concentric tubes). Both of these are typically a few nanometres in diameter and several micrometres to centimetres long. CNTs have assumed an important role in the context of nanomaterials, because of their novel chemical and physical properties. They are mechanically very strong (their Young's modulus is over 1 terapascal, making CNTs as stiff as diamond), flexible (about their axis), and can conduct electricity extremely well (the helicity of the graphene sheet determines whether the CNT is a semiconductor or metallic). All of these remarkable properties give CNTs a range of potential applications: for example, in reinforced composites, sensors, nanoelectronics and display devices. Watch an animation of various nanotubes and a fullerene (buckyball): CNTs are now available commercially in limited quantities. They can be grown by several techniques. However, the selective and uniform production of CNTs with specific dimensions and physical properties is yet to be achieved. The potential similarity in size and shape between CNTs and asbestos fibres has led to concerns about their safety.

Inorganic Nanotubes

Inorganic nanotubes and inorganic fullerene-like materials based on layered compounds such as molybdenum disulphide were discovered shortly after CNTs. They have excellent tribological (lubricating) properties, resistance to shockwave impact, catalytic reactivity, and high capacity for hydrogen and lithium storage, which suggest a range of promising applications. Oxide-based nanotubes (such as titanium dioxide) are being explored for their applications in catalysis, photocatalysis and energy storage.

Nanowires

Nanowires are ultrafine wires or linear arrays of dots, formed by self-assembly. They can be made from a wide range of materials. Semiconductor nanowires made of silicon, gallium nitride and indium phosphide have demonstrated remarkable optical, electronic and magnetic characteristics (for example, silica nanowires can bend light around very tight corners). Nanowires have potential applications in high-density data storage, either as magnetic read heads or as patterned storage media, and electronic and opto-electronic nanodevices, for metallic interconnects of quantum devices and nanodevices. The preparation of these nanowires relies on sophisticated growth techniques, which include selfassembly processes, where atoms arrange themselves naturally on stepped surfaces, chemical vapour deposition (CVD) onto patterned substrates, electroplating or molecular beam epitaxy (MBE). The 'molecular beams' are typically from thermally evaporated elemental sources.

Biopolymers

The variability and site recognition of biopolymers, such as DNA molecules, offer a wide range of opportunities for the self-organization of wire nanostructures into much more complex patterns. The DNA backbones may then, for example, be coated in metal. They also offer opportunities to link nano- and biotechnology in, for example, biocompatible sensors and small, simple motors. Such self-assembly of organic backbone nanostructures is often controlled by weak interactions, such as hydrogen bonds, hydrophobic, or van der Waals interactions (generally in aqueous environments) and hence requires quite different synthesis strategies to CNTs, for example. The combination of one-dimensional nanostructures consisting of biopolymers and inorganic compounds opens up a number of scientific and technological opportunities.

NANOTECHNOLOGY APPLICATIONS

The ability to observe nano-sized materials has opened up a world of possibilities in a variety of industries and scientific endeavours. Because nanotechnology is essentially a set of techniques that allow manipulation of properties at a very small scale, it can have many applications.

Drug delivery. Today, most harmful side effects of treatments such as chemotherapy are a result of drug delivery techniques that don't pinpoint their intended target cells accurately. Researchers at Harvard and MIT have been able to attach special RNA strands, measuring about 10 nm in diameter, to nanoparticles and fill the nanoparticles with a chemotherapy drug. These RNA strands are attracted to cancer cells. When the nanoparticle encounters a cancer cell it adheres to it and releases the drug into the cancer cell. This directed technique of drug delivery has great potential for treating cancer patients while producing less side harmful affects than those produced by conventional chemotherapy.

Fabrics. The properties of familiar materials are being changed by manufacturers who are adding nano-sized components to conventional materials to improve performance. For example, some clothing manufacturers are making water and stain repellent clothing using nano-sized whiskers in the fabric that cause water to bead up on the surface.

Reactivity of Materials. The properties of many conventional materials change when formed as nano-sized particles (nanoparticles). This is generally because nanoparticles have a greater surface area per weight than larger particles; they are therefore more reactive to some other molecules. For example studies have show that nanoparticles of iron can be effective in the cleanup of chemicals in ground-water because they react more efficiently to those chemicals than larger iron particles.

Strength of Materials. Nano-sized particles of carbon, (for example nanotubes and bucky balls) are extremely strong. Nanotubes and bucky balls are composed of only carbon and their strength comes from special characteristics of the bonds between carbon atoms. One proposed application that illustrates the strength of nanosized particles of carbon is the manufacture of t-shirt weight bullet proof vests made out of carbon nanotubes.

Micro/Nano ElectroMechanical Systems. The ability to create gears, mirrors, sensor elements, as well as electronic circuitry in silicon surfaces allows the manufacture of miniature sensors such as those used to activate the airbags in your car. This technique is called MEMS (Micro-ElectroMechanical Systems). The MEMS technique results in close integration of the mechanical mechanism with the necessary electronic circuit on a single silicon chip, similar to the technique used to produce computer chips. Using MEMS to produce a device reduces both the cost and size of the product, compared to similar devices made with conventional techniques.

MEMS is a stepping stone to NEMS or Nano-ElectroMechanical Systems. NEMS products are being made by a few companies, and will take over as the standard once manufacturers make the investment in the equipment needed to produce nano-sized features.

THE NANOTECHNOLOGY DEBATE

There are numerous different points of view about the nanotechnology. These differences start with the definition of nanotechnology. Some define it as any activity that involves manipulating materials between one nanometer and 100 nanometers. However the original definition of nanotechnology involved building machines at the molecular scale and involves the manipulation of materials on an atomic (about two-tenths of a nanometer) scale. The debate continues with varying opinions about exactly what nanotechnology can achieve. Some researchers believe nanotechnology can be used to significantly extend the human lifespan or produce replicator-like devices that can create almost anything from simple raw materials.

The third major area of debate concerns the timeframe of nanotechnology-related advances. Will nanotechnology have a significant impact on our day-to-day lives in a decade or two, or will many of these promised advances take considerably longer to become realities? Finally, all the opinions about what nanotechnology can help us achieve echo with ethical challenges. If nanotechnology helps us to increase our lifespans or produce manufactured goods from inexpensive raw materials, what is the moral imperative about making such technology available to all? Is there sufficient understanding or regulation of nanotech based materials to minimize possible harm to us or our environment? Only time will tell how nanotechnology will affect our lives, but browsing through the topics on the navigation bar to the left or on our Nanotechnology Applications page will help you understand the possibilities and anticipate the future.

NANOMACHINES DEVICES

Nanomachines are devices build from individual atoms. Some researchers believe that nanomachines will one day be able to enter living cells to fight disease. They also hope to one day build nanomachines that will be able to rearrange atoms in order to construct new objects. If they succeed, nanomachines could be used to literally turn dirt into food and perhaps eliminate poverty.

What are Nanomachines?

Namomachines are extremely small machines and is also called as nanite. The dimensions of a nanomachine are measured in nanometers. In simple words, a nanomachine refers to a system that is made up of tiny component or nanoscale components. Nanotechnology seeks to develop nanomachines in such a way that they can handle complex tasks independently.

As the terminology implies, nanomachines are extremely small devices. Their size is measured in nanometers (a nanometer is about 1 billionth of a meter) and they are built from individual atoms. During the 1980's and 1990's, futurist and visionary K. Eric Drexler popularized the potential of nanomachines. For Drexler, the ultimate goal of nanomachine technology is the production of the 'assembler'. The assembler is a nanomachine designed to manipulate matter at the atomic level. It will be built with extremely small 'pincers' (as small as a chain of atoms) which will be used to move atoms from existing molecules into new structures. The idea is that the assembler will be able to rearrange atoms from raw material in order to produce useful items. In theory, one could shovel dirt into a vat and wait patiently for a team of nanomachine assemblers to convert the dirt into an apple, a chair, or even a computer. The machines in the vat would have a molecular schematic of the object to be built encoded in their 'memory'. They would then systematically rearrange the atoms contained in the dirt to produce the desired item.

Figure: This is a representation of a nanomachine. The coloured balls represent the individual atoms that comprise the machine.

Another ambition of nanotechnology is to design nanomachines that can make copies of themselves. The thought is that if a machine can rearrange atoms in order to build new materials, it should also be able to build copies of itself. If this goal is achieved, products produced by nanomachines will be extremely inexpensive. This is because the technology (once perfected) will be self-replicating and will not require specific materials, which might be rare and therefore cost money. Arthur C. Clarke has predicted that nanotechnology will herald an end to conventional monetary systems.

If scientists manage to build nanomachines that can rearrange atoms, a world of exciting possibilities will open up. Purpose designed nanomachines could be used to provide breakthrough treatments for many diseases. Medical nanomachines programmed to recognize and disassemble cancerous cells could be injected into the bloodstream of cancer suffers, thus providing a quick and effective treatment for all types of cancer. Nanomachines could be used to repair damaged tissue and bones. They could even be used to strengthen bones and muscle tissue by building molecular support structures by reassembling nearby tissue. With the ability to manipulate human cells at the atomic level, medical science will rapidly devise treatments for most human illnesses. And since nanomachines will be designed to make copies of themselves, these treatments will be inexpensive and available to the entire population.

Food shortages and starvation will be a thing of the past if nanotechnology is perfected. Nanomachines will be able to turn any material into food, and this food could be used to feed millions of people world wide. Again, since the technology is self replicating, food produced by nanomachines will be low cost and available to all.

As well as food, nanomachines will be able to build other items to satisfy the demands of our growing population of consumers. Clothing, houses, cars, televisions, and computers will be readily available at virtually no cost. Furthermore, there will be no concern about the garbage produced by the new consumerist society because nanomachines will convert it all back into new consumable goods.

Environmental problems such as ozone depletion and global warming could be solved with nanotechnology. Swarms of nanomachines could be released into the upper atmosphere. Once there, they could systematically destroy the ozone depleting chlorofluorocarbons (CFCs) and build new ozone molecules out of water (H_2O) and carbon dioxide (CO_2). Ozone (O_3) is built out of 3 oxygen atoms, and since water and carbondioxide both contain oxygen, the atmosphere contains a plentiful supply of oxygen atoms. While the ozone construction teams are at work in the upper atmosphere, teams of specialized nanomachines could be employed to destroy the excess CO_2 in the lower atmosphere. CO_2 is a heat trapping gas, which has been identified as one of the major contributors to global warming. Removing excess CO_2

could help halt global warming and bring the planet's ecosystem back into balance. This will benefit all species on Earth.

The perfection of nanotechnology and the production of nanomachines could herald a new age for humanity. Starvation, illness, and environmental problems could quickly come to an end. But how realistic are the goals of nanotechnology? Will it ever be possible to produce machines the size of atoms? And if so, how feasible is it to build nanomachines that can build objects from the atom up? Is it possible for nanomachines to build copies of themselves? Before we get carried away with the promises of nanotechnology, we should take a look at some of the problems that are yet to be solved.

Challenges to Overcome

An important challenge to conquer is one of engineering. How can we physically build machines out of atoms? Rearranging atoms into new shapes is essentially building new molecules (nanomachines are sometimes called 'molecular machines') and this is no easy task. Using contemporary technology to rearrange atoms has been said to be analogous to assembling LEGO blocks while wearing boxing gloves. It is virtually impossible to snap individual atoms together. All we can do is crudely push large piles of them together and hope for the best. Scientists hope that once this initial challenge is overcome, nanomachines will usher in a new age of molecular engineering and previous problems will be a thing of the past. The new nanomachines will allow scientists to take off the boxing gloves and accurately snap together individual atoms to build virtually any molecule (within the laws of physics, of course).

This is nice in principle, but the question of how to build the *first* nanomachines remains. Nanotechnologists think that it will be impossible to build the first nanomachines by using large scale equipment. Although progress is being made in the miniaturization of integrated circuits and in the ultra-fine finishing of high quality optical components, the large scale technology being used doesn't let us take off the boxing gloves. There is a limit to how far down these machines can go. Super smooth lens polishing is one thing, but moving individual atoms is something else all together. Nanotechnologists need to get the boxing gloves off before they can build the first nanomachines.

One way to work without boxing gloves is to patiently experiment with chemical synthesis. The idea is to build molecules of increasing complexity by allowing atoms to assemble or rearrange in natural ways. When molecules are mixed, they naturally form new molecules. Through extensive experimentation, more control can be gained over how molecules are formed. In time, it is conceivable that chemists will be able to position individual atoms by using a range of techniques developed in chemical synthesis. One of these techniques might involve the removal and relocation of hydrogen atoms. This technique could be developed with knowledge of how hydrogen atoms interact with other atoms. For example, it is known that the *propynyl* radical C3H3 (its made out of 3 carbon atoms and 3 hydrogen atoms) is 'attracted' to hydrogen. It is also known that this radical has two ends. At one end there is a highly reactive radical, while at the other end there is stable carbon. This feature means that chemists may be able to synthesize a larger molecule with the *propynyl* radical at one end (the rest of the molecule would be built from the stable carbon end). If this larger molecule was held on a positioning device, it could be used to extract hydrogen from a range of different molecules by passing them by the reactive radical.

Chemical synthesis is talented. In computer simulations, molecularly stable gears and cogs have been formed through chemical synthesis.

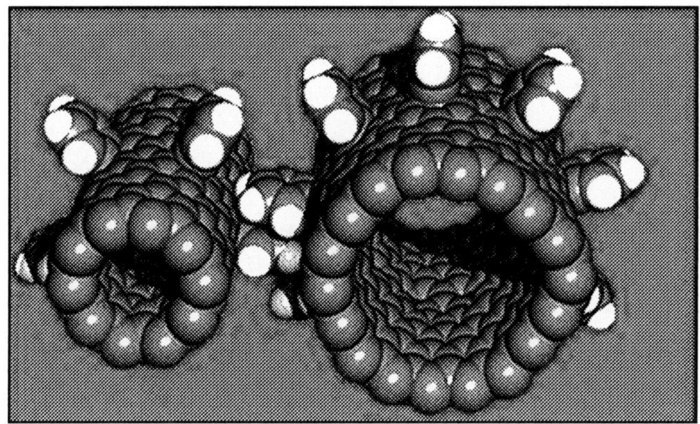

Figure: A representation of nanogears made from graphitetubes billionths of a meter wide.

If chemists and engineers succeed in building nanomachines the hope is that these machines will be able to build a whole range of new molecules from the atom up. If all goes well, scientists will never have to move atoms round while wearing boxing gloves and the lengthy experimental process of chemical synthesis will no longer be required. But will it be that easy?

In order to create new molecules, a nanomachine has to somehow 'grab' individual atoms with its pincers and move them into new positions or attach them to other molecules. This seems to be quite simple, but as George M. Whitesides (2001) points out, there are serious problems that need to be overcome. Consider, for example, the fact that a nanomachine's pincers will be made out of several atoms and will therefore be larger than the individual atoms that it needs to move around. This means that the intricacy and accuracy of the nanomachine's movement will be severely limited. It will be clumsy. Assembling atoms would be like trying to piece together a mechanical wristwatch with your fingers rather than small tweezers. Another problem arises from the fact that individual atoms are compelled to 'attach' to other atoms. Some atomic bonds can be extremely strong (especially with carbon atoms) so pulling them apart will require large amounts of energy. Furthermore, since carbon atoms attach to just about anything it seems likely that they will bond to the nanomachine's pincers after they've been pried away from their original molecules. The only way to remove them could be to move them to molecules that they are more strongly attracted to. But then there is the possibility that the entire nanomachine will stick to the molecule. The situation is analogous to trying to build a wristwatch with magnetized tweezers and screwdrivers. It can't be done because the individual components stick to the tools.

Drexler et al (2001) brush aside these problems. They suggest that such concerns arise from a misunderstanding of how nanomachines work. For example, the idea that nanomachines use 'pincers' to move objects around is nothing more than a poor metaphor. In reality, nanomachines might contain an active tip, which is no larger than the atom it is designed to manipulate. So Whitesides' concerns about the size of a nanomachine's pincers are easily answered. However, his concerns about the bonding of carbon atoms to nanomachines seem more difficult to answer. Drexler attempts to bury the problem by citing theoretical work done with the hydrogen extraction tool and by referring to experimental work done with hydrogen atoms. He doesn't directly address concerns about manipulating carbon atoms. This is important, because carbon is one of the most common atoms found on Earth and will no doubt be involved if nanomachines are used to build new molecules. Progress made with hydrogen might not translate easily to future work on carbon atoms.

Drexler does, however, talk about some very promising work by Wilson Ho and Hyojune Lee. In an experiment, Ho and Lee;

"...used an STM tip first to locate two carbon monoxide (CO) molecules and one iron (Fe) atom adsorbed on a silver surface in vacuum at 13 K. Next, they lowered the tip over one CO molecule and increased the voltage and current flow of the instrument to pick up the molecule; then they moved the tip-bound molecule over the surface-bound Fe atom and reversed the current flow, causing the CO molecule to covalently bond to the Fe atom, forming an iron carbonyl Fe(CO) molecule on the surface. Finally, the researchers repeated the procedure, returning to the exact site of the first Fe(CO) and adding a second CO molecule to the Fe(CO), forming a molecule of $Fe(CO)_2$*, which in subsequent images of the surface appeared as a tiny "rabbit ears" structure, covalently bound to the silver surface. Ho's group has also demonstrated single-atom hydrogen abstraction experimentally, using an STM".*

This type of work will hopefully lead to more complex manipulation of atoms, and this could result in the development of tools that successfully 'pick and place' carbon atoms.

As our technological capacities develop, the promise of nanomachine technology becomes more of a reality. We may one day see the successful creation of nanomachine assemblers. These machines could end hunger and bring in a new age of advancement for humanity. Nanotechnology offers us big promises in a small package. However, the advantages it promises do not come for free. They come with some very big risks.

Big Risks Come in Small Packages

Cutting edge technology can obtain a while to catch on in the commercial world. However, there is one place in which it catches on very quickly: The Military! During humanity's history, technological research has moved fastest when there is a potential military application. The danger is that this trend will continue with nanotechnology. Imagine the possible uses of nanomachines in warfare. Self replicating nanomachines designed to target and destroy organic material could be released over enemy territory reducing the population to dust within a matter of hours. If these machines were designed to destroy each other after (say) 24 hours, the enemy's country would be left empty and safe to be invaded by military forces. Biological warfare would be a thing of the past since nanomachine warfare would be so much safer (well, for the 'good guys' anyway).

The only way to prevent this use of nanomachines would be through international agreements. Unfortunately, not all countries are willing to sign such agreements. And those who *do* sign might be tempted to develop the technology in secret—just incase the enemy is doing the same thing. Perhaps the most we could hope for would be a stalemate situation like the one between the United States and the U.S.S.R during the cold war. If both sides have the technology, they might be too nervous to use it, since they know that the other side will retaliate.

A more serious danger of nanomachine technology involves the ability to self replicate. Imagine that a nanomachine has the ability to make a copy of itself by rearranging the atoms contained in any nearby matter. Since it is producing an exact copy of itself, it is likely that the 'offspring' machine will be able to replicate. This is, after all, the way in which nanotechnologists intend to keep the cost of nanomachines down.

So now we have 2 nanomachines that can replicate. One more cycle will produce 2 more, which leaves a total of 4.

4 becomes 8.

8 becomes 16.

16 becomes 32, and so on.

After only 27 generations we would have over 134 million nanomachines on our hands. Since they are molecular, this doesn't seem like a big number. But the number could keep growing. After 39 generations there would be over 549 billion nanomachines on the planet. The point is obvious. Without a way of controlling the reproduction of nanomachines, the planet is in danger of being over run. Furthermore, since the nanomachines are using the planet's resources as raw material with which to replicate, the danger is that the planet could eventually be *transformed* into a seething mass of nanomachines.

George Whitesides (2001) responds to this problem by pointing out that Earth has already been ravaged by molecular machines—namely, biological cells. This is true. Earth was a much different place 3.5 billion years ago before the emergence of life. Self replicating cells have, over 3.5 billion years, completely transformed the planet. They have changed the planet from a world of inorganic minerals with a CO2 rich atmosphere, to a world that is perfect for biological life.

But this fact doesn't counteract the danger in creating replicating nanomachines. In fact, Whitesides has reminded us that it *is* possible for molecular machines to replicate exponentially and transform the planet. If self replicating nanomachines get out of control, then they could alter the planet to such an extent that it is no longer suitable for biological life.

A possible solution to the problem is to limit the replicating abilities of nanomachines. For example, a mechanism could be developed by which new nanomachines are tagged with a number. This number could represent their generation. So, a nanomachine labelled 'gen 2' would produce offspring labelled 'gen 3', and their offspring would be labelled 'gen 4'. The replicating algorithm could be designed to only function if the generation number is less than 4. Also, nanomachines with a generation number higher than 1 could be encoded with a function that limits the number of reproductive cycles they can execute. By building in these safeguards, we may be able to control the population of nanomachines while at the same time allowing the existence of a number necessary to facilitate some of the advantages mentioned earlier.

However, these safeguards may not be enough. The biological world has shown us that evolution occurs and cannot be stopped. The same may be true of the nano-world. Consider the idea that each time a nanomachine makes a copy of itself, there is a possibility that an error could be made during the copying process. Such errors could be very small—perhaps no larger than a single 'bit' of information. Now imagine what would happen if an error occurred while a nanomachine was building its offspring's copying mechanism. To be more specific, imagine that a single 'bit' error occurred when encoding the function that limits the machine's replicative abilities. So, instead of checking that the machine's generation number is less than 4, it checks to see that it is less than 40. When this error is passed on to the machine's offspring, they will reproduce providing their generation number is less than 40. Since the error will be passed on to each subsequent generation, there will be a substantial explosion in nanomachine population. A single error could have the potential to send the nanomachine population out of control. And the more reproducing nanomachines there are, the greater the chance of another error occurring in at least one of them. The only way to avoid the problem of uncontrollable replication is to avoid building self-replicating nanomachines.

It may be true that self-replicating machines are the only way to ensure a cheap supply of nanomachines, but the potential risks outweigh the benefits. If nanomachines are built individually in labs, they will still be useful to cure disease and they will still be potentially useful for rearranging molecules to build new objects such as food. The only drawback is that none of it will come for free. Someone will still have to pay for the construction of the machines, and this means that their products will have to be paid for by consumers. So, poverty will not be eradicated. However, it could be that producing food with nanomachines is faster and cheaper than conventional means, which will mean that poverty may be eased a bit. Furthermore, if governments are willing to invest in the technology, nanomachines may be able to be used to fix some of our environmental problems by repairing the damage we've done to the atmosphere. So the research is worth continuing.

STRUCTURE OF ATOMIC CLUSTERS

In the macroscopic world, the physical properties of a fabric are independent of a sample's size. However, when sample dimensions are made sufficiently small, the properties of a cluster of atoms must ultimately depart from those in the bulk, and evolve as a function of size. As discussed in a recent issue of *Scientific American*, this transition region, where the 'homogeneous' bulk picture gives way to an increasingly 'atomistic' description, is an area of intense current research interest. The determination of the structure of nanometre-size clusters of atoms is a central problem underpinning all activity in this field.

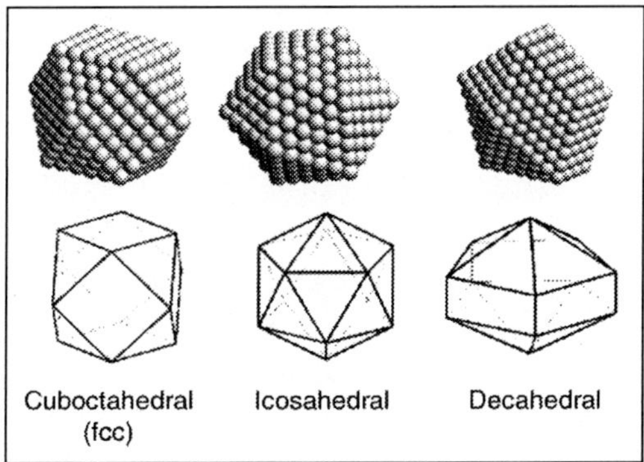

Figure: This image shows the three typical structures that are observed for small particles of materials whose bulk structure is fac-centred-cubic.

In bulk crystals, structure is characterised by the periodicity of a lattice, whereas in a cluster of a few hundred atoms, translational symmetry is not required. In such clusters, the high proportion of atoms at the surface make a crucial contribution to the particle's total energy. Energetically favourable patterns of atoms on a cluster's surface can force a rearrangement in the interior. Hence clusters can, *and do*, adopt new atomic arrangements, which may be forbidden in the bulk, *e.g.* in most materials that are face-centred-cubic (fcc) in the bulk, *small* clusters form with axes of fivefold symmetry (icosahedral and decahedral structures) and irregular atomic spacing.

Fundamental Questions

A large numeral of fundamental questions need to be answered: How do these structures nucleate and grow? How do clusters transform from one structure to another as successive atoms are added during growth? At what size does the bulk structure prevail? Given that atomic structure plays a critical role in determining all cluster properties, it is remarkable that these questions remain unanswered – the inherent difficulties in both the theoretical and experimental determination of cluster structure have been major obstacles. As a result, much structural information has been obtained *indirectly*, from experimental observations of other cluster properties, which has led to incomplete and sometimes ambiguous interpretation.

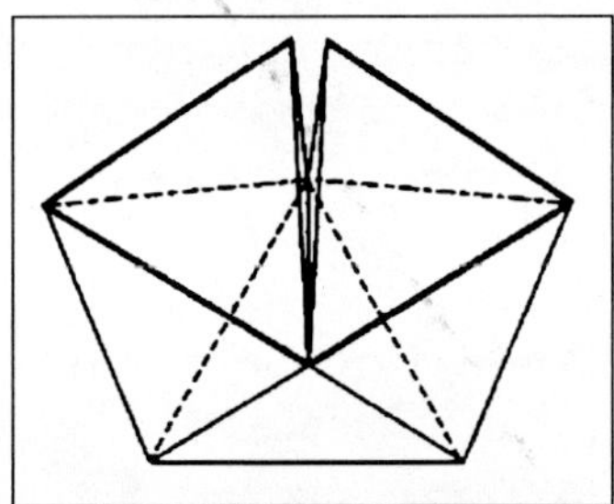

5 tetrahedra can almost, BUT NOT QUITE, form a particle with 5 fold symmetry (an icosahedron). Icosahedral atomic clusters are therefore strained.

Electron Diffraction

Electron diffraction from a beam of clusters is a powerful technique for the *direct* study of cluster structure. In such experiments, observations are made on unsupported clusters under vacuum, free of interaction with a substrate, or from surface contamination, which might otherwise alter the delicate balance of energies determining structure. The technique was demonstrated beautifully in a series of classic studies of rare gas clusters by a group at Orsay. A similar apparatus has been developed by Blair Hall (now at IRL in Wellington, and a collaborator on this project) *specifically* for the study of metal clusters, and will soon arrive at the University of Canterbury.

This equipment has a proven track record of success. Preceding structural studies of metal clusters have concentrated primarily on elements whose bulk atomic arrangement is fcc.

The primary objective of our current work is to investigate the structures of clusters of a range of non-fcc materials.

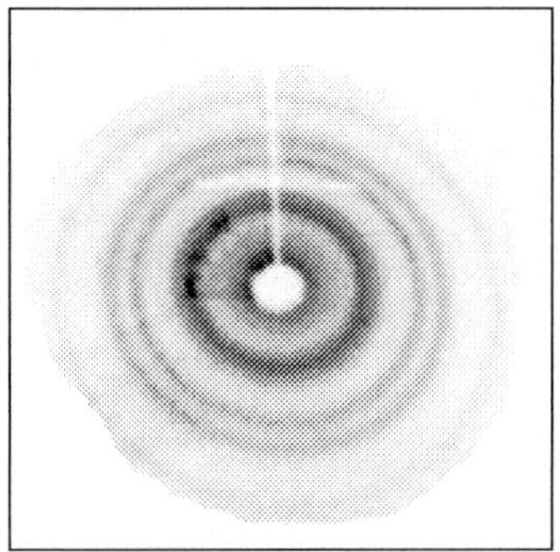

This is the first electron diffraction pattern obtained in the atomic cluster research laboratory. The sample is a thin film of polycrystalline gold and each diffraction ring is due to scattering from a specific plane of atoms (*e.g.* [100] or [111]).

This diffraction prototype was observed on a phosphor screen. Diffraction patterns for clusters are much weaker and must be detected with a sensitive electronic detector (a CCD chip, or in our new apparatus, a Reticon Linear Diode Array).

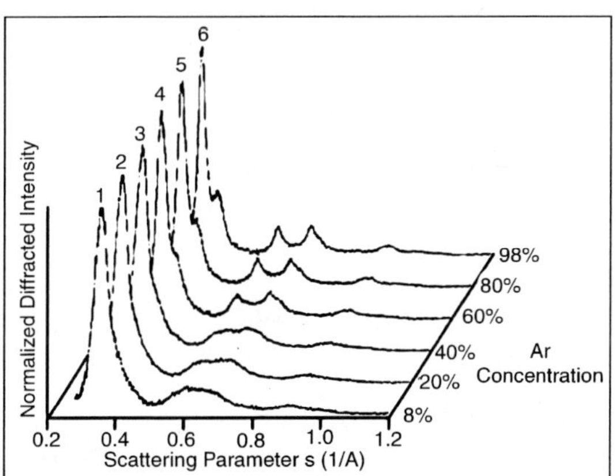

A set of diffraction patterns from lead clusters. These patterns are cross-sections along a diameter of the powder diffraction pattern immediately above.

As the argon concentration in the source chamber increases the particle size increases and more well developed structure is observed in the diffraction patterns. Pattern 1 is similar to that from a simple geometrical model of an icosahedral particle, while pattern 6 is similar to calculated diffraction patterns for decahedral particles.

Experimental

A shoot of the inert gas aggregation source used to generate atomic clusters. Although the image is not very clear, the cluster source chamber can be seen at bottom left and the column of a converted scanning electron microscope is on top of the main vacuum chamber. The converted SEM is used to produce a stream of electrons which are scattered from the clusters - the scattered intensity allows us to determine the structure of the clusters. The turbo pumps at right are used to achieve the low pressures needed in the vacuum chamber.

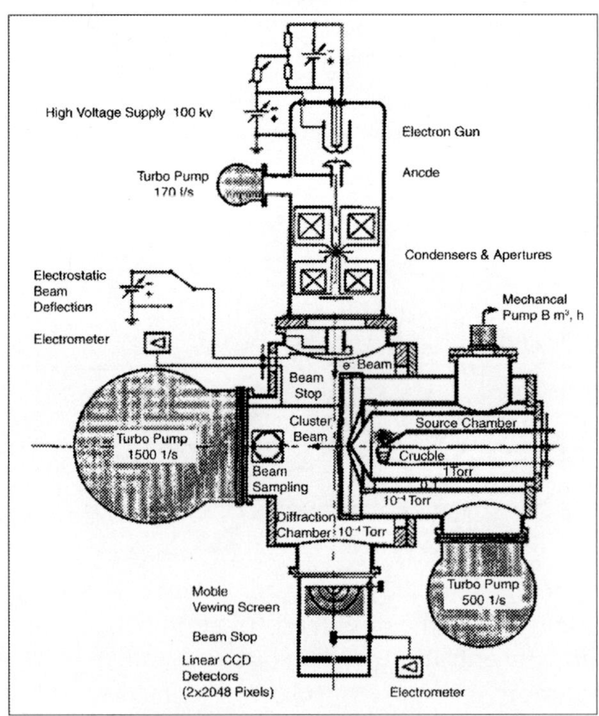

Comprehensive drawing of the inert gas aggregation source used to generate atomic clusters, together with the high vacuum and electron diffraction apparatus.

Rear view of the cluster production and electron scattering apparatus. Again the column of the converted scanning electron microscope can be seen on top of the main vacuum chamber.

A HYBRID NANO-ENERGY HARVESTER

The device harnesses both sunlight and mechanical energy.

- Thursday, April 9, 2009
- By Katherine Bourzac.

Nanoscale generators can turn ambient mechanical energy—vibrations, fluid flow, and even biological movement—into a power source. Now researchers have combined a nanogenerator with a solar cell to create an integrated mechanical- and solar-energy-harvesting device. This hybrid generator is the first of its kind and might be used, for instance, to power airplane sensors by capturing sunlight as well as engine vibrations.

Nanogenerators typically use piezoelectric nanowires—hairlike zinc oxide structures that generate an electrical potential when mechanically stressed—to produce small amounts of power. The first such devices were made by Zhong Lin Wang, a professor at Georgia Tech and director of the institute's Centre for Nanostructure Characterization. Wang hopes that nanogenerators will one day eliminate the need for batteries in implantable medical sensors, and will eventually generate enough power to charge up larger personal electronics.

Compared with solar cells, nanogenerators are still a relatively inefficient way of harvesting energy, says Wang, but "sometimes solar energy isn't available." So he collaborated with Xudong Wang, an assistant professor of materials science and engineering at the University of Wisconsin-Madison, to make the new hybrid device.

It combines two previously developed technologies, both of which rely on zinc oxide nanowires, in a layered silicon substrate. The top layer consists of a thin-film solar cell embedded with dye-coated zinc oxide nanowires. The large surface area of the nanowires boosts the device's light absorption, a design based on work by Peidong Yang, a professor of chemistry at the University of California, Berkeley. The bottom layer contains Wang's nanogenerator. On the underside of the silicon is a jagged array of polymer-coated zinc oxide nanowires in a toothlike arrangement. When the device is exposed to vibrations, these "teeth" scrape against an underlying array of vertically aligned zinc oxide nanowires, creating an electrical potential.

The solar cell and the nanogenerator are electrically associated by the silicon substrate itself, which acts as both the anode of the solar cell and the cathode of the nanogenerator. It is possible to string together large groups of solar cells and nanogenerators, but having them integrated in a single system takes up less space and is also more energy efficient. The prototype device can generate 0.6 volts of solar power and 10 millivolts of piezoelectric power. While the prototype device had only one nanogenerator, Wang expects to increase the power output by creating devices with multiple layers of nanogenerators. He says that a likely first application of these devices might be in sensor-laden military aircraft. The U.S. Air Force recently issued a call for research funding proposals related to hybrid energy-scavenging devices.

Charles Lieber, a professor of chemistry at Harvard University, says that Wang's device is "creative" and is, to his knowledge, the first hybrid nanoscale device capable of harvesting two types of energy. "That is particularly important, given that one is light active, while the other can work in the dark," says Lieber. He expects Wang's work to inspire other researchers to focus on hybrid nanogenerator devices, as well as on devices that combine nanogenerators with "complementary nano-enabled power storage."

WHY NANO-ENERGY?

We believe that nanotechnology provides our best hope for a new, cleaner and more efficient energy system, one that will form the backbone of our future energy industry.

Nanotechnology is science at the smallest scale, and allows us to manipulate the fundamental building blocks of all matter – atoms. By designing and tweaking materials at this level scientists can create super materials with incredible new chemical, electrical and physical properties –

properties never before thought possible. Nanotechnology is by now being used to produce stain-resistant fabrics, odour-eating socks even fire-resistant paints, but its real potential is in the field of energy. By re-engineering the way we generate, store, transmit and use energy at the nano-scale, we will be able to create energy technologies that are far cleaner, more efficient and therefore cheaper, than any of the fossil fuel or alternative energy technologies around today.

And now is the time to invest. Nanotechnology is a relatively new science – the first major breakthroughs took place in the mid 1980's and since then work has focused on developing the tools and equipment needed to work at the nano-scale. But now industrial applications are starting to multiply and the industry is set to explode over the next few years. Nano-energy in particular has received little attention to-date but is increasingly talked about as the most significant area for development. We believe that over the next 4-5 years, the sort of comprehensive nano-energy portfolio we are developing will prove to be exceptionally valuable.

MOLECULAR NANOTECHNOLOGY

Molecular Nanotechnology (MNT) is nanotechnology using "molecular manufacturing", an anticipated technology based on positionally-controlled mechanosynthesis guided by molecular machine systems. It involves combining physical principles demonstrated by chemistry, other nanotechnologies, and the molecular machinery of life with the systems engineering principles found in modern macroscale factories. Its most well-known exposition is in the books of K. Eric Drexler.

Ralph Merkle has compared today's chemistry (in contrast to mechanosynthesis) to an attempt to build interesting Lego brick constructions while wearing boxing gloves. Because conventional chemistry has no tools that allow us to place a particular molecule in a particular place (so that it bonds in a predictable way), we must work with randomly moving molecules. As a result, when we cause a particular chemical reaction, we frequently get a mix of several different product species. We must often follow up after the reaction with a physical filtering process to extract the species we actually wanted, with the other species discarded as waste. Nanotechnology could therefore offer much cleaner manufacturing processes than today's bulk technology offers.

FORESIGHTS

Confused Terminology

Drexler has noted that some British writers have applied the term "nanotechnology" to microscale technologies like MEMS. This prompted Drexler to use the term "molecular nanotechnology".

Recently, the term "nanotechnology" is applied to the currently available fine-scale chemistry or materials science or molecular engineering. That would include "nanotech" suntan lotion and "nanotech" stain-resistant pants. This has prompted Drexler to use the term "zettatechnology" to refer to molecular manufacturing and its products. The "zetta" prefix is a reference to number of atoms in macro-sized product, unlike the Nano prefix for the number of subdivisions of a meter. Such a product is likely to have around a sextillion (10^{21}) distinct atomic parts.

HYPOTHETICAL APPLICATIONS AND CAPABILITIES

Smart Materials and Nanosensors

One submission of nanotechnology is the development of so-called smart materials. This term refers to any sort of material designed and engineered at the nanometre scale to perform a specific task, and encompasses a wide variety of possible commercial applications. One example is materials designed to respond differently to various molecules; such a capability could lead, for example, to artificial drugs which would recognize and render inert specific viruses. Another is the idea of self-healing structures, which would repair small tears in a surface naturally in the same way as self-sealing tires or human skin; and while this technology is relatively new, it is already seeing commercial application in various engineering plastics.

A nanosensor would resemble a smart material, involving a small component within a larger machine that would react to its environment and change in some fundamental, intentional way. As a very simple example: a photosensor could passively measure the incident light and discharge its absorbed energy as electricity when the light passes above or below a specified threshold, sending a signal to a larger machine. Such a sensor would cost less and use less power than a conventional sensor, and yet function usefully in all the same applications — for example, turning on parking lot lights when it gets dark. While smart materials and nanosensors both exemplify useful applications of nanotechnology, they pale in comparison with the complexity of the technology most popularly associated with the term: the replicating nanorobot.

Replicating Nanobots

Nanofacturing is popularly linked with the idea of swarms of coordinated nanoscale robots working together, as proposed by Drexler in his 1986 popular discussions of the subject. In theory, nanobots could construct more nanobots.

However, critics doubt the feasibility of controllable self-replicating nanobots: they cite the possibility of mutations removing any control and favouring reproduction of mutant pathogenic variations. Advocates counter that bacteria are (of necessity) evolved to evolve, while nanobot mutation can be actively prevented by common error-correcting techniques. Similar ideas are advocated in the Foresight Guidelines on Molecular Nanotechnology.

Recent technical proposals for nanofactories do not include self-replicating nanobots, and recent ethical guidelines prohibit self-replication.

Medical Nanorobots

One of the most important applications of molecular nanotechnology will be medical nanorobotics or nanomedicine. The ability to design, build, and deploy large numbers of medical nanorobots will make possible the rapid elimination of disease and the reliable and relatively painless recovery from physical trauma. Medical nanorobots will also make possible the convenient correction of genetic defects, and can help to ensure a greatly expanded healthspan. More controversially, medical nanorobots could be used to augment natural human capabilities. However, mechanical medical nanodevices will not be allowed (or designed) to self-replicate inside the human body, nor will medical nanorobots have any need for self-replication themselves since they will be manufactured exclusively in carefully regulated nanofactories.

Utility Fog

Another proposed application of nanotechnology involves utility fog—in which a cloud of networked microscopic robots (simpler than assemblers) changes its shape and properties to form macroscopic objects and tools in accordance with software commands. Rather than modify the current practices of consuming material goods in different forms, utility fog would simply replace most physical objects.

HYPOTHETICAL SOCIAL IMPACTS

In spite of the current early developmental status of nanotechnology, much concern surrounds its anticipated impact on economics and on law.

Some conjecture that nanotechnology would elicit a strong public-opinion backlash, as has occurred recently around genetically modified plants and the prospect of human cloning. Whatever the exact effects, nanotechnology would probably upset existing economic structures, as it should reduce the scarcity of manufactured goods and make many more goods (such as food and health aids) manufacturable.

The majority futurists and all economists believe that future citizens of a nanotechnological society would still need money, in the form of unforgeable digital cash or physical specie. They might use such money to buy goods and services that are unique, or limited within the solar system. These might include: matter, energy, information, real estate, design services, entertainment services, legal services, fame, political power, or the attention of other people to your political/religious/philosophical message. Furthermore, futurists must consider war, even between prosperous states, and non-economic goals.

A number of resources will remain limited, because unique physical objects are limited (a plot of land in the real Jerusalem, mining rights to the larger near-earth asteroids) or because they depend on the goodwill of a particular person (the love of a famous person, a painting from a famous artist). Demand will always exceed supply for some things, and a political economy may continue to exist in any case. Whether the interest in these limited resources will diminish with the advent of virtual reality, where they can be easily substituted, is yet unclear, although the only reason why it might not is hypothetical irrational preference towards "real thing".

Risks

Further than the fantasy scenarios, nanotechnology has daunting risks. It enables cheaper and more destructive conventional weapons. Also, nanotechnology permits weapons of mass destruction that self-replicate, as viruses and cancer cells do when attacking the human body. Commentators generally agree that humankind should permit self-replication only under very controlled or "inherently safe" conditions.

A fright exists that nanomechanical robots, if designed to self-replicate using naturally occurring materials (a difficult task), could consume the entire planet in their hunger for raw materials, or simply crowd out natural life, out-competing it for energy (as happened historically when blue-green algae appeared and outcompeted earlier life forms). Some commentators sometimes refer to this situation as the "grey goo" or "ecophagy" scenario. K. Eric Drexler considers an accidental "grey goo" scenario extremely unlikely. The "grey goo" scenario begs the Tree Sap Answer: what chances exist that one's car could spontaneously mutate into a wild car, run off-road and live in the forest off tree sap? In light of these dangers, the Foresight Institute (founded by K. Eric Drexler to prepare for the arrival of future technologies) has drafted a set of guidelines for the ethical development of nanotechnology. These include the banning of free-foraging self-replicating pseudo-organisms on the Earth's surface, at least, and possibly in other places.

Electron Microscope

The electron microscope is a category of microscope that uses electrons to create an image of the target.

TYPES OF ELECTRON MICROSCOPY

The electron microscope is a category of microscope that uses a beam of electrons to create an image of the specimen. It is capable of much higher magnifications and has a greater resolving power than a light microscope. They are large, expensive pieces of equipment, generally standing alone in a small, specially designed room and requiring trained personnel to operate them.

THE HISTORY OF ELECTRON MICROSCOPY

By the centre of the 19th century, microscopists had accepted that it was simply not possible to resolve structures of less than half a micrometre with a light microscope because of the Abbe's formula, but the development of the cathode ray tube was literally about to change the way they looked at things; by using electrons instead of light! Hertz (1857-94) suggested that cathode rays were a form of wave motion and Weichert, in 1899, found that these rays could be concentrated into a small spot by the use of an axial magnetic field produced by a long solenoid. But it was not until 1926 that Busch showed theoretically that a short solenoid converges a beam of electrons in the same way that glass can converge the light of the sun, that a direct comparison was made between light and electron beams. Busch should probably therefore be known as the father of electron optics.

In 1931 the German engineers Ernst Ruska and Maximillion Knoll succeeded in magnifying and electron image. This was, in retrospect, the moment of the invention of the electron microscope but the first prototype was actually built by Ruska in 1933 and was capable of resolving to 50 nm. Although it was primitive and not really fit for practical use, Ruska was recognised some 50 years later by the award of a Nobel Prize. The first commercially available electron microscope was built in England by Metropolitan Vickers for Imperial College, London, and was called the EM1, though it never surpassed the resolution of a good optical microscope. The early electron microscopes did not excite the optical microscopists because the electron beam, which had a very high current density, was concentrated into a very small area and was very hot and therefore charred any non-metallic specimens that were examined. When it was found that you could successfully examine biological specimens in the electron microscope after treating them with osmium and cutting very thin slices of the sample, the electron microscope began to appear as a viable proposition.

At the University of Toronto, in 1938, Eli Franklin Burton and students Cecil Hall, James Hillier and Albert Prebus constructed the first electron microscope in the New World. This was an effective, high-resolution instrument, the design of which eventually led to what was to become known as the RCA (Radio Corporation of America) range of very successful microscopes.

Unfortunately, the outbreak of the Second World War in 1939 held back their further development somewhat, but within 20 years of the end of the war routine commercial electron microscopes were capable of 1 nm resolution.

USE OF ELECTRON MICROSCOPES

All electron microscopes use electromagnetic and/or electrostatic lenses to control the path of electrons. Glass lenses, used in light microscopes, have no effect on the electron beam. The basic design of an electromagnetic lens is a solenoid (a coil of wire around the outside of a tube) through which one can pass a current, thereby inducing an electromagnetic field. The electron beam passes through the centre of such solenoids on its way down the column of the electron microscope towards the sample. Electrons are very sensitive to magnetic fields and can therefore be controlled by changing the current through the lenses.

The faster the electrons travel, the shorter their wavelength. The resolving power of a microscope is directly related to the wavelength of the irradiation used to form an image. Reducing wavelength increases resolution. Therefore, the resolution of the microscope is increased if the accelerating voltage of the electron beam is increased. The accelerating voltage of the beam is quoted in kilovolts (kV). It is now possible to purchase a 1,000kV electron microscope, though this is not commonly found.

Although modern electron microscopes can magnify objects up to about two million times, they are still based upon Ruska's prototype and the correlation between wavelength and resolution. The electron microscope is an integral part of many laboratories such as The John Innes Centre. Researchers can use it to examine biological materials (such as microorganisms and cells), a variety of large molecules, medical biopsy samples, metals and crystalline structures, and the characteristics of various surfaces. Nowadays, electron microscopes have many other uses outside research. They can be used as part of a production line, such as in the fabrication of silicon chips, or within forensics laboratories for looking at samples such as gunshot residues. In the arena of fault diagnosis and quality control, they can be used to look for stress lines in engine parts or simply to check the ratio of air to solids in ice cream!

TRANSMISSION ELECTRON MICROSCOPE (TEM)

The original shape of electron microscopy, Transmission electron microscopy (TEM) involves a high voltage electron beam emitted by a cathode and formed by magnetic lenses. The electron beam that has been partially transmitted through the very thin (and so semitransparent for electrons) specimen carries information about the structure of the specimen. The spatial variation in this information (the "image") is then magnified by a series of magnetic lenses until it is recorded by hitting a fluorescent screen, photographic plate, or light sensitive sensor such as a CCD (charge-coupled device) camera. The image detected by the CCD may be displayed in real time on a monitor or computer.

Transmission electron microscopes produce two-dimensional, black and white images. Declaration of the TEM is also limited by spherical and chromatic aberration, but a new generation of aberration correctors has been able to overcome or limit these aberrations.

Software correction of spherical aberration has allowed the production of images with sufficient resolution to show carbon atoms in diamond separated by only 0.089 nm and atoms in silicon at 0.078 nm at magnifications of 50 million times.

The ability to determine the positions of atoms within materials has made the TEM an indispensable tool for nanotechnologies research and development in many fields, including heterogeneous catalysis and the development of semiconductor devices for electronics and photonics. In the life sciences, it is still mainly the specimen preparation which limits the resolution of what we can see in the electron microscope, rather than the microscope itself.

JIC have a high voltage (200kV) TEM, which was installed in 2008. There are two digital cameras on it, one is higher resolution than the other, so that the need for developing and printing film has been negated. Our TEM is designed for use with biological samples and is capable of resolving to better than 1nm. It is also capable of 3-D tomography which involves taking a succession of images whilst tilting the specimens through increasing angles, which can then be combined to form a three-dimensional image of the specimen.

SCANNING ELECTRON MICROSCOPE (SEM)

Different the TEM, where the electrons in the primary beam are transmitted through the sample, the Scanning Electron Microscope (SEM) produces images by detecting secondary electrons which are emitted from the surface due to excitation by the primary electron beam. In the SEM, the electron beam is scanned across the surface of the sample in a raster pattern, with detectors building up an image by mapping the detected signals with beam position.

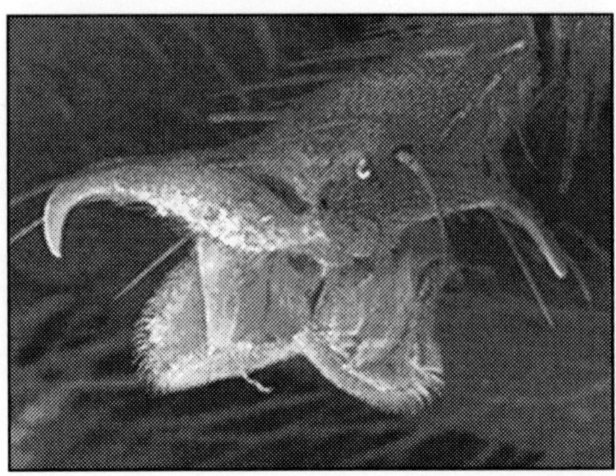

Figure: SEM image of a fly's foot taken at JIC in 2006.

TEM resolution is about an order of magnitude better than the SEM resolution. Our TEM can easily resolve details of 0.2 nm. Our two SEMs at JIC are both relatively recent acquisitions and are high-resolution instruments capable of about 2 nm resolution on biological samples. Because the SEM image relies on electron interactions at the surface rather than transmission it is able to image bulk samples and has a much greater depth of view, and so can produce images that are a good representation of the 3D structure of the sample. SEM images are therefore considered to provide us with 3D, topographical information about the sample surface but will still always be only in black and white. In the SEM, use much lower accelerating voltages to prevent beam penetration into the sample since what we require is generation of the secondary electrons from the true surface structure of a sample. Therefore, it is common to use low KV, in the range 1-5 kV for biological samples, even though our SEMs are capable of up to 30 kV. At JIC we currently have two SEMs, both with high-resolution capabilities, digital imaging facilities and cryo-systems which enable them to be used for looking at frozen-hydrated specimens.

SAMPLE PREPARATION

Materials to be view in an electron microscope generally require processing to produce a suitable sample. This is mainly because the whole of the inside of an electron microscope is under high vacuum in order to enable the electron beam to travel in straight lines. The technique required varies depending on the specimen, the analysis required and the type of microscope:

Cryofixation - freezing a specimen rapidly, typically to liquid nitrogen temperatures or below, that the water forms ice. This preserves the specimen in a snapshot of its solution state with the minimal of artefacts. An entire field called cryo-electron microscopy has branched from this technique. With the development of cryo-electron microscopy, it is now possible to observe virtually any biological specimen close to its native state.

Fixation - a general term used to describe the process of preserving a sample at a moment in time and to prevent further deterioration so that it appears as close as possible to what it would be like in the living state, although it is now dead. In chemical fixation for electron microscopy, glutaraldehyde is often used to crosslink protein molecules and osmium tetroxide to preserve lipids.

Dehydration - removing water from the samples. The water is generally replaced with organic solvents such as ethanol or acetone as a stepping stone towards total drying for SEM specimens or infiltration with resin and subsequent embedding for TEM specimens.

Embedding - infiltration of the tissue with wax (for light microscopy) or a resin (for electron microscopy) such as araldite or LR White, which can then be polymerised into a hardened block for subsequent sectioning.

Sectioning - the production of thin slices of the specimen. For light microscopy, the sections can be a few micrometres thick but for electron microscopy they must be very thin so that they are semitransparent to electrons, typically around 90 nm. These ultrathin sections for electron microscopy are cut on an ultramicrotome with a glass or diamond knife. Glass knives can easily be made in the laboratory and are much cheaper than diamond, but they blunt very quickly and therefore need replacing frequently.

Staining - uses heavy metals such as lead and uranium to scatter imaging electrons and thus give contrast between different structures, since many (especially biological) materials are nearly "transparent" to the electron beam. By staining the samples with heavy metals, we add electron density to it which results in there being more interactions between the electrons in the primary beam and those of the sample, which in turn provides us with contrast in the resultant image. In biology, specimens can be stained "en bloc" before embedding and/or later, directly after sectioning, by brief exposure of the sections to solutions of the heavy metal stains.

Freeze-fracture and freeze-etch - a preparation technique particularly useful for examining lipid membranes and their incorporated proteins in "face on" view. The fresh tissue or cell suspension is frozen rapidly (cryofixed), then fractured by simply breaking or by using a microtome while maintained at liquid nitrogen temperature. The cold, fractured surface is generally "etched" by increasing the temperature to about -95°C for a few minutes to let some surface ice sublime to reveal microscopic details. For the SEM, the sample is now ready for imaging. For the TEM, it can then be rotary-shadowed with evaporated platinum at low angle (typically about 6°) in a high vacuum evaporator. A second coat of carbon, evaporated perpendicular to the average surface plane is generally performed to improve stability of the replica coating. The specimen is returned to room temperature and pressure, and then the extremely fragile "shadowed" metal replica of the fracture surface is released from the underlying biological material by careful chemical digestion with acids, hypochlorite solution or SDS detergent. The floating replica is thoroughly washed from residual chemicals, carefully picked up on an EM grid, dried then viewed in the TEM.

Sputter Coating - an ultrathin coating of electrically-conducting material, deposited by low vacuum coating of the sample. This is done to prevent charging of the specimen which would occur because of the accumulation of static electric fields due to the electron irradiation required during imaging. It also increases the amount of secondary electrons that can be detected from the surface of the sample in the SEM and therefore increases the signal to noise ratio. Such coatings include gold, gold/palladium, platinum, chromium etc.

DISADVANTAGES OF ELECTRON MICROSCOPY

Electron microscopes are very exclusive to buy and maintain. They are dynamic rather than static in their operation: requiring extremely stable high voltage supplies, extremely stable currents to each electromagnetic coil/lens, continuously-pumped high/ultra-high vacuum systems and a cooling water supply circulation through the lenses and pumps. As they are very sensitive to vibration and external magnetic fields, microscopes aimed at achieving high resolutions must be housed in buildings with special services.

A significant amount of training is required in order to operate an electron microscope successfully and electron microscopy is considered a specialised skill.

The samples have to be viewed in a vacuum, as the molecules that make up air would scatter the electrons. This means that the samples need to be specially prepared by sometimes lengthy and difficult techniques to withstand the environment inside an electron microscope. Recent advances have allowed some hydrated samples to be imaged using an environmental scanning electron microscope, but the applications for this type of imaging are still limited.

ATOMIC FORCE MICROSCOPY

The Atomic Force Microscope was developed to conquer a basic drawback with STM - that it can only image conducting or semiconducting surfaces. The AFM, however, has the advantage of imaging almost any type of surface, including polymers, ceramics, composites, glass, and biological samples.

Binnig, Quate, and Gerber invented the Atomic Force Microscope in 1985. Their original AFM consisted of a diamond shard attached to a strip of gold foil. The diamond tip contacted the surface directly, with the interatomic van der Waals forces providing the interaction mechanism. Detection of the cantilever's vertical movement was done with a second tip - an STM placed above the cantilever.

AFM Probe Deflection

Today, most AFMs use a laser wide smile deflection system, introduced by Meyer and Amer, where a laser is reflected from the back of the reflective AFM lever and onto a position-sensitive detector. AFM tips and cantilevers are microfabricated from Si or Si_3N_4. Typical tip radius is from a few to 10s of nm.

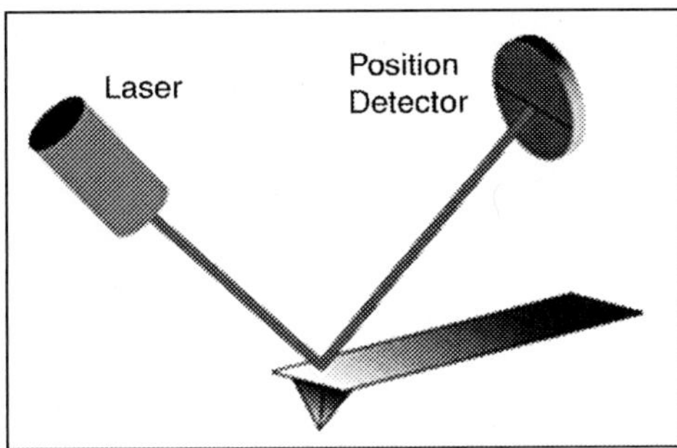

Figure: Beam deflection system, using a laser and photodector to measure the beam position.

Measuring Forces

Because the atomic force microscope relies on the forces between the tip and sample, knowing these forces is important for proper imaging. The force is not measured directly, but calculated by measuring the deflection of the lever, and knowing the stiffness of the cantilever. Hook's law gives F = -kz, where F is the force, k is the stiffness of the lever, and z is the distance the lever is bent.

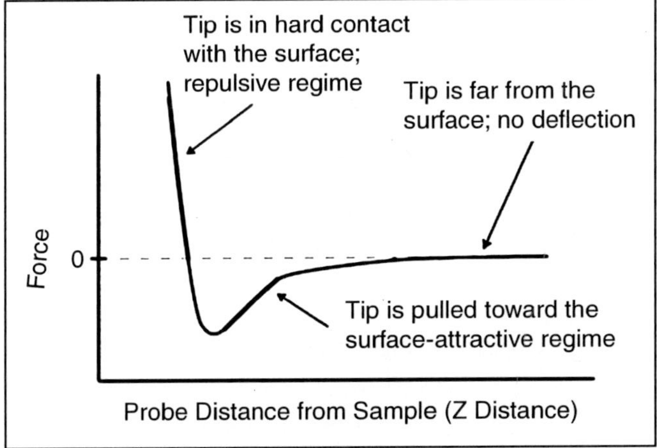

AFM Modes of Operation

Because of AFM's adaptability, it has been applied to a large number of research topics. The Atomic Force Microscope has also gone through many modifications for specific application requirements.

Contact Mode

The first and foremost mode of operation, contact mode is widely used. As the tip is raster-scanned across the surface, it is deflected as it moves over the surface corrugation. In constant force mode, the tip is constantly adjusted to maintain a constant deflection, and therefore constant height above the surface. It is this adjustment that is displayed as data. However, the ability to track the surface in this manner is limited by the feedback circuit. Sometimes the tip is allowed to scan without this adjustment, and one measures only the deflection. This is useful for small, high-speed atomic resolution scans, and is known as variable-deflection mode.

Because the tip is in hard contact with the surface, the stiffness of the lever needs to be less that the effective spring constant holding atoms together, which is on the order of 1 - 10 nN/nm. Most contact mode levers have a spring constant of < 1 N/m.

Lateral Force Microscopy

LFM measures frictional forces on a surface. By measuring the "twist" of the cantilever, rather than merely its deflection, one can qualitatively determine areas of higher and lower friction.

Noncontact Mode

Noncontact mode belongs to a family of AC modes, which refers to the use of an oscillating cantilever. A stiff cantilever is oscillated in the attractive regime, meaning that the tip is quite close to the sample, but not touching it (hence, "noncontact"). The forces between the tip and sample are quite low, on the order of pN (10^{-12} N). The detection scheme is based on measuring changes to the resonant frequency or amplitude of the cantilever.

Dynamic Force/Intermittant-contact/"tapping mode" AFM. Commonly referred to as "tapping mode" it is also referred to as intermittent-contact or the more general term Dynamic Force Mode (DFM).

A stiff cantilever is oscillated closer to the sample than in noncontact mode. Part of the oscillation extends into the repulsive regime, so the tip intermittently touches or "taps" the surface. Very stiff cantilevers are typically used, as tips can get "stuck" in the water contamination layer.

The advantage of tapping the surface is improved lateral resolution on soft samples. Lateral forces such as drag, common in contact mode, are virtually eliminated. For poorly adsorbed specimens on a substrate surface the advantage is clearly seen.

Force Modulation

Force modulation refers to a technique used to probe properties of materials through sample/tip interactions. The tip (or sample) is oscillated at a high frequency and pushed into the repulsive regime. The slope of the force-distance curve is measured which is correlated to the sample's elasticity. The data can be acquired along with topography, which allows comparison of both height and material properties.

Phase Imaging

In Phase mode imaging, the phase shift of the oscillating cantilever relative to the driving signal is measured.

This phase shift can be correlated with specific material properties that effect the tip/sample interaction. The phase shift can be used to differentiate areas on a sample with such differing properties as friction, adhesion, and viscoelasticity. The techniques is used simultaneously with DFM mode, so topography can be measured as well.

GENERAL CONCEPT AND DEFINING CHARACTERISTICS OF AFM

Scanned-proximity search microscopes provide very high resolution images of various sample properties: The atomic force microscope is one of about two dozen types of scanned-proximity probe microscopes.

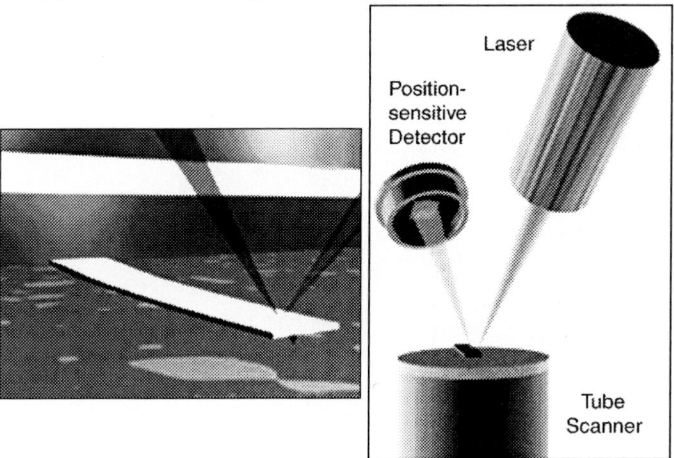

Figure : Concept of AFM and the optical lever: (left) a cantilever touching a sample; (right) the optical lever. Scale drawing; the tube scanner measures 24 mm in diameter, while the cantilever is 100 m long.

All of these microscopes work by measuring a local property - such as height, optical absorption, or magnetism - with a probe or "tip" placed very close to the sample. The small probe-sample separation (on the order of the instrument's resolution) makes it possible to take measurements over a small area.

To acquire an image the microscope raster-scans the probe over the sample while measuring the local property in question. The resulting image resembles an image on a television screen in that both consist of many rows or lines of information placed one above the other. Unlike traditional microscopes, scanned-probe systems do not use lenses, so the size of the probe rather than diffraction effects generally limit their resolution.

AFM operates by measure attractive or repulsive forces between a tip and the sample. In its repulsive "contact" mode, the instrument lightly touches a tip at the end of a leaf spring or "cantilever" to the sample.

As a raster-scan drags the tip over the sample, some sort of detection apparatus measures the vertical deflection of the cantilever, which indicates the local sample height.

Thus, in contact mode the AFM measures hard-sphere repulsion forces between the tip and sample. In noncontact mode, the AFM derives topographic images from measurements of attractive forces; the tip does not touch the sample.

Because it does not allow the imaging of samples under water. AFMs can achieve a resolution of 10 pm, and unlike electron microscopes, can image samples in air and under liquids. In principle, AFM resembles the record player as well as the stylus profilometer. However, AFM incorporates a number of refinements that enable it to achieve atomic-scale resolution:

- Sensitive detection
- Flexible cantilevers
- Sharp tips
- High-resolution tip-sample positioning
- Force feedback.

BASIC PRINCIPLES OF AFM

Scanning force microscopy is based on the survival of a separation-dependency force between any two bodies. It is the force between the tip and the substrate that is present at close separations.

Typically, pyramidal silicon nitride tips are used, which have a radius of curvature on the order of $100A^{0.}$ These are made by etching process that removes silicon from the substrate, leaving an etched or sharpened tip behind. The force is detected by placing the top on a flexible cantilever that deflects proportionally to the exerted force.

The deflection is then measured by some convenient procedure, such as laser deflection or some other device. Actually, the main innovation may be seen as being a copy of the principle behind the record player.

MODES OF OPERATION

Contact Mode

In contact AFM, the tip is in perpetual contact with the sample. The tip is attached to the end of a cantilever with a low spring constant, lower than the effective spring constant holding the atoms of most solid samples together.

As the scanner gently traces the tip across the sample (or the sample under the tip), the contact force causes the cantilever to bend and the Z-feedback loop works to maintain a constant cantilever deflection.

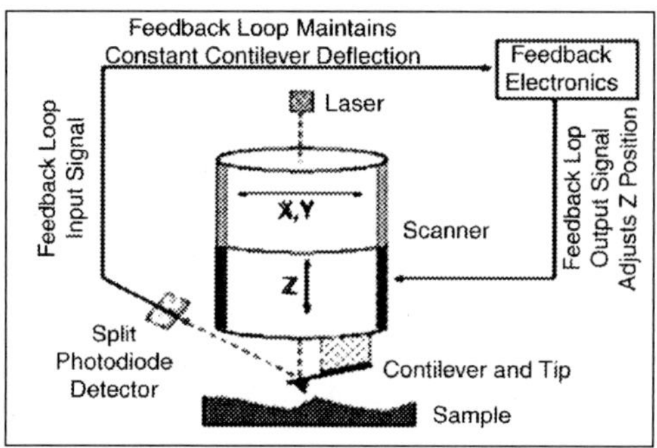

Conductive AFM Option (C-AFM)

Conductive Atomic Force Microscopy (CAFM) is a secondary imaging mode derived from contact AFM that characterizes conductivity variations across medium- to low-conducting and semiconducting materials. CAFM performs general-purpose measurements, and has a current range of 2 pA to 1A. CAFM employs a conductive probe tip.

Typically, a DC bias is applied to the tip, and the sample is held at ground potential. While the z feedback signal is used to generate a normal contact AFM topography image, the current passing between the tip and sample is measured to generate the conductive AFM image.

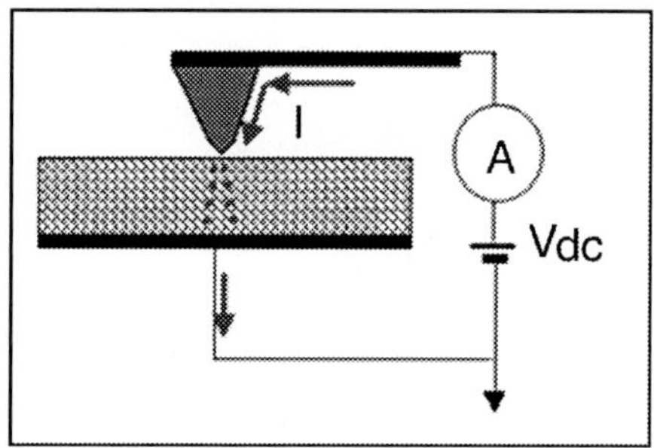

Figure: Current flow in conductive AFM (CAFM).

Lateral Force Microscopy (LFM)

Lateral Force Microscopy (LFM) is a secondary get in touch with AFM mode that detects and maps relative differences in the frictional forces between the probe tip and the sample surface. In LFM, the scanning is always perpendicular to the long axis of the cantilever. Forces on the cantilever that are parallel to the plane of the sample surface cause twisting of the cantilever around its long axis. This twisting is measured by a quad-cell Position Sensitive PhotoDetector (PSPD), as with TRmode.

AFM Tip Lateral Movement in LFM

Twisting of the cantilever usually arises from two sources: changes in surface friction and changes in topography. In the first case, the tip may experience greater friction as it traverses some areas, causing the cantilever to twist more. In the second case, the cantilever may twist when it encounters edges of topographical features. To separate one effect from the other, usually three signals are collected simultaneously: the trace and retrace LFM signals, and the AFM height (topography) signal.

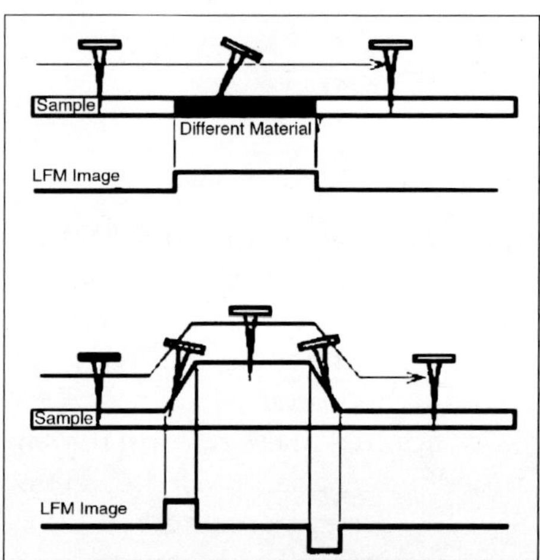

LFM applications comprise identifying transitions between different components in polymer blends and composites, identifying contaminants on surfaces, delineating coverage by coatings, and chemical force microscopy (CFM) using probe tips functionalized for specific chemical or biological species.

Magnetic Force Microscopy (MFM)

Magnetic Force Microscopy (MFM) is a secondary imaging mode derived from TappingMode mode that maps magnetic force gradient above the sample surface. This is performed through a patented two-pass technique, LiftMode. LiftMode separately measures topography and another selected property (magnetic force, electric force, etc.) using the topographical information to track the probe tip at a constant height (Lift Height) above the sample surface during the second pass.

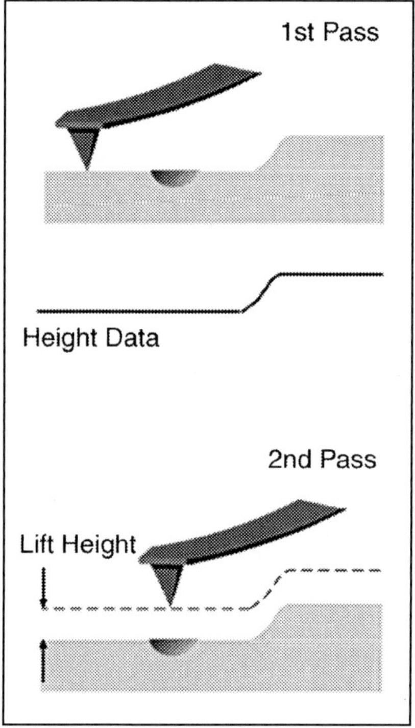

Lift Mode AFM.

Scanning Tunnelling Microscopy (STM)

Scanning Tunnelling Microscopy (STM) measures topography of outside electronic states using a tunnelling current that is dependent on the separation between the probe tip and a sample surface. STM is typically performed on conductive and semiconductive surfaces. Common applications consist of atomic resolution imaging, electrochemical STM, Scanning Tunnelling Spectroscopy (STS), and low current imaging of less conductive samples.

TappingMode

TappingMode AFM, the most commonly used of all AFM modes, that maps topography by lightly tapping the surface with an oscillating probe tip. The cantilever's oscillation amplitude changes with sample surface topography, and the topography image is obtained by monitoring these changes and closing the z feedback loop to minimize them.

TappingMode has become an important AFM technique, as it overcomes some of the limitations of both contact and non-contact AFM. By eliminating lateral forces that can damage soft samples and reduce image resolution, TappingMode allows routine imaging of samples once considered impossible to image with AFM, especially in contact mode. Another major advantage of TappingMode is related to limitations that can arise due to the thin layer of liquid that forms on most sample surfaces in an ambient imaging environment, *i.e.,* in air or some other gas. The amplitude of the cantilever oscillation in TappingMode is typically on the order of a few 10's of nanometers, which ensures that the tip does not get stuck in this liquid layer. The amplitude used in non-contact AFM is much smaller, as different forces are being measured. As a result, the non-contact tip often gets stuck in the liquid layer unless the scan is performed at a very slow speed.

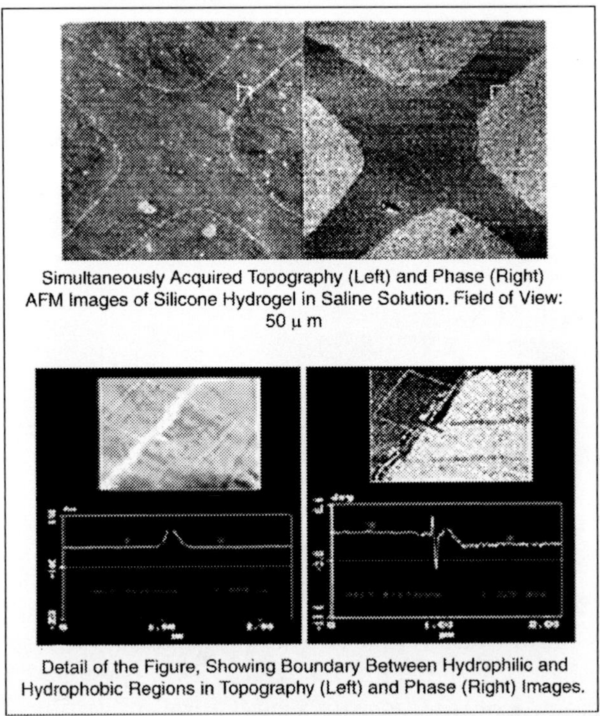

Simultaneously Acquired Topography (Left) and Phase (Right) AFM Images of Silicone Hydrogel in Saline Solution. Field of View: 50 μ m

Detail of the Figure, Showing Boundary Between Hydrophilic and Hydrophobic Regions in Topography (Left) and Phase (Right) Images.

Phase imaging is a secondary imaging mode derived from TappingMode that goes beyond topographical data to detect variations in composition, adhesion, friction, viscoelasticity, and other properties, including electric and magnetic. Applications include contaminant identification, mapping of components in composite materials, differentiating regions of high and low surface adhesion or hardness and regions of different electrical or magnetic properties. Phase imaging is the mapping of the phase lag between the periodic signal that drives the cantilever and the oscillations of the cantilever. Changes in the phase lag often indicate changes in the properties of the sample surface.

The system's feedback loop operates in the usual manner, using changes in the cantilever's oscillation amplitude to map sample topography. The phase lag is monitored while the topographic image is being taken so that images of topography and material properties can be collected simultaneously.

This shape shows simultaneously acquired topography (left) and phase (right) AFM images of silicone hydrogel in saline solution. The four outer areas were exposed to a sequence of chemical processing steps. The central cross-like region was masked and so protected from the processing steps and hence retained its hydrophobicity.

In the phase image (right), a marked phase shift is clearly seen across the boundaries. However, the hydrophilic and hydrophobic regions show no topographic contrast (left). The phase image is clearly providing material property contrast on this well-defined experimental hydrogel surface.

The phase signal is sensitive to both short- and long-range tip-sample interactions. Short-range interactions include adhesive forces and visco-elastic forces; long-range include electric fields and magnetic fields.

Phase imaging is a key element in the use of a number of scanning probe techniques, including Magnetic Force Microscopy (MFM), Electric Force Microscopy (EFM) and Scanning Capacitance Microscopy (SCM). This technique of detection provides a more sensitive measurement than other detection techniques, such as amplitude detection.

AFM CANTILEVER DEFLECTION MEASUREMENT

Laser light from a solid state diode is reflected off the reverse of the cantilever and collected by a position sensitive detector (PSD) consisting of two closely spaced photodiodes whose output signal is collected by a differential amplifier. Angular displacement of the cantilever results in one photodiode collecting more light than the other photodiode, producing an output signal (the difference between the photodiode signals normalized by their sum) which is proportional to the deflection of the cantilever. It detects cantilever deflections <10 nm (thermal noise limited). A long beam path (several centimetres) amplifies changes in beam angle.

Force Spectroscopy

Another major application of AFM (besides imaging) is force spectroscopy, the direct measurement of tip-sample interaction forces as a function of the gap between the tip and sample (the result of this measurement is called a force-distance curve). For this technique, the AFM tip is extended towards and retracted from the surface as the deflection of the cantilever is monitored as a function of piezoelectric displacement. These measurements have been used to measure nanoscale contacts, atomic bonding, van der Waals forces, and Casimir forces, dissolution forces in liquids and single molecule stretching and rupture forces. Furthermore, AFM was used to measure, in an aqueous environment, the dispersion force due to polymer adsorbed on the substrate. Forces of the order of a few piconewtons can now be routinely measured with a vertical distance resolution of better than 0.1 nanometers. Force spectroscopy can be performed with either static or dynamic modes. In dynamic modes, information about the cantilever vibration is monitored in addition to the static deflection.

Problems with the technique include no direct measurement of the tip-sample separation and the common need for low stiffness cantilevers which tend to 'snap' to the surface. The snap-in can be reduced by measuring in liquids or by using stiffer cantilevers, but in the latter case a more sensitive deflection sensor is needed. By applying a small dither to the tip, the stiffness (force gradient) of the bond can be measured as well.

Identification of Individual Surface Atoms

The AFM can be used to image and manipulate atoms and structures on a variety of surfaces. The atom at the apex of the tip "senses" individual atoms on the underlying surface when it forms incipient chemical bonds with each atom. Because these chemical interactions subtly alter the tip's vibration frequency, they can be detected and mapped. This principle was used to distinguish between atoms of silicon, tin and lead on an alloy surface, by comparing these 'atomic fingerprints' to values obtained from large-scale density functional theory (DFT) simulations.

The trick is to first measure these forces precisely for each type of atom expected in the sample, and then to compare with forces given by DFT simulations. The team found that the tip interacted most strongly with silicon atoms, and interacted 23% and 41% less strongly with tin and lead atoms, respectively. Thus, each different type of atom can be identified in the matrix as the tip is moved across the surface.

ADVANTAGES AND DISADVANTAGES

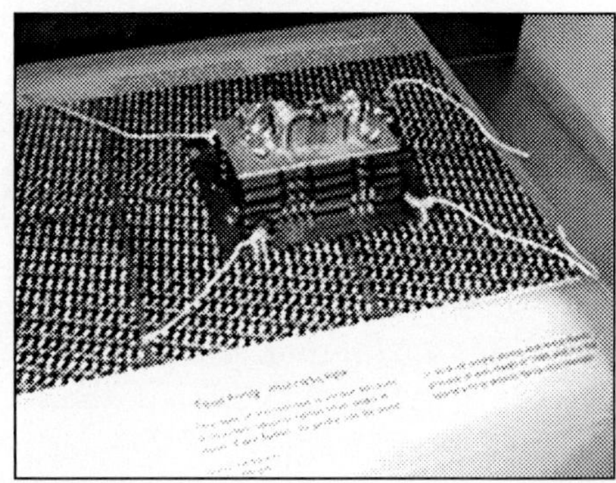

Figure: The first atomic force microscope

Just like any other tool, an AFM's usefulness has limitations. When determining whether or not analysing a sample with an AFM is appropriate, there are various advantages and disadvantages that must be considered.

Advantages

AFM has several advantages over the scanning electron microscope (SEM). Unlike the electron microscope which provides a two-dimensional projection or a two-dimensional image of a sample, the AFM provides a three-dimensional surface profile. Additionally, samples viewed by AFM do not require any special treatments (such as metal/carbon coatings) that would irreversibly change or damage the sample, and does not typically suffer from charging artifacts in the final image. While an electron microscope needs an expensive vacuum environment for proper operation, most AFM modes can work perfectly well in ambient air or even a liquid environment. This makes it possible to study biological macromolecules and even living organisms. In principle, AFM can provide higher resolution than SEM. The atomic resolution in ultra-high vacuum (UHV) and, more recently, in liquid environments. High resolution AFM is comparable in resolution to scanning tunnelling microscopy and transmission electron microscopy. AFM can also be combined with a variety of optical microscopy techniques, further expanding its applicability. Combined AFM-optical instruments have been applied primarily in the biological sciences but has also found a niche in some materials applications, especially those involving photovoltaics research.

Disadvantages

A disadvantage of AFM compared with the scanning electron microscope (SEM) is the single scan image size. In one pass, the SEM can image an area on the order of square millimetres with a depth of field on the order of millimetres, whereas the AFM can only image a maximum height on the order of 10-20 micrometers and a maximum scanning area of about 150 × 150 micrometers. One technique of improving the scanned area size for AFM is by using parallel probes in a fashion similar to that of millipede data storage.

The scanning speed of an AFM is also a limitation. Traditionally, an AFM cannot scan images as fast as a SEM, requiring several minutes for a typical scan, while a SEM is capable of scanning at near real-time, although at relatively low quality. The relatively slow rate of scanning during AFM imaging often leads to thermal drift in the image making the AFM microscope less suited for measuring accurate distances between topographical features on the image. However, several fast-acting designs were suggested to increase microscope scanning productivity including what is being termed videoAFM (reasonable quality images are being obtained with videoAFM at video rate: faster than the average SEM). To eliminate image distortions induced by thermal drift, several techniques have been introduced.

AFM images can in addition be affected by nonlinearity, hysteresis, and creep of the piezoelectric material and cross-talk between the x, y, z axes that may require software enhancement and filtering. Such filtering could "flatten" out real topographical features. However, newer AFMs utilize real-time correction software (for example, feature-oriented scanning) or closed-loop scanners which practically eliminate these problems. Some AFMs also use separated orthogonal scanners (as opposed to a single tube) which also serve to eliminate part of the cross-talk problems. As with any other imaging technique, there is the possibility of image artifacts, which could be induced by an unsuitable tip, a poor operating environment, or even by the sample itself. As depicted on the right. These image artifacts are unavoidable however, their occurrence and effect on results can be reduced through various techniques. Artifacts resulting from a too coarse tip can be caused for example by inappropriate handling or de-facto collisions with the sample by either scanning too fast or having an unreasonably rough surface causing actual wearing of the tip.

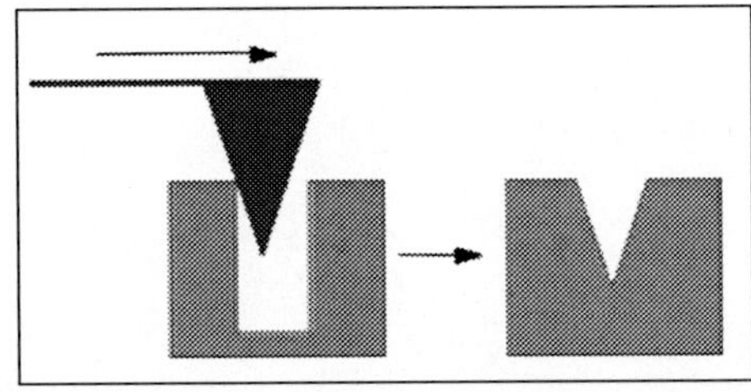

Figure: AFM artifact, steep sample topography

Due to the scenery of AFM probes, they cannot normally measure steep walls or overhangs. Specially made cantilevers and AFMs can be used to modulate the probe sideways as well as up and down (as with dynamic contact and non-contact modes) to measure sidewalls, at the cost of more expensive cantilevers, lower lateral resolution and additional artifacts.

Piezoelectric Scanners

AFM scanners are made from piezoelectric material, which expands and contracts proportionally to an applied voltage. Whether they elongate or contract depends upon the polarity of the voltage applied. The scanner is constructed by combining independently operated piezo electrodes for X, Y, and Z into a single tube, forming a scanner which can manipulate samples and probes with extreme precision in 3 dimensions. Independent stacks of piezos can be used instead of a tube, resulting in decoupled X, Y, and Z movement.

Scanners are characterized by their sensitivity which is the ratio of piezo movement to piezo voltage, *i.e.*, by how much the piezo material extends or contracts per applied volt. Because of differences in material or size, the sensitivity varies from scanner to scanner. Sensitivity varies non-linearly with respect to scan size. Piezo scanners exhibit more sensitivity at the end than at the beginning of a scan. This causes the forward and reverse scans to behave differently and display hysteresis between the two scan directions. This can be corrected by applying a non-linear voltage to the piezo electrodes to cause linear scanner movement and calibrating the scanner accordingly. One disadvantage of this approach is that it requires re-calibration because the precise non-linear voltage needed to correct non-linear movement will change as the piezo ages. This problem can be circumvented by adding a linear sensor to the sample stage or piezo stage to detect the true movement of the piezo. Deviations from ideal movement can be detected by the sensor and corrections applied to the piezo drive signal to correct for non-linear piezo movement. This design is known as a 'closed loop' AFM. Non-sensored piezo AFMs are referred to as 'open loop' AFMs.

The sensitivity of piezoelectric materials decreases exponentially with time. This causes most of the change in sensitivity to occur in the initial stages of the scanner's life. Piezoelectric scanners are run for approximately 48 hours before they are shipped from the factory so that they are past the point where they may have large changes in sensitivity. As the scanner ages, the sensitivity will change less with time and the scanner would seldom require recalibration, though various manufacturer manuals recommend monthly to semi-monthly calibration of open loop AFMs.

SCANNING TUNNELING MICROSCOPY

The development of the family of scanning search microscopes starts with the original invention of the STM in 1981. Gerd Binnig and Heinrich Rohrer developed the first working STM while working at IBM Zurich Research Laboratories in Switzerland. This instrument would later win Binnig and Rohrer the Nobel prize in physics in 1986.

The STM works by scanning a very sharp metal wire tip over a surface. By bringing the tip very close to the surface, and by applying an electrical voltage to the tip or sample, we can image the surface at an extremely small scale – down to resolving individual atoms.

The STM is based on several principles. One is the quantum mechanical effect of tunnelling. It is this effect that allows us to "see" the surface. Another principle is the piezoelectric effect. It is this effect that allows us to precisely scan the tip with angstrom-level control. Lastly, a feedback loop is required, which monitors the tunnelling current and coordinates the current and the positioning of the tip.

Tunnelling

Tunnelling is a quantum mechanical effect. A tunnelling current occurs when electrons move through a barrier that they classically shouldn't be able to move though. In classical terms, if you don't have enough energy to move "over" a barrier, you won't. However, in the quantum mechanical world, electrons have wavelike properties. These waves don't end abruptly at a wall or barrier, but taper off quite quickly. If the barrier is thin enough, the probability function may extend into the next region, though the barrier! Because of the small probability of an electron being on the other side of the barrier, given enough electrons, some will indeed move through and appear on the other side. When an electron moves though the barrier in this fashion, it is called tunnelling.

Quantum mechanics tells us that electrons have both wave and particle like properties. Tunnelling is an effect of the wavelike nature. The top image shows us that when an electron (the wave) hits a barrier, the wave doesn't abruptly end, but tapers off very quickly - exponentially.

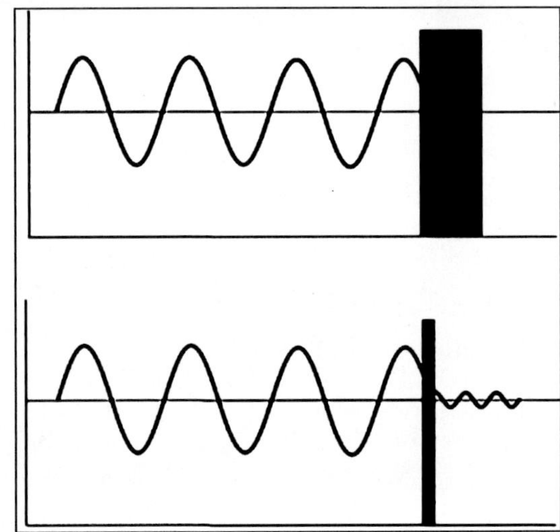

The bottom image shows the scenario if the barrier is quite thin (about a nanometer). Part of the wave does get through, and therefore some electrons may appear on the other side of the barrier..

Because of the sharp decompose of the probability function through the barrier, the number of electrons that will actually do this is very dependent upon the thickness of the barrier. The actual current through the barrier drops off exponentially with the barrier thickness.

To extend this description to the STM: The starting point of the electron is either the tip or sample (depending on the setup of the instrument). The barrier is the gap (air, vacuum, liquid), and the second region is the "other side" – tip or sample, again, depending on the experimental setup. By monitoring the current through the gap, we have very good control of the tip-sample distance.

Piezoelectric Materials

The piezoelectric effect was discovered by Pierre Curie in 1880. The effect is created by squeezing the sides of certain crystals, such as quartz or barium titanate. The result is the creation of opposite charges on the sides. The effect can be reversed as well; by applying a voltage across a piezoelectric crystal, it will elongate or compress. These materials are used to scan the tip in an STM, and most other scanning probe techniques. A typical piezoelectric material used in STMs is PZT (Lead Zirconium Titanate).

Electronics and the Feedback Loop

Obviously, you need electronics to measure the current, scan the tip, and translate this information into a form that we can use. A feedback loop constantly monitors the tunnelling current and makes adjustments to the tip to maintain a constant tunnelling current. These adjustments are recorded by the computer and presented as an image in the STM software. Such a setup is called a "constant current" image. In addition, for very flat surfaces, the feedback loop can be turned off and only the current is displayed. This is a "constant height" image.

The Result

The STM is capable of acquiring remarkable images on the most extreme scale, easily resolving atomic structure in the right environments.

The Quantum Corral

This STM picture shows the direct of standing-wave patterns in the local density of states of the Cu(111) surface. These spatial oscillations are quantum-mechanical interference patterns caused by scattering of the two-dimensional electron gas off the Fe adatoms and point defects.

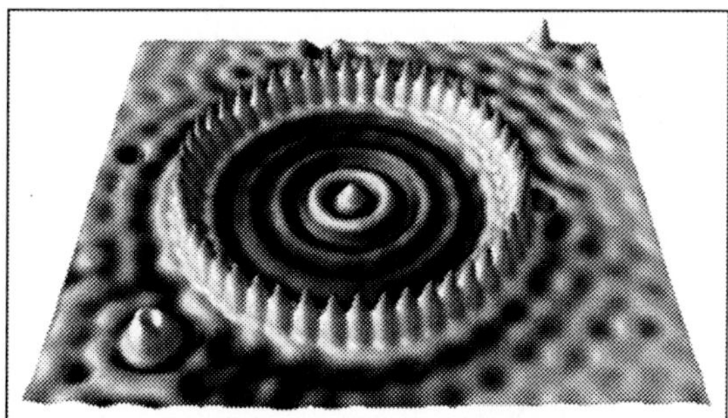

Figure : Courtesy of International Business Machines Corporation. Unauthorized use not permitted

REFERENCES

- Bhatt, R.C. : *Nanostructure and Nanoculture,* Swastik Pub, Delhi, 2009.
- Boucher P. M.: *"Nanotechnology : Legal Aspects."* CRC Press Boca Raton, 2008.
- Parthasarathy, B.K. : *Nanoscience and Nanotechnology,* Isha Books, Delhi, 2007.

3

Fundamentals of Nanoelectronics

Developments in integrated circuits and storage space devices used in computers have proceeded at an exponential rate: at present it takes 2-3 years for each successive halving of the component size. Information storage has followed a similar trend in miniaturization of the size of the bits of magnetized material used in hard disks. However, these technologies have fundamental limits, below which the devices no longer function in a predictable manner. For instance, the oxide layers used in complementary metal oxide semiconductor (CMOS) devices are becoming so thin that they conduct electricity in a quantum-mechanical manner by electron tunnelling. In 1998 it was estimated that microelectronics and magnetic storage technologies would reach their ultimate limits within 10-30 years. Projections for very large scale integration (VLSI) predict that a single chip will accommodate 90 million transistors with a feature size of 70 nanometers and a clock speed of 900 MHz by the year 2010. Currently, many critical dimensions in semiconductor devices are in the 100-nm range, with some insulating layers being tens of nanometers thick.

There is intensive effort to drive miniaturization even further. Miniaturization all the way down to the level of individual atoms and molecules would enable the fabrication of highly dense, fast, and energy-efficient devices. Energy dissipation in current devices can approach 100 W. As devices become more dense, the electronic elements must be designed to be more efficient. Another potential advantage of using individual molecules is that electronic circuitry could be prefabricated using chemical synthesis. This approach would permit techniques such as self-assembly to be used in fabrication, whereby molecules diffuse and dock onto specific connections.

The term nanoelectronics refers to electronic devices in which dimensions are in the range of atoms up to 100 nm. Nanoelectronics is regarded as the successor to microelectronics because it is capable of extending miniaturization further toward the ultimate limit of individual atoms and molecules. The first applications will probably be in the military sector. The implementation schemes and device architectures for nanoelectronics may involve conceptually different approaches to future computational devices, such as DNA (deoxyribonucleic acid) computing and quantum computing, where atoms could act as quantum logic gates. Both of these concepts rely on controlling individual atoms or molecules, and may also be regarded as being in the realm of nanoelectronics. Currently there are no established mass-production techniques for commercial nanoelectronic devices, nor has a well-defined fabrication strategy that can be scaled for massive integration of components been established. Whereas for microelectronic devices the transistor forms a basic building block, it is not clear what the basic (three-terminal) element of a nanoelectronic device will be. Likewise, a well-defined architecture, required to process data, has yet to be established.

INFLUENCE OF QUANTUM MECHANICS

Nanoelectronics research is currently looking not only for the successor to CMOS processing but also for a replacement for the transistor itself. On the scale of 10-nm dimensions, components have a wavelength comparable to that of an electron at the Fermi energy. The confinement and coherence of the electron gives rise to gross deviations from the classical charge transport found in conventional devices. Quantum-mechanical laws become increasingly dominant on the nanoscale, and it is probable that nanoelectronics will operate on quantum principles.

Single-electron Devices

One popular approach has been to use small conducting islands in which electrons are confined and quantized in a definite state. These islands are typically connected to electrodes by thin tunnelling barriers. Quantum dots, resonant tunnel devices, and single-electron transistors are examples of devices that use this basic concept, albeit in different ways. Single-electron transistors are an example of a three-terminal device in which the charge of a single electron is sufficient to switch the source-to-drain current. The tiny energy required to drive single-electron transistor devices makes this approach very appealing. Nevertheless, a variety of drawbacks and obstacles limit the application of such devices in solid-state nanoelectronics, for example, their sensitivity to small fluctuations in voltage and background charges, which tend to accumulate in semiconductors. Such problems suggest that single-electron devices may be used for storage rather than logic functions. Currently, such devices operate at cryogenic temperatures, although there are a few examples of room temperature operation.

Fabrication Schemes

Owing to the dimensions in which nanoelectronics operates, there are two probable approaches to the fabrication of nanoelectronic devices: top down and bottom up.

Top-down Approaches

Top-down is the addition of established techniques of engineering and microelectronics processing that relies on a patterning process often described in terms of depositing, patterning, and etching layers of material to define the circuitry and active elements. Top-down approaches rely on control of damage, and as the structures become smaller the defects make device operation increasingly problematic. Moreover, as these technologies are pushed to smaller sizes, the cost increases drastically, tolerances become more difficult to maintain, and the engineering laws of scaling become inapplicable. To scale such technologies to the atomic and molecular level, the use of softer techniques with atomic tolerances based on bottom-up approaches will eventually become mandatory.

Present top-down fabrication is limited by the wavelength of the light used in the photolithography process, which illuminates proximity masks containing the pattern to be transferred to the substrate. To overcome this limit, shorter wavelengths are required, such as in extreme ultraviolet lithography. X-ray lithography is currently the leading technology in the drive to replace photolithography as a large-scale production tool because it uses masks, which are suited to high-volume production. Major challenges are mask fabrication and placement accuracy. Electron-beam lithography uses a finely focused beam of high-energy electrons to define a pattern. However, because the beam must be scanned to define the pattern, the technique in its present form is considered too slow for anything but the production of masks. Proposals to employ arrays of miniaturized electron guns are under investigation to alleviate this problem.

Bottom-up Approach

This term describes device fabrication on an atom-by-atom foundation. Molecules, which are prefabricated arrangements of atoms in a functional form, are also appealing for bottom-up fabrication. The advantage of the bottom-up approach lies in the design and chemical synthesis of functional molecules by the billions, which can then be assembled into nanoelectronic devices. The assembly of these perfect molecular units is more problematic, but two principal techniques for engineering atoms and molecules currently exist.

One approach involves self-assembly, whereby specific intermolecular forces allow molecules to arrange themselves into more complex structures. Supramolecular chemistry is the art of designing molecules that arrange themselves into functional entities held together by intermolecular forces. Success has also been achieved in patterning surfaces using elastomeric stamps containing self-assembling molecules that transfer themselves to a surface upon contact. The development of the scanning tunnelling microscope and related scanning probe microscopy tools such as the scanning force microscope provide another approach to bottom-up fabrication. In scanning tunnelling microscopy, a sharp tip, a needle of conical shape typically terminating in a single atom, is used to reposition, chemically change, or deposit individual atoms. It is an extension of mechanical assembly techniques on the atomic scale. Scanning tunnelling microscopy is the highest-resolution member of the family of scanning probe microscopy techniques, which all use sharp tips in different modes of operation. A major drawback of scanning probe microscopy is that it is a serial process. To overcome this, a system employing 1000 scanning tips has been realised using micromechanics. It is anticipated that several thousand tips working in unison can be fabricated. These small devices also allow patterning to be conducted at higher speeds because of the higher mechanical frequencies of the components.

Clear evidence that scanning probe microscopy techniques are an ultimate engineering tool has been provided by demonstrations of the ability to finely control and assemble individual atoms at low temperatures, around -270°C (-454°F). The low temperature is necessary to keep the atoms from becoming thermally agitated and diffusing away. This capability has also been extended to room temperature in experiments in which molecules were repositioned on surfaces.

Atomic and Molecular Devices

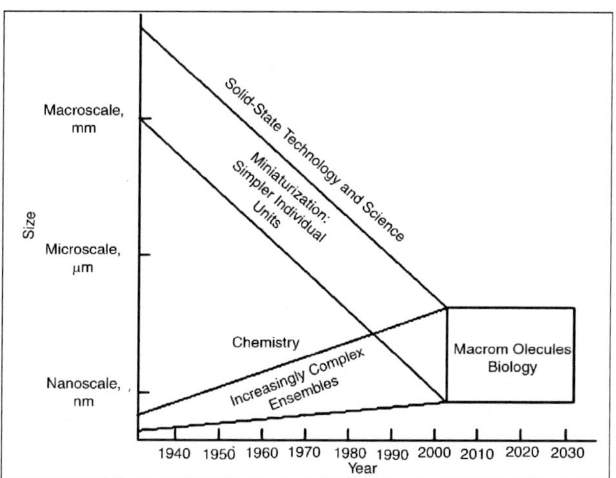

Figure: Developments in solid-state technology and chemistry. Miniaturization in technology is based on ever smaller and simpler components. Macromolecular chemistry involves a bottom-up approach, in which units of atoms are connected to form structures of ever-increasing complexity. Understanding and control of biology is also approaching the scale of individual atoms and molecules (for example, in drug design), and so converges with chemistry and microelectronics. Concepts in all three fields may be used in the design, fabrication, and operational principles of nanoelectronic devices.

An abacus with individual molecules as beads having a diameter of less than 1 nm has been constructed. The "finger" to move these beads is the ultrafine tungsten tip of a scanning tunnelling microscope, which also renders the result of such a "calculation" visible when operated in imaging mode. Stable rows of ten C_{60} (buckminsterfullerene) molecules were formed along steps just one atom high on a copper surface. These steps act as "rails," comparable to the earliest form of the abacus, which had grooves instead of rods to keep the beads in line. Individual molecules were then approached by the scanning tunnelling microscope tip and pushed back and forth in a precisely controlled way to count from 0 to 10. This work is a further step in bottom-up fabrication. It points the way to the assembly of more complex structures molecule by molecule, as nature does, and thus breaks ground for entirely new fabrication technologies in nanoelectronics.

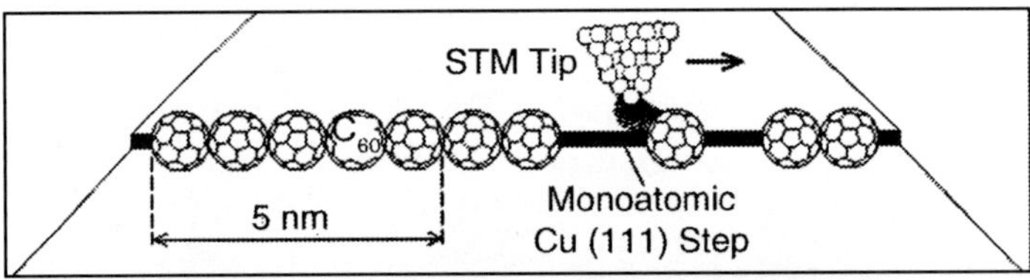

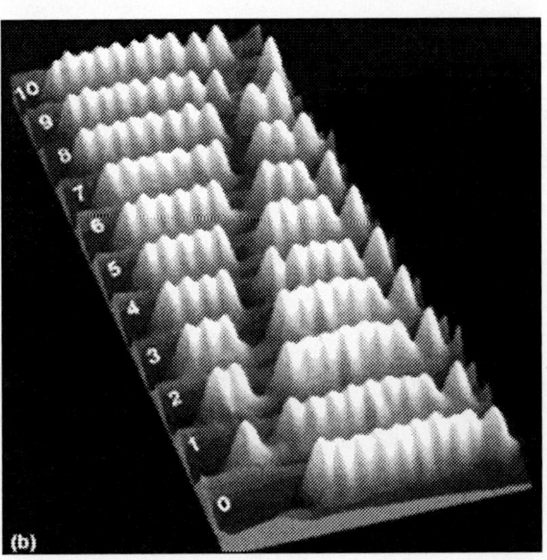

Figure : Atomic and molecular devices. Abacus with individual molecules that are moved by the tip of a scanning tunnelling microscope (STM). (*a*) Repositioning of a C_{60} molecule adsorbed on Cu(111) step sites along the step direction. (*b*) Sequence of STM images of 10 C_{60} molecules adsorbed on the lower step, plotted in a pseudo-three-dimensional illuminated representation. Using the tip as a nanoactuator, the numbers 0 to 10 have been sequentially computed and read using moleculesascounters.

MOLECULAR AMPLIFIERS AND TRANSISTORS

In an extension of electrical transport studies of C_{60} molecules that involved electrically contacting individual molecules, an experimental electronic device with a tiny active part was demonstrated. It consists of a single C_{60} molecule, also known as a bucky ball, actuated by the ultrasharp tip of a scanning tunnelling microscope, and represents a fully functional electromechanical amplifier. Applying a mechanical force to a single bucky ball using a scanning tunnelling microscope tip can change its electrical conductance continuously and reversibly. By gently squeezing the bucky ball by lowering the tip only one-tenth of a nanometer, the molecular structure is deformed slightly, which in turn changes its electrical properties. The resistance of the squeezed bucky ball is 100 times lower than that of the uncompressed molecule, which allows electrons to tunnel more easily through the molecule. A small voltage (10 mV) applied to a piezoelectric element of the tip to squeeze the molecule results in a fivefold voltage gain in the device.

By externally deforming the molecule, the highest occupied and lowest unoccupied molecular orbital (HOMO and LUMO) levels (which are the equivalent of the valence and conduction bands in a conventional semiconductor) are broadened and shifted. This effect increases the weight of the tails of those levels at the electrode's Fermi level and therefore accounts for the increased conductance (transparency) of the molecule. (Electron transport at small bias voltage is still linear with voltage in this region.) One interesting consequence of the molecule's compression is that normally degenerate molecular levels (such as the fivefold HOMO and the threefold LUMO of a C_{60} molecule) are split in energy, which explains the five and three peaks of these orbitals that appear on the conductance spectrum under compression. The next step in further decreasing the size of the device is to replace the macroscopic piezo-actuated electromechanical gate (the scanning tunnelling microscope tip) with a smaller component. A micromechanical electrode actuated by electrostatic, thermomechanical, or other means might be used to vary the pressure on a molecule, thereby reducing the overall dimensions of the complete amplifier.

In another approach, researchers fabricated a transistor using nanotubes. Carbon nanotubes are molecularly perfect tubes with nanometer-scale diameters formed from carbon atoms. They are electrical conductors and have attracted much attention as candidates for molecular computers. Currently, carbon nanotubes are limited in their application to nanoelectronics by their variety of forms and types. They lack the uniformity of molecules synthesized by conventional chemical techniques, which can be purified to trillions of exactly identical copies. However, their good electrical characteristics do make them attractive candidates for nanoelectronics. Much research is being conducted to further develop techniques to fabricate better tubes in larger quantities.

Atomic and Molecular Switches

The first atomic switch consisted of a xenon atom that was switched back and forth between a scanning tunnelling microscope tip and a surface at low temperature. Associated with the motion of the atom was a change in the electric current flowing in the junction. This approach to the ultimate miniaturization of electronic components encouraged researchers to propose a number of devices.

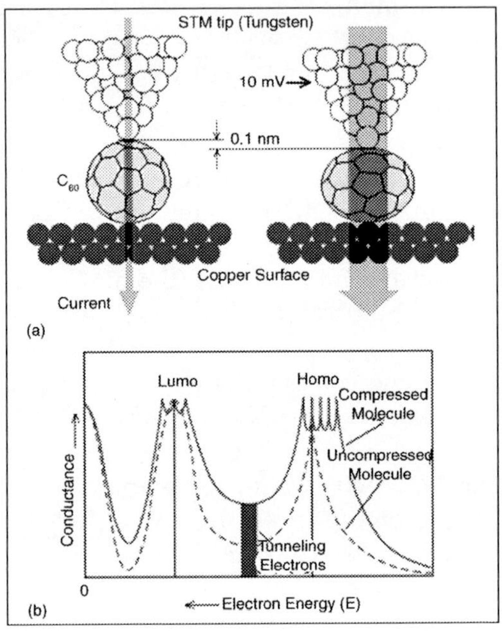

Figure: Effects of compressing the cage of a C_{60} molecule with the tip apex of a scanning tunnelling microscope (STM). (*a*) Compression of 0.1 nm, producing a two-order-of-magnitude increase in the current intensity through the molecule. (*b*) Comparison of conductance spectra of compressed and uncompressed molecules, showing how external deformation causes broadening and shifting of the highest occupied and lowest unoccupied molecular orbital (HOMO and LUMO) levels. The vertical band indicates the path of the tunnelling electrons from the Fermi level of the tip.

In a simulated atomic relay based on this principle, a single atom in an atomic wire was switched in and out of registry by a third electrode, also made of an atomic wire. An atomic wire consists of single atoms lined up in a row. Such structures have been created in the laboratory using, for example, a surface with atomic grooves. A cheap and reliable technique to fabricate massive arrays of such wires just where they are needed is the focus of much research. A molecular relay based on similar concepts but with a differently shaped molecule has been proposed.

Molecular switches and other types of devices have been chemically synthesized from molecules that are mechanically interlinked. Although such devices are appealing in terms of their size, the fabrication of a reliable switch represents a major challenge in the quest for molecular-scale computers.

PROGRAMMABLE LOGIC DEVICE

A programmable logic device or PLD is an electronic component used to build reconfigurable digital circuits. Unlike a logic gate, which has a fixed function, a PLD has an undefined function at the time of manufacture. Before the PLD can be used in a circuit it must be programmed, that is, reconfigured.

Using a ROM as a PLD

Before PLDs were invented, read-only memory (ROM) chips were used to create arbitrary combinational logic functions of a number of inputs. Consider a ROM with m inputs (the address lines) and n outputs (the data lines). When used as a memory, the ROM contains 2^m words of n bits each. Now imagine that the inputs are driven not by an m-bit address, but by m independent logic signals. Theoretically, there are 2^{2m} possible Boolean functions of these m signals, but the structure of the ROM allows just n of these functions to be produced at the output pins. The ROM therefore becomes equivalent to n separate logic circuits, each of which generates a chosen function of the m inputs.

The advantage of using a ROM in this way is that any imaginable function of the m inputs can be made to appear at any of the n outputs, making this the most general-purpose combinational logic device available. Also, PROMs (programmable ROMs), EPROMs (ultraviolet-erasable PROMs) and EEPROMs (electrically erasable PROMs) are available that can be programmed using a standard PROM programmer without requiring specialised hardware or software. However, there are several disadvantages:

- they are usually much slower than dedicated logic circuits,
- they cannot necessarily provide safe "covers" for asynchronous logic transitions so the PROM's outputs may glitch as the inputs switch,
- they consume more power,
- they are often more expensive than programmable logic, especially if high speed is required.

Since most ROMs do not have input or output registers, they cannot be used stand-alone for sequential logic. An external TTL register was often used for sequential designs such as state machines. Common EPROMs, for example the 2716, are still sometimes used in this way by hobby circuit designers, who often have some lying around. This use is sometimes called a 'poor man's PAL'.

Early Programmable Logic

In 1969, Motorola offered the XC157, a mask-programmed gate array with 12 gates and 30 uncommitted input/output pins. In 1970, Texas Instruments developed a mask-programmable IC based on the IBM read-only associative memory or ROAM. This device, the TMS2000, was programmed by altering the metal layer during the production of the IC. The TMS2000 had up to 17 inputs and 18 outputs with 8 JK flip flop for memory. TI coined the term Programmable Logic Array for this device.

In 1971, General Electric Company (GE) was developing a programmable logic device based on the new PROM technology. This experimental device improved on IBM's ROAM by allowing multilevel logic. Intel had just introduced the floating-gate UV erasable PROM so the researcher at GE incorporated that technology. The GE device was the first erasable PLD ever developed, predating the Altera EPLD by over a decade. GE obtained several early patents on programmable logic devices.

In 1973 National Semiconductor introduced a mask-programmable PLA device (DM7575) with 14 inputs and 8 outputs with no memory registers. This was more popular than the TI part but cost of making the metal mask limited its use. The device is significant because it was the basis for the field programmable logic array produced by Signetics in 1975, the 82S100. (Intersil actually beat Signetics to market but poor yield doomed their part.)

In 1974 GE entered into an agreement with Monolithic Memories to develop a mask- programmable logic device incorporating the GE innovations. The device was named the 'Programmable Associative Logic Array' or PALA. The MMI 5760 was completed in 1976 and could implement multilevel or sequential circuits of over 100 gates. The device was supported by a GE design environment where Boolean equations would be converted to mask patterns for configuring the device. The part was never brought to market.

PLA

In 1970, Texas Instruments developed a mask-programmable IC based on the IBM read-only associative memory or ROAM. This device, the TMS2000, was programmed by altering the metal layer during the production of the IC. The TMS2000 had up to 17 inputs and 18 outputs with 8 JK flip flop for memory. TI coined the term Programmable Logic Array for this device.

A programmable logic array (PLA) has a programmable AND gate array, which links to a programmable OR gate array, which can then be conditionally complemented to produce an output.

Programmable Array Logic

MMI introduced a breakthrough device in 1978, the Programmable Array Logic or PAL. The architecture was simpler than that of Signetics FPLA because it omitted the programmable OR array. This made the parts faster, smaller and cheaper. They were available in 20 pin 300 mil DIP packages while the FPLAs came in 28 pin 600 mil packages. The PAL Handbook demystified the design process. The PALASM design software (PAL Assembler) converted the engineers' Boolean equations into the fuse pattern required to program the part. The PAL devices were soon second-sourced by National Semiconductor, Texas Instruments and AMD.

After MMI succeeded with the 20-pin PAL parts, AMD introduced the 24-pin 22V10 PAL with additional features. After buying out MMI (1987), AMD spun off a consolidated operation as Vantis, and that business was acquired by Lattice Semiconductor in 1999.

Generic Array Logic

Figure: Lattice GAL 16V8 and 20V8

An innovation of the PAL was the generic array logic device, or GAL, invented by Lattice Semiconductor in 1985. This device has the same logical properties as the PAL but can be erased and reprogrammed. The GAL is very useful in the prototyping stage of a design, when any bugs in the logic can be corrected by reprogramming. GALs are programmed and reprogrammed using a PAL programmer, or by using the in-circuit programming technique on supporting chips. Lattice GALs combine CMOS and electrically erasable (E^2) floating gate technology for a high-speed, low-power logic device.

A similar device called a PEEL (programmable electrically erasable logic) was introduced by the International CMOS Technology (ICT) corporation.

Complex Programmable Logic Device

PALs and GALs are obtainable only in small sizes, equivalent to a few hundred logic gates. For bigger logic circuits, complex PLDs or CPLDs can be used. These contain the equivalent of several PALs linked by programmable interconnections, all in one integrated circuit. CPLDs can replace thousands, or even hundreds of thousands, of logic gates. Some CPLDs are programmed using a PAL programmer, but this technique becomes inconvenient for devices with hundreds of pins. A second technique of programming is to solder the device to its printed circuit board, then feed it with a serial data stream from a personal computer. The CPLD contains a circuit that decodes the data stream and configures the CPLD to perform its specified logic function.

Field-programmable Gate Array

While PALs were busy developing into GALs and CPLDs, a separate stream of development was happening. This type of device is based on gate array technology and is called the field-programmable gate array (FPGA). Early examples of FPGAs are the 82s100 array, and 82S105 sequencer, by Signetics, introduced in the late 1970s. The 82S100 was an array of AND terms. The 82S105 also had flip flop functions. FPGAs use a grid of logic gates, and once stored, the data doesn't change, similar to that of an ordinary gate array. The term "field-programmable" means the device is programmed by the customer, not the manufacturer.

FPGAs are usually programmed after being soldered down to the circuit board, in a manner similar to that of larger CPLDs. In most larger FPGAs the configuration is volatile, and must be re-loaded into the device whenever power is applied or different functionality is required. Configuration is typically stored in a configuration PROM or EEPROM. EEPROM versions may be in-system programmable (typically via JTAG). The difference between FPGAs and CPLDs is that FPGAs are internally based on Look-up tables (LUTs) whereas CPLDs form the logic functions with sea-of-gates (*e.g.* sum of products). CPLDs are meant for simpler designs while FPGAs are meant for more complex designs. In general, CPLDs are a good choice for wide combinational logic applications, whereas FPGAs are more suitable for large state machines (*i.e.* microprocessors).

Other Variants

At present, much interest exists in reconfigurable systems. These are microprocessor circuits that contain some fixed functions and other functions that can be altered by code running on the processor. Designing self-altering systems requires engineers to learn new techniques, and that new software tools be developed. PLDs are being sold now that contain a microprocessor with a fixed function (the so-called *core*) surrounded by programmable logic. These devices let designers concentrate on adding new features to designs without having to worry about making the microprocessor work.

How PLDs Retain their Configuration

A PLD is a combination of a logic device and a memory device. The memory is used to store the pattern that was given to the chip during programming. Most of the techniques for storing data in an integrated circuit have been adapted for use in PLDs. These include:

- Silicon antifuses
- SRAM
- EPROM or EEPROM cells
- Flash memory.

Silicon antifuses are the storage elements used in the PAL, the first type of PLD. These are connections that are made by applying a voltage across a modified area of silicon inside the chip. They are called antifuses because they work in the opposite way to normal fuses, which begin life as connections until they are broken by an electric current. SRAM, or static RAM, is a volatile type of memory, meaning that its contents are lost each time the power is switched off. SRAM-based PLDs therefore have to be programmed every time the circuit is switched on. This is usually done automatically by another part of the circuit.

An EPROM cell is a MOS (metal-oxide-semiconductor) transistor that can be switched on by trapping an electric charge permanently on its gate electrode. This is done by a PAL programmer. The charge remains for many years and can only be removed by exposing the chip to strong ultraviolet light in a device called an EPROM eraser. Flash memory is non-volatile, retaining its contents even when the power is switched off. It can be erased and reprogrammed as required. This makes it useful for PLD memory.

As of 2005, most CPLDs are electrically programmable and erasable, and non-volatile. This is because they are too small to justify the inconvenience of programming internal SRAM cells every time they start up, and EPROM cells are more expensive due to their ceramic package with a quartz window.

PLD Programming Languages

Many PAL programming devices believe input in a standard file format, commonly referred to as 'JEDEC files'. They are analogous to software compilers. The languages used as source code for logic compilers are called hardware description languages, or HDLs.

PALASM and ABEL are frequently used for low-complexity devices, while Verilog and VHDL are popular higher-level description languages for more complex devices. The more limited ABEL is often used for historical reasons, but for new designs VHDL is more popular, even for low-complexity designs.

PLD Programming Devices

A device programmer is used to transfer the boolean logic pattern into the programmable device. In the early days of programmable logic, every PLD manufacturer also produced a specialized device programmer for its family of logic devices. Later, universal device programmers came onto the market that supported several logic device families from different manufacturers. Today's device programmers usually can program common PLDs (mostly PAL/GAL equivalents) from all existing manufacturers.

Common file formats used to store the boolean logic pattern (fuses) are JEDEC, Altera POF (Programmable Object File), or Xilinx BITstream.

THE FUNDAMENTAL PHYSICAL LIMITS OF COMPUTATION

A computation, whether it is performed by electronic machinery, on an abacus or in a biological system such as the brain, is a physical process. It is subject to the same questions that apply to other physical processes: How much energy must be expended to perform a particular computation? How long must it take? How large must the computing device be? In other words, what are the physical limits of the process of computation?

So far it has been easier to ask these questions than to answer them. To the extent that we have found limits, they are terribly far away from the real limits of modern technology. We cannot profess, therefore, to be guiding the technologist or the engineer. What we are doing is really more fundamental. We are looking for general laws that must govern all information processing, no matter how it is accomplished. Any limits we find must be based solely on fundamental physical principles, not on whatever technology we may currently be using.

There are precedents for this kind of fundamental examination. In the 1940's Claude E. Shannon of the Bell Telephone Laboratories found there are limits on the amount of information that can be transmitted through a noisy channel; these limits apply no matter how the message is encoded into a signal. Shannon's work represents the birth of modern information science. Earlier, in the mid- and late 19th century, physicists attempting to determine the fundamental limits on the efficiency of steam engines had created the science of thermodynamics. In about 1960 one of us (Landauer) and John Swanson at IBM began attempting to apply the same type of analysis to the process of computing. Since the mid-1970's a growing number of other workers at other institutions have entered this field.

In our analysis of the physical limits of computation we use the term "information" in the technical sense of information theory. In this sense information is destroyed whenever two previously distinct situations become indistinguishable. In physical systems without friction, information can never be destroyed; whenever information is destroyed, some amount of energy must be dissipated (converted into heat). As an example, imagine two easily distinguishable physical situations, such as a rubber ball held either one meter or two meters off the ground. If the ball is dropped, it will bounce. If there is no friction and the ball is perfectly elastic, an observer will always be able to tell what state the ball started out in (that is, what its initial height was) because a ball dropped from two meters will bounce higher than a ball dropped from one meter. If there is friction, however, the ball will dissipate a small amount of energy with each bounce, until it eventually stops bouncing and comes to rest on the ground. It will then be impossible to determine what the ball's initial state was; a ball dropped from two meters will be identical with a ball dropped from one meter. Information will have been lost as a result of energy dissipation.

Here is another example of information destruction: the expression 2 + 2 contains more information than the expression = 4. If all we know is that we have added two numbers to yield 4, then we do not know whether we have added 1 + 3, 2 + 2, 0 + 4 or some other pair of numbers. Since the output is implicit in the input, no computation ever generates information

In fact, computation as it is currently carried out depends on many operations that destroy information. The so-called *and* gate is a device with two input lines, each of which may be set at 1 or 0, and one output, whose value depends on the value of the inputs.

If both inputs are 1, the output will be 1. If one of the inputs is 0 or if both are 0, the output will also be 0. Any time the gate's output is a 0 we lose information, because we do not know which of three possible states the input lines were in (0 and 1, 1 and 0, or 0 and 0). In fact, any logic gate that has more input than output lines inevitably discards information, because we cannot deduce the input from the output. Whenever we use such a "logically irreversible" gate, we dissipate energy into the environment. Erasing a bit of memory, another operation that is frequently used in computing, is also fundamentally dissipative; when we erase a bit, we lose all information about that bit's previous state.

Are irreversible logic gates and erasures essential to computation? If they are, any computation we perform has to dissipate some minimum amount of energy.

This conclusion has since been demonstrated in several models; the easiest of these to describe are based on so-called reversible logic elements such as the Fredkin gate, named for Edward Fredkin of the Massachusetts Institute of Technology. The Fredkin gate has three input lines and three outputs. The input on one line, which is called the control channel, is fed unchanged through the gate. If the control channel is set at 0, the input on the other two lines also passes through unchanged. If the control line is a 1, however, the outputs of the other two lines are switched: the input of one line becomes the output of the other and *vice-versa*. The Fredkin gate does not discard any information; the input can always be deduced from the output.

Fredkin has exposed that any logic device required in a computer can be implemented by an appropriate arrangement of Fredkin gates. To make the computation work, certain input lines of some of the Fredkin gates must be preset at particular values.

Fredkin gates have more output lines than the gates they are made to simulate. In the process of computing, what seem to be "garbage bits," bits of information that have no apparent use, are therefore generated. These bits must somehow be cleared out of the computer if we are to use it again, but if we erase them, it will cost us all the energy dissipation we have been trying to avoid.

Actually these bits have a most important use. Once we have copied down the result of our computation, which will reside in the normal output bits, we simply run the computer in reverse. That is, we enter the "garbage bits" and output bits that were produced by the computer's normal operation as "input" into the "back end" of the computer. This is possible because each of the logic gates in the computer is itself reversible. Running the computer in reverse discards no information, and so it need not dissipate any energy. Eventually the computer will be left exactly as it was before the computation began. Hence it is possible to complete a "computing cycle" to run a computer and then to return it to its original state-without dissipating any energy.

So far we have discussed a set of logic operations, not a physical device. It is not hard, however, to imagine a physical device that operates as a Fredkin gate. In this device the information channels are represented by pipes. A bit of information is represented by the presence or absence of a ball in a particular section of pipe; the presence of a ball signifies a 1 and the absence of a ball signifies a 0.

The control line is represented by a narrow section of pipe that is split lengthwise down the middle. When a ball enters the split segment of pipe, it pushes the two halves of the pipe apart, actuating a switching device. The switching device channels any input balls that may be in the other two pipes: when a ball is present in the control line, any ball that enters an input pipe is automatically redirected to the other pipe. To ensure that the switch is closed when no control ball is present, there are springs that hold the two halves of the split pipe together. A ball entering the split pipe must expend energy when it compresses the springs, but this energy is not lost; it can be recovered when the control ball leaves the split pipe and the springs expand.

All the balls are linked together and pushed forward by one mechanism, so that they move in synchrony; otherwise we could not ensure that the various input and controlling balls would arrive at a logic gate together. In a sense the forward progress of the computation is really motion along a single degree of freedom, like the motion of two wheels rigidly attached to one axle. Once the computation is done we push all the balls backward, undoing all the operations and returning the computer to its initial state.

If the entire assembly is immersed in an ideal viscous fluid, then the frictional forces that act on the balls will be proportional to their velocity; there will be no static friction. The frictional force will therefore be very weak if we are content to move the balls slowly. In any mechanical system the energy that must be expended to work against friction is equal to the product of the frictional force and the distance through which the system travels. If we move the balls through the Fredkin gates at a low speed, then the energy expended (the product of force and distance) will be very small, because the frictional force depends directly on the balls' speed. Infact, we can expend as little energy as we wish, simply by taking a long time to carry out the operation. There is thus no minimum amount of energy that must be expended in order to perform any given computation.

The energy lost to friction in this model will be very small if the machine is operated very slowly. Is it possible to design a more idealized machine that could compute without any friction? Or is friction essential to the computing process? Fredkin, together with Tommaso Toffoli and others at M.I.T., has shown that it is not.

They demonstrated that it is possible to do computation by firing ideal, frictionless billiard balls at one another. In the billiard-ball model perfect reflecting "mirrors," surfaces that redirect the balls' motion, are arranged in such a way that the movement of the balls across a table emulates the movement of bits of information through logic gates. As before, the presence of a ball in a particular part of the computer signifies a 1, whereas the absence of a ball signifies a 0. If two balls arrive simultaneously at a logic gate, they will collide and their paths will change; their new paths represent the output of the gate. Fredkin, Toffoli and others have described arrangements of mirrors that correspond to different types of logic gate, and they have shown that billiard-ball models can be built to simulate any logic element that is necessary for computing.

To start the computation we fire a billiard ball into the computer wherever we wish to input a 1. The balls must enter the machine simultaneously. Since they are perfectly elastic, they do not lose energy when they collide; they will emerge from the computer with the same amount of kinetic energy we gave them at the beginning.

In operation a billiard-ball computer produces "garbage bits," just as a computer built of Fredkin gates does. After the computer has reached an answer we reflect the billiard balls back into it, undoing the computation. They will come out of the machine exactly where we sent them in, and at the same speed. The mechanism that launched them into the computer can then be used to absorb their kinetic energy. Once again we will have performed a computation and returned the computer to its initial state without dissipating energy.

The billiard-ball computer has one major flaw: it is extremely sensitive to slight errors. If a ball is aimed slightly incorrectly or if a mirror is tilted at a slightly wrong angle, the balls' trajectories will go astray. One or more balls will deviate from their intended paths, and in due course errors will combine to invalidate the entire computation. Even if perfectly elastic and frictionless billiard balls could be manufactured, the small amount of random thermal motion in the molecules they are made of would be enough to cause errors after a few dozen collisions.

Of course we could install some kind of corrective device that would return any errant billiard ball to its desired path, but then we would be obliterating information about the ball's earlier history. For example, we might be discarding information about the extent to which a mirror is tilted incorrectly. Discarding information, even to correct an error, can be done only in a system in which there is friction and loss of energy. Any correctional device must therefore dissipate some energy.

Many of the difficulties inherent in the billiard-ball computer can be made less extreme if microscopic or submicroscopic particles, such as electrons, are used in place of billiard balls. As Wojciech H. Zurek, who is now at the Los Alamos National Laboratory, has pointed out, quantum laws, which can restrict particles to a few states of motion, could eliminate the possibility that a particle might go astray by a small amount.

Although the conversation so far has been based primarily on classical dynamics, several investigators have proposed other reversible computers that are based on quantum-mechanical principles. Such computers, first proposed by Paul Benioff of the Argonne National Laboratory and refined by others, notably Richard P. Feynman of the California Institute of Technology, have so far been described only in the most abstract terms. Essentially the particles in these computers would be arranged so that the quantum-mechanical rules governing their interaction would be precisely analogous to the rules describing the predicted outputs of various reversible logic gates. For example, suppose a particle's spin can have only two possible values: up (corresponding to a binary 1) and down (corresponding to a 0). The interactions between particle spins can be prescribed in such a way that the value of one particle's spin changes depending on the spin of nearby particles; the spin of the particle would correspond to one of the outputs of a logic gate.

So far this discussion has concentrated on information processing. A computer must store information as well as process it. The interaction between storage and processing is best described in terms of a device called a Turing machine, for Alan M. Turing, who first proposed such a machine in 1936. A Turing machine can perform any computation that can be performed by a modern computer. One of us (Bennett) has shown that it is possible to build a reversible Turing machine: a Turing machine that does not discard information and can therefore be run with as small an expenditure of energy as the user wishes.

A Turing machine has several components. There is a tape, divided into discrete frames or segments, each of which is marked with a 0 or a 1; these bits represent the input. A "read/write head" moves along the tape. The head has several functions. It can read one bit of the tape at a time, it can print one bit onto the tape and it can shift its position by one segment to the left or right. In order to remember from one cycle to the next what it is doing, the head mechanism has a number of distinct "states"; each state constitutes. In each cycle the head reads the bit on the segment it at present occupies; then it prints a new bit onto the tape, changes its internal state and moves one segment to the left or right. The bit it prints, the state it changes into and the direction in which it moves are determined by a fixed set of transition rules. Each rule specifies a particular set of actions. Which rule the machine follows is determined by the state of the head and the value of the bit that it reads from the tape. For example, one rule might be: "If the head is in state *A* and is sitting on a segment of tape that is printed with a 0, it should change that bit to a 1, change its state to state *B* and move one segment to the right." It may happen that the transition rule instructs the machine not to change its internal state, not to print a new bit onto the tape or to halt its operation. Not all Turing machines are reversible, but a reversible Turing machine can be built to perform any possible computation.

The reversible Turing-machine models have an advantage over such machines as the frictionless billiard-ball computer. In the billiard-ball computer random thermal motion causes unavoidable errors. Reversible Turing-machine models actually exploit random thermal motion: they are constructed in such a way that thermal motion itself, with the assistance of a very weak driving force, moves the machine from one state to the next. The progress of the computation resembles the motion of an ion (a charged particle) suspended in a solution that is held in a weak electric field. The ion's motion, as seen over a short period of time, appears to be random; it is nearly as likely to move in one direction as in another. The applied force of the electric field, however, gives the net motion a preferred direction: the ion is a little likelier to move in one direction than in the other.

It may at first seem inconceivable that a purposeful sequence of operations, such as a computation, could be achieved in an apparatus whose direction of motion at any one time is nearly random. This style of operation is quite common, however, in the microscopic world of chemical reactions. There the trial-and-error action of Brownian motion, or random thermal motion, suffices to bring reactant molecules into contact, to orient and bend them into the specific conformation required for them to react, and to separate the product molecules after the reaction. All chemical reactions are in principle reversible: the same Brownian motion that accomplishes the forward reaction sometimes brings product molecules together and pushes them backward through the transition. In a state of equilibrium a backward reaction is just as likely to occur as a forward one.

In order to keep a reaction moving in the forward direction, we must supply reactant molecules and remove product molecules; in effect, we must provide a small driving force. When the driving force is very small, the reaction will take nearly as many backward steps as forward ones, but on the average it will move forward. In order to provide the driving force we must expend energy, but as in our ball-and-pipe realization of the Fredkin gate the total amount of energy can be as small as we wish; if we are willing to allow a long time for an operation, there is no minimum amount of energy that must be expended. The reason is that the total energy dissipated depends on the number of forward steps divided by the number of backward steps. (It is actually proportional to the logarithm of this ratio, but as the ratio increases or decreases so does its logarithm.) The slower the reaction moves forward, the smaller the ratio will be.

We can see how a Brownian Turing machine strength work by examining a Brownian tape-copying machine that already exists in nature: RNA polymerase, the enzyme that helps to construct RNA copies of the DNA constituting a gene. A single strand of DNA is much like the tape of a Turing machine. At each position along the strand there is one of four "bases": adenine, guanine, cytosine or thymine (abbreviated *A, G, C and I)*. RNA is a similar chain like molecule whose four bases, adenine, guanine, cytosine and uracil *(A. G. C and U)* bind to "complementary" DNA bases.

The RNA polymerase catalyses this pairing reaction. The DNA helix is normally surrounded by a solution containing a large number of nucleoside triphosphate molecules, each consisting of an RNA base linked to a sugar and a tail of three phosphate groups. The RNA-polymerase enzyme selects from the solution a single RNA base that is complementary to the base about to be copied on the DNA strand. It then attaches the new base to the end of the growing RNA strand and releases two of the phosphates into the surrounding solution as a free pyrophosphate ion. Then the enzyme shifts forward one notch along the strand of DNA in preparation for attaching the next RNA base. The result is a strand of RNA that is complementary to the template strand of DNA. Without RNA polymerase this set of reactions would occur very slowly, and there would be little guarantee that the RNA and DNA molecules would be complementary.

The reactions are reversible: sometimes the enzyme takes up a free pyrophosphate ion, combines it with the last base on the RNA strand and releases the resulting nucleoside triphosphate molecule into the surrounding solution, meanwhile backing up one notch along the DNA strand. At equilibrium, forward and backward steps would occur with equal frequency; normally other metabolic processes drive the reaction forward by removing pyrophosphate and supplying the four kinds of nucleoside triphosphate.

In the laboratory the speed with which RNA polymerase acts can be varied by adjusting the concentrations of the reactants. As the concentrations are brought closer to equilibrium the enzyme works more slowly and dissipates less energy to copy a given section of DNA, because the ratio of forward to backward steps is smaller.

Although RNA polymerase merely copies information without processing it, it is relatively easy to imagine how a hypothetical chemical

Turing machine might work. The tape is a single long backbone molecule to which two types of base, representing the binary 0 and 1, attach at periodic sites. A small additional molecule is attached to the 0 or 1 group at one site along the chain. The position of this additional molecule represents the position of the Turing machine's head. There are several different types of "head molecule," each type representing a different machine state.

The machine's transition rules are represented by enzymes. Each enzyme is capable of catalysing one particular reaction. The way these enzymes work is best demonstrated by an example.

Suppose the head molecule is type *A* (indicating that the machine is in state *A*) and is attached to a 0 base. Also suppose the following transition rule applies: "When the head is in state *A* and reads a 0, change the 0 to a 1, change state to *B* and move to 'the right." A molecule of the enzyme representing this rule has a site that fits a type-*A* head molecule attached to a 1 base. It also has one site that fits a 0 base and one site that fits a *B* head.

To achieve the transition, the enzyme molecule first approaches the tape molecule at a location 'just to the right of the base on which the *A* head resides. Then it detaches from the tape both the head molecule and the 0 base to which the head was attached, putting in their place a 1 base. Next it attaches a *B* head to the base that is to the right of the 1 base it has just added to the tape. At this point the transition is complete. The head's original site is changed from a 0 to a 1, the head molecule is now a type *B*, and it is attached to the base that is one notch to the right of the previous head position.

During the operation of a Brownian Turing machine the tape would have to be immersed in a solution containing many enzyme molecules, as well as extra O's, 1 's, *A*'s and *B*'s. To drive the reaction forward there would have to be some other reaction that cleaned the enzyme molecules of detached heads and bases. The concentrations of the reactants that clean the enzyme molecules represent the force that drives the Turing machine forward. Again we can expend as little energy as we wish simply by driving the machine forward very slowly. The enzymatic Turing machine would not be error-free. Occasionally a reaction that is not catalyzed by any enzyme might occur; for example, a 0 base could spontaneously detach itself from the backbone molecule and a 1 base could be attached in its place. Similar errors do indeed occur during RNA synthesis.

In principle it would be probable to eliminate errors by building a Brownian Turing machine out of rigid, frictionless clockwork. The clockwork Turing machine involves less idealization than the billiard-ball computer but more than the enzymatic Turing machine. On the one hand, its parts need not be manufactured to perfect tolerances, as the billiard balls would have to be; the parts fit loosely together, and the machine can operate even in the presence of a large amount of thermal noise. Still, its parts must be perfectly rigid and free of static friction, properties not found in any macroscopic body.

Because the machine's parts fit together loosely, they are held in place not by friction but by grooves or notches in neighbouring parts. Although each part of the machine is free to jiggle a little, like the pieces of a well-worn wood puzzle, the machine as a whole can only follow one "computational path." That is, the machine's parts interlock in such a way that at any time the machine can make only two kinds of large-scale motion: the motion corresponding to a forward computational step and that corresponding to a backward step. The computer makes such transitions only as the accidental result of the random thermal motion of its parts biased by the weak external force. It is nearly as likely to proceed backward along the computational path, undoing the most recent transition, as it is to proceed forward. A small force, provided externally, drives the computation forward. This force can again be as small as we wish, and so there is no minimum amount of energy that must be expended in order to run a Brownian clockwork Turing machine.

According to classical thermodynamics, then, there is no minimum amount of energy required in order to perform a computation. Is the classical thermodynamical analysis in conflict with quantum theory? After all, the quantum-mechanical uncertainty principle states there must be an inverse relation between our uncertainty about how long a process takes and our uncertainty about how much energy the process involves. Some investigators have suggested that any switching process occurring in a short period of time must involve a minimum expenditure of energy.

In fact the uncertainty standard does not require any minimum energy expenditure for a fast switching event. The uncertainty principle would be applicable only if we attempted to measure the precise time at which the event took place. Even in quantum mechanics extremely fast events can take place without any loss of energy. Our confidence that quantum mechanics allows computing without any minimum expenditure is bolstered when we remember that Benioff and others have developed models of reversible quantum-mechanical computers, which dissipate no energy and obey the laws of quantum mechanics. Thus the uncertainty principle does not seem to place a fundamental limit on the process of computation; neither does classical thermodynamics.

For example, do elementary logic operations require some minimum amount of time? What is the smallest possible gadgetry that could accomplish such operations? Because scales of size and time are connected by the velocity of light, it is likely that these two questions have related answers. We may not be able to find these answers, however, until it is determined whether or not there is, some ultimate graininess in the universal scales of time and length.

At the other extreme, how large can we make a computer memory? How many particles in the universe can we bring and keep together for that purpose? The maximum possible size of a computer memory limits the precision with which we can calculate. It will limit, for example, the number of decimal places to which we can calculate pi. The inevitable deterioration processes that occur in real computers pose another, perhaps related, question: Can deterioration, at least in principle, be reduced to any desired degree, or does it impose a limit on the maximum length of time we shall be able to devote to any one calculation? That is, are there certain calculations that cannot be completed before the computer's hardware decays into uselessness?

Such questions really concern limitations on the physical execution of mathematical operations. Physical laws, on which the answers must ultimately be based, are themselves expressed in terms of such mathematical operations. Thus we are asking about the ultimate form in which the laws of physics can be applied, given the constraints imposed by the universe that the laws themselves describe.

INFORMATION PROCESSING AND SYSTEMS

Individuals, groups, organizations, and computers are all capable of performing information-processing tasks, though some entities perform certain information tasks better than others. Recent advances combine people and technology to achieve even more effective information-processing systems. Ultimately, the way information is processed determines whether we succeed or fail, make or lose money, understand or misunderstand. The consequences of how we process information are profound.

INFORMATION PROCESSING

Individuals procedure information, groups and teams process information, and computer systems process information. In most cases, processing information leads to a decision. The success or failure of that decision can usually be directly tied to how the information was processed. To truly capitalize on the power that is tied to your information, you must use it correctly.

INFORMATION SYSTEMS

An information system (IS) is the group of components including hardware, software, people, procedures, and data that work together to produce meaningful information for individuals and organizations. Information systems may not involve computer systems at all. For example, some employees maintain calendar information on bulletin boards in their office using paper and pens. The purpose of this manually operated information system is to capture, organize, and communicate scheduling information. However, more and more information tasks are being automated using computerized information systems.

Functions of Information Systems

Information systems process data using a set of interrelated functions. In the most basic sense, information systems accept, store, and process data, and finally produce results in the form of information. Information systems carry out and are defined by four primary functions: input functions, storage functions, processing functions, and output functions. This set of functions applies to information systems, which may be composed of multiple computers in a network or a single computer.

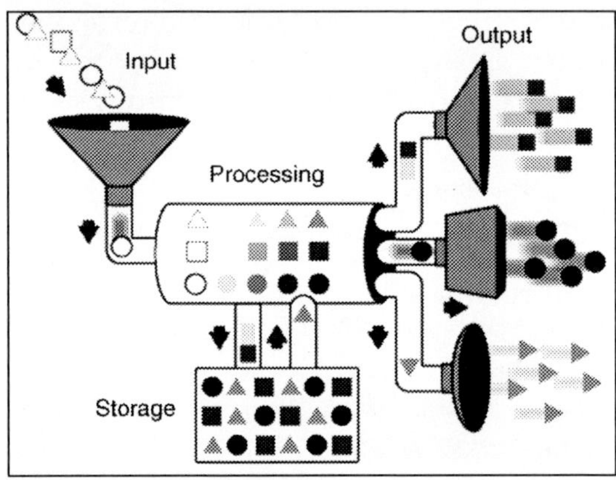

The data that is fed into the system is called input data. Input data is accepted into the system by the input function and is immediately processed by the processing function or stored by the storage function as stored data for processing later on. The processing function accepts data from the storage function and processes it by manipulating it. The output function produces the results of the processing function and sends it out of the system as output information.

THE FUNCTIONS OF IS IN PRACTICE AND THE PURPOSES OF IS

The Functions of IS in Practice

Let's now follow an invented set of data through an imaginary information system. Input data from 100 people participating in a drug test for a major pharmaceutical company are entered into the company's information system (input data). Of the 100 participants in the study, half took the experimental medication while the other half took sugar pills. The input function accepts the drug test input data for both groups of participants and the storage function records it as stored data. When the time comes to analyse the data, the storage function retrieves the stored data about each set of patients and sends it to the processing function to compare one group against the other. The output function accepts the analysed data from the processing function and sends the results of the experiment, or information, out to users in the company. Company researchers may use the information to determine the effectiveness of the new medication and also understand its side effects.

The Purposes of IS

Information systems take many forms and serve a wide array of business and research purposes. Here are a few simple, yet powerful examples. One common type of business information system is an inventory control system. Inventory represents what a company has in stock for a given product. If a furniture retail store has 32 king beds on hand for sale to customers, one could also say it has 32 king beds in stock. Each style of furniture is called an item in the store's inventory.

An inventory control system keeps track of information about the items that the business stocks. Every time the furniture store sells an item or removes it from stock, the quantity of that item decreases. The inventory control system is used to track increases and decreases of furniture as it moves into or out of the store.

SCANNING, SELECTING, CAPTURING, AND STORING

Scanning refers to the development of searching for information, whether it is in one's internal and/or external environment. It involves observation and looking for sources of information. The effectiveness of information processing depends to a large degree on strategies used for seeking information and the resources available to conduct the search. The more people search for information, the more raw material they have for making decisions. A limited or narrow search might miss critical information needed for decision-making purposes. A broad search may take considerable time and increase the challenge of drawing conclusions from thousands of facts. Information technology also performs the task of scanning. Computer software applications scan databases for information based on the needs of the user.

Selecting refers to what people actually decide to pay attention to as a source or sources of information. While scanning involves looking for sources of information, selecting consists of selecting a source and using it for the purpose of collecting information. What people actually pay attention to may dictate the effectiveness of information processing. Some top executives tend to pay attention to information sources inside their organization, while others tend to pay attention to sources outside the organization. Each of these strategies has its strengths and weaknesses.

Once a source (or sources) of information is selected for processing purposes, the next step is to capture information. Obtaining information is the act of gathering it at its point of origin. This information-processing task is performed by input computer technologies such as the keyboard or a bar code reader. From an individual perspective, your eyes and ears capture information. Information-processing systems often require a mechanism for storing information so it can be used and/or reused later on. Storage devices include the human brain, notes, libraries, hard drives in a computer, and computer databases. Organizations store information in policy manuals and in their cultural practices, traditions, and habits.

INFORMATION-PROCESSING TASKS

There are seven generic information-processing tasks: Scanning, Selecting, Capturing, Storing, Creating, Communicating, and Implementing. In general, this set of tasks is performed in sequential fashion beginning with scanning and ending with implementation. In reality, however, these tasks may be performed out of sequence. In some cases a person or group may even skip a step. The information-processing framework presented below is an attempt to categorize and describe the entire spectrum of information-processing tasks performed by individuals, groups, and organizations. These tasks are presented in a sequence that tends to occur naturally in organizations and other social situations.

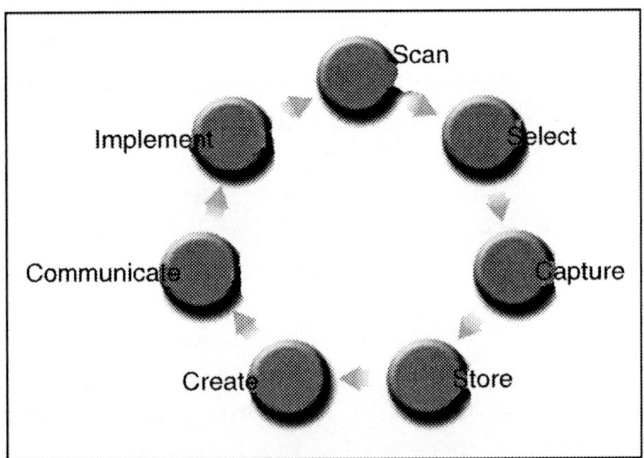

Each task serves a unique, but important role in the overall information-processing model. Ultimately, each task affects the end result of the information-processing effort. This is reflected in the performance measures we use to determine effectiveness and efficiency, such as satisfaction, return on investment, error rates, and employee turnover. In summary, each of these tasks may profoundly affect the overall effectiveness of information processing in organizations. Before we continue with an in-depth examination of information processing, the following example is used to briefly illustrate each of the seven information-processing tasks.

One day, John decides it's time buy a new car. This is a big investment for John, and, as you would suspect, he wants to construct the best possible decision. John begins searching (scanning) for information that will help him make the right decision. He begins searching the Internet, talking to friends and car dealers, and purchases an automobile buyers guide at the local bookstore. With so many information sources and car models available, John decides he must narrow his search and chooses to focus on mid-sized sport utility vehicles (SUVs) and the Internet and a buyers guide he recently purchased for fact-finding purposes (selecting). John documents his research findings in an electronic spreadsheet (capturing) where he lists the performance characteristics and attributes of various mid-sized SUVs. This spreadsheet is maintained on the hard drive of his home computer (storing). John is pressured by his wife to make a decision the following week. The next day, John examines the spreadsheet information, interprets it, and produces an overall score for each SUV he is considering buying (creating). John communicates his findings to his wife (communicating) and together they make a final decision. The next day they buy their new car (implementing), and John and his wife live happily ever after. John has performed all seven of the generic information-processing tasks to reach his final decision.

Creating, Communicating, and Implementing

Once information has been captured, it must now be interpreted and/or processed. Creating this interpretation involves the cognitive process of drawing a conclusion as to the meaning of information. This can be a very subjective process performed by decision makers. Meaning attached to information is often the result of the categories or labels that people use. Information can be interpreted as positive or negative, reliable or unreliable, relevant or irrelevant, as a threat or an opportunity. These are just a few of the possibilities. How one interprets information influences the actions one will take in response. This is key to understanding how our perceptions influence our behaviour, and how the same information might lead to different types of actions, depending on who is interpreting it. This task also involves processing information to obtain new information. Many computer software packages exist to analyse data and produce new forms of information for the user.

Communicating involves sending information to another location or other people. Information can be communicated via telephone, formal written letters, e-mail, personal contact, or through other channels. A variety of communication media is available for transmitting or receiving information. The media we may choose from varies in terms of media richness, which is the medium's capacity for carrying multiple information cues (*e.g.*, verbal, facial expressions, and voice inflections) and providing rapid feedback. Face-to-face discussion and telephone conversations have a higher degree of media richness than formal written letters or e-mail. Information technology sends and receives information and data using modems, satellites, and pagers. Once again, both people and machines perform communication-based information processing tasks. Implementation refers to the process of acting on information. Sometimes the greatest challenge is to act appropriately based on the information one has interpreted. Certain individuals, groups, and organizations do this well, while others do not. One may act quickly or slowly, with determination or reluctance, efficiently or inefficiently, immediately or sometime in the future.

The actions one may take usually involve change. Change can range from small-scale forms, such as changes in procedures, to larger-scale forms, such as revisions in overall corporate strategy. How one translates information into action often determines the end results or performance outcomes. Performance outcomes may include feelings of satisfaction, increased profits, increased market share, change in behaviour, customer satisfaction, and many others.

REVERSIBLE COMPUTATION

A computation is the abstract representation of a physical process in terms of *states,* and transitions between states or *events*. The definition of possible states and events is formulated in a *computation model,* such as the Turing machine or the finite automaton. For example, a Turing machine state is the complete sequence of symbols on its tape plus the head's position and internal symbol, and an event is the motion between two successive states, defined deterministically as a combination of read, write, move left and move right elementary motions. In order to perform a computation, a robust mapping is first established between a computation model and a physical system, meaning that states and events in the model are used to label states and events observed in the system, and that the choosen correspondence is sufficiently stable in respect to various kinds of perturbations.

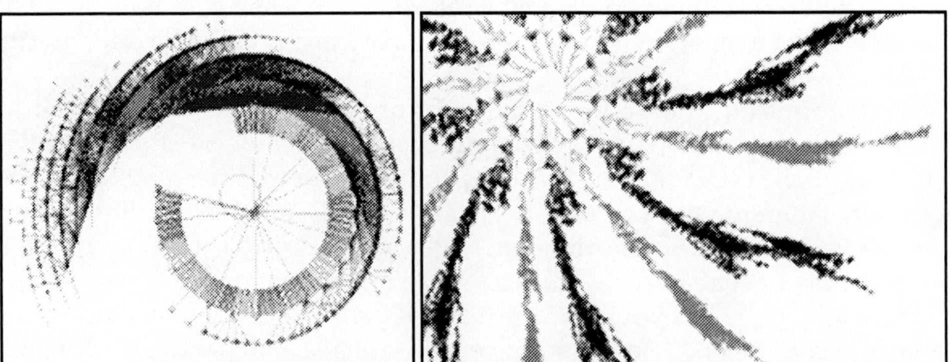

Figure : Irreversible computational dynamics

The system is then prepared in an initial state and is allowed to evolve through a path of events within the space of states, until it eventually reaches a state labelled as final. The discretized dynamics of the computational space may be represented with a directed graph, where nodes are possible states of the system and edges are events transforming a state into another.

Logical Reversibility

A function is said reversible if, given its output, it is always possible to determine back its input, which is the case when there is a one-to-one relationship between input and output states. If the space of states is finite, such a function is a permutation. Logical reversibility implies conservation of information.

When several input states are mapped onto the same output state, then the function is irreversible, since it is impossible by only knowing the final state to find back the initial state. In boolean algebra, NOT is reversible, while SET TO ONE is irreversible. Two-argument boolean functions like AND, OR, XOR are also irreversible, since they map 2^2 input states into 2^1 output states so that information is lost in the merging of paths, like shown in the following graph of a NAND computation, whose reverse evolution is no longer deterministic.

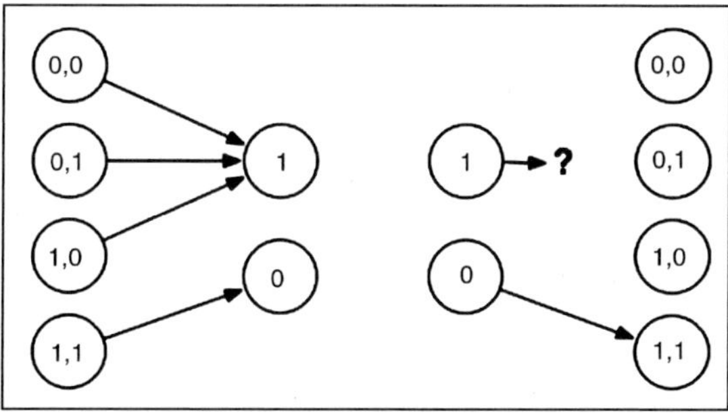

Figure: The right side tries to depict the inverse mapping to the left side

Physical Reversibility

Known laws of physics are reversible. This is the case both of classical mechanics, based on lagrangian/hamiltonian dynamics, and of standard quantum mechanics, where closed systems evolve by unitary transformations, which are objective and invertible. As a consequence, when a physical system performs an irreversible computation, the computation model's mapping indicates that the computing system cannot stay closed.

More precisely, since an irreversible computation reduces the space of physical information-bearing states, then their entropy must decrease by increasing the entropy of the non-information bearing states, representing the thermal part of the system.

In 1961 Landauer deliberate this thermodynamical argument, and proposed the following principle: if a physical system performs a logically irreversible classical computation, then it must increase the entropy of the environment with an absolute minimum of heat release of $kT \times \ln(2)$ per lost bit (where k is Boltzmann's constant and T the temperature, ie. about 3×10^{-21} joules at room temperature), which emphasizes two facts:

- The logical irreversibility of a computation implies the physical irreversibility of the system performing it ("information is physical");
- Logically reversible computations may be at least in principle intrinsically nondissipative.

Reversible Embedding of Irreversible Computations

Landauer further noticed that any irreversible computation may be transformed into a reversible one by embedding it into a larger computation where no information is lost, *e.g.* by replicating every output in the input ('sources') and every input in the output ('sinks').

For example, the NAND irreversible function seen above may be embedded in the following bijection, also known as Toffoli gate (the original function is indicated in red):

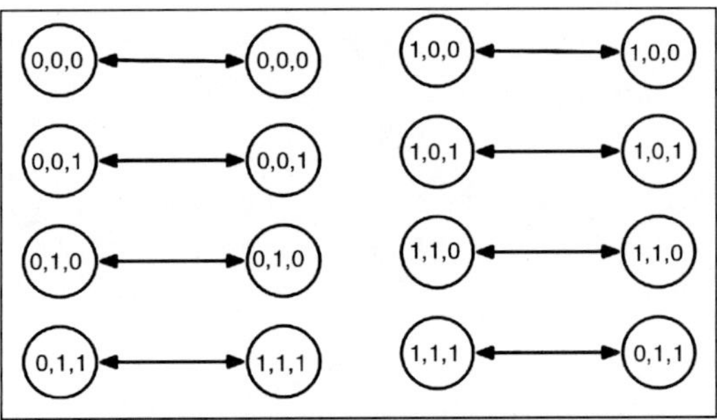

The additional bits of information, like Ariadne's threads, ensure that any computational path may be reversed: they are the garbage of the forward path and the program of the backwards path. Instead of losing them in the environment, they are kept in the controlled computational space.

Figure: Some railroad switches are reversible

Toffoli gates are universal reversible logic primitives, meaning that any reversible function may be constructed in terms of Toffoli gates. The Fredkin gate is another example of universal reversible logic primitive. It exchanges its two inputs depending on the state of a third control input, thus allowing to embed any computation into a conditional routing of paths carrying conserved signals.

Reversible Computation Models

The billiard-ball representation, invented by Fredkin and Toffoli, was one of the first computation models focusing on implementation with reversible physical components. Based on the laws of classical mechanics, it is equivalent to the formalism of kinetic theory of perfect gases. The presence of moving rigid spheres at specified points are defined as 1's, their absence as 0's. Interactions by means of right-angle collisions allow to construct various logic primitives, like for example the following 2-input, 3-output universal gate due to Feynman, who also proposed with Ressler a billiard-ball version of the Fredkin gate.

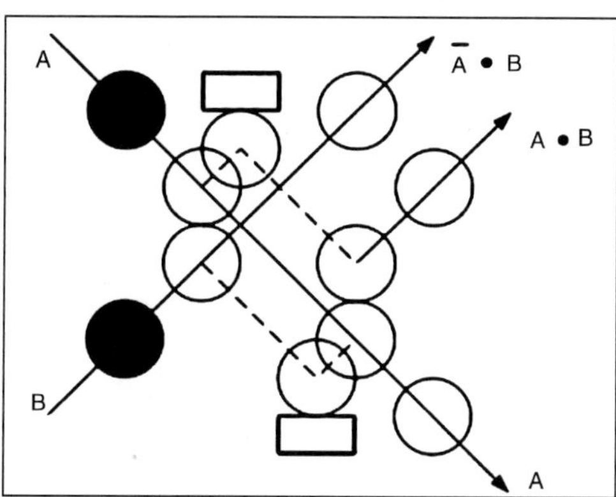

Figure: Feynman switch gate, B detects A without affecting its path

In practice, these computing spheres would be very unreliable, as instability arising from arbitrarily small perturbations would quickly generate chaotic deviations, producing an output saturated with errors. The errors may be corrected (for example, by adding potentials to stabilize the paths), however the error correction process is itself irreversible and dissipative - since it has to erase the erroneous information. Hence, error correction appears to be the only aspect of computation defining a lower bound to energy dissipation.

A stabler approach is that of the Brownian computation model, where thermal noise is on the contrary allowed to freely interact with a computing system near equilibrium. Potential energy barriers define the paths of a computational space, where the system walks randomly until it eventually reaches a final state. RNA polymerase, the enzyme involved in DNA transcription, is an example of a brownian logically reversible tape-copying machine. The DNA replication process also follows a similar mechanism, but adds a logically irreversible error-correcting step.

Lecerf-Bennett Reversal

The embedding technique is however insufficient to build a physically reversible universal computer, since the growing amount of information needing to be replicated for each event would saturate any finite memory. Then, computation would come to an end - unless the memory would be irreversibly erased, but then dissipation would have been merely postponed, and not avoided. This seemed to rule out the opportunity of useful reversible computing machines, until a remarkable solution was found by Bennett (earlier work by Lecerf anticipated its formal technique), showing that it is possible at least in principle to perform an unlimited amount of computation without any energy dissipation.

The reversible system shall compute the embedding function twice: the first time "forwards" to obtain and save the computation result, and the second time "backwards", as a mirror-image computation of the inverse function, de-computing the first step and returning the closed system to its initial state.

Figure: M. C. Escher, Swans (1956).

Logical Irreversibility and Maxwell's Demon

In 1867 Maxwell devised a thought experiment involving a finite microscopic "demon" capable of observing the motion of individual molecules. This demon guards a small hole separating two containers, filled with gas at the same temperature. When a molecule approaches, the demon checks its speed and then opens or closes a shutter, so as slower molecules always go in one container (cooling it), and faster molecules go in the other (heating it), in apparent violation of the 2nd law of thermodynamics.

A first important step toward the solution of this controversial paradox was taken in 1929 by Szilard who, after avoiding dualistic traps by substituting the intelligent demon with a simple machine, suggested that proper accounting of entropy is restored in the process of measuring the molecule position. This explanation became the standard one until 1981, when Bennett showed that the fundamentally dissipative step is surprisingly not the measurement (which can be done reversibly) but the logically irreversible erasure of demon's memory, to make room for new measurements.

Reversibility in Quantum Computation

Quantum computation takes advantage of the physical effects of superposition and entanglement, leading to a qualitatively new computation paradigm. In quantum mechanical computation models, all events occur by unitary transformations, so that all quantum gates are reversible.

Quantum systems are fewer susceptible to certains kinds of errors affecting classical computations, since their discrete spectrum prevents trajectories from becoming chaotic, so that, for example, a quantum "billiard ball model" is more reliable than its classical counterpart. However, quantum systems are also affected by new sources of error, as a consequence of interactions with the environment, such as the loss of quantum coherence. It is possible to correct generic quantum errors up to a limit, so as to reconstruct an error-free quantum state, at the price of performing an irreversible quantum erasure of the erroneous quantum information.

LIMITING POWER CONSUMPTION

Our techniques influence the fact that modern memory devices have multiple low-power states that retain the stored data. Each power state consumes a dissimilar amount of power, whereas the transitions between states involve different energy and performance overheads. The power states, their power consumptions, and transition overheads of RDRA Memory Chips, each of which can be transitioned independently. Memory accesses can only occur in active state, even though the data is retained even in power down state.

The thought behind our techniques is to have the memory controllerad just the power state of the memory devices so that their overall power consumption does not exceed the budget. We assume that the budget is pre-defined by the user or the manufacturer of the system containing the memory. Obviously, the budget has to be high enough that at least one device can be accessed at any time. Our techniques reserve enough power for active devices to be accessed. Thus, adhering to the power budget means that, when a memory device that is not in active state needs to be accessed, the controller may need to change the state of one or more other devices. *The main challenge is to design techniques to guide the memory controller in selecting power states, so that it can avoid exceeding the budget while minimizing transition overheads.* In particular, when the power budget is relatively low, some applications may suffer performance degradations; nevertheless, it is important to minimize these degradations by intelligently selecting power states.

Because the memory subsystem is a major energy consumer in several environments, *another important challenge for the state-selection techniques is to adhere to the power budget while enabling as much memory energy conservation as possible without excessive performance degradation.* Thus, instead of keeping the power consumption just below the budget, the techniques can reduce power consumption further, as long as the resulting performance degradation is acceptable. Again, we assume that the user or the manufacturer of the system pre-defines a maximum acceptable performance degradation resulting from energy conservation.

In the next four subsections, we introduce our techniques. The unifying principle behind the techniques is that they represent different approaches to solving the well-known Multi-Choice Knapsack problem (MCKP). The Knapsack technique is the optimal *static* approach, whereas the others are dynamic heuristics that differ in how they form and traverse a list of recently accessed memory devices. The fifth subsection describes extensions to the techniques that enable energy conservation without excessive performance degradation. The last sub section discusses the complexity and overheads our techniques impose on the memory controller.

Knapsack

This technique is based on the observation that the goal of limiting power consumption to a pre-established budget is equivalent to the MCKP. The budget represents the knapsack's capacity, whereas each memory device and potential power state represent an object. Objects are grouped by memory device, so that each group contains objects representing all the potential power states of the device.

The weight of each object is its power consumption, whereas the object's cost is the performance overhead of transitioning from its power state to active state. The goal is to pick one object from each set, so that the potential performance degradation is minimized under the constraint that the power budget is not exceeded. Typically, the optimal solution is the one in which the most memory devices can be in active state.

Based on this formulation, our Knapsack technique computes the optimal configuration of power states for a given power budget. Specifically, the configuration determines the number of devices that should be in each state. For example, assuming the RDRAM information of memory chips, the optimal configuration would be 1 chip in active state and 3 chips in nap state for a power budget of 1399 m Watts. (1399 m Watts = 25% of the range between the lowest possible budget, 1176 m Watts, and the highest possible budget, 2067 m Watts, plus the lowest budget.

However, out of these 1399 m Watts, the difference between the accessing and active power consumptions, 867 m Watts, is reserved to allow a chip to be activated. Thus, the actual 25% power budget for 4 RDRAM chips is 532 = 1399 - 867 m Watts. Henceforth, the absolute values we list for the power budgets already exclude the 867 m Watts.) For a budget of 755 m Watts (50% of the same range mentioned above plus 1176 m Watts minus 867 m Watts), the best configuration would be 2 chips in active state and 2 chips in nap state.

The computation of the optimal configuration is performed offline, so that the memory controller can be initialized with the configuration information. Although the MCKP is NP-hard, the number of memory devices is typically small enough that even a brute force solution is feasible. For example, our executions for 16 memory devices take only a few minutes to explore all the possible configurations. However, when the number of devices is moderate to large, a heuristic algorithm (and likely more search time) is required. For now, we use brute force.

Regardless of how the optimal configuration is computed, the initialization of the memory controller assigns the states described in the configuration to each device randomly. To guarantee that the power budget is not exceeded at run time, Knapsack manages power states dynamically as follows.

When the memory device to be accessed, the "target" device, is already in active state, no action is taken; it can be accessed without transitioning any other devices. When it is in a low-power state, an active device is selected to transition to the current power state of the target device. After this transition occurs, the target device can be activated and accessed. This approach maintains the invariant that the number of devices in each state is always as determined by solving the MCKP offline. To account for the locality of accesses across the different devices, Knapsack selects the active device to be transitioned to a low-power state using an LRU queue. Specifically, the LRU active device is selected as the victim.

The Knapsack technique. We assume 4 RDRAM memory chips and a power budget of 755mWatts. The figure shows one box per chip, listing the chip number on the left and its current power state on the right (A = active, S = standby, N = nap, and P = power down). When chip 2, currently in nap state, needs to be accessed: chips 0, 3, and 1 are in nap, active, and active state, respectively.

Chips 3 and 1 are on the LRU queue of active chips. This is one of the optimal configurations for this number of chips and power budget. Because the total power consumption is 660 m Watts at this point, simply activating chip 2 would violate the budget. To remain at an optimal configuration after chip 2 is activated, Knapsack changes the state of the LRU active chip (chip 3) to that of chip 2 and then allows the access to proceed.

The main problem with Knapsack is that it is only feasible when the number of devices is small enough that a heuristic algorithm can produce a close-to-optimal solution within a reasonable amount of time. Furthermore, every time a change in power budget is desired, Knapsack involves a re-computation of the configuration. Because re-computing the configuration may be time-consuming when the number of devices is relatively large, we next describe three techniques that do not rely on finding an optimal or close-to optimal configuration: LRU-Greedy, LRU-Smooth, and LRU-Ordered. The techniques leverage an LRU queue of memory devices. The main difference between them is the way each one traverses the LRU queue and which devices are included in it.

REFERENCES

- Cerofolini, G. F.: *Nanoscale Devices,* Springer-Verlag, Berlin, 2009.
- Kaushik, N L : *Nanotechnology and Micromachines,* Pearl Books, Delhi, 2007.
- Kumar, H.D. : *Material Science: Nanotechnology and Applications,* I.K. International, Delhi, 2011.
- Pope M and Sweenberg : *Electronic Processes in Organic Crystals and Polymers,* Oxford: Oxford University Press, 1999.
- Randy Frank: *Understanding Smart Sensors,* Artech House, Delhi, 2000.
- Suri, Shalini : *Nano Technology : Basic Science to Emerging Technology,* APH, Delhi, 2006.
- Tai-Ran Hsu: *MEMS and Microsystems: Design and Manufacture,* McGraw-Hill, London, 2001.

4

Instrumentation Techniques Utilization for Nanoscale Electronics

The main instrumentation techniques utilized at nanoscale to observe, manipulate, and measure nanosized components and devices. Unlike instrumentation techniques at other scales, which are extremely specialized, the nanoscale instrumentation has multiple purposes. For example, using an atomic force microscope (AFM) it is possible to manipulate nano-objects, to measure the distribution of the electrical resistivity or the electromagnetic field over a certain surface, to perform surface topography and lithographical processes, both with A resolution. Of course, a lot of instrumentation techniques, such as Raman or luminescence test equipments, is common for different device scales; these must however be properly changed in order to work with material samples, components, or devices that have at least one dimension of less than a few nm.

The instrumentation precise for nanoelectronics, which is mainly based on the principles of micro-electro-mechanical systems (MEMS) and nano-electro-mechanical systems (NEMS). These systems are the first to be treated in this chapter, but not with the aim to provide a full description of MEMS/NEMS and their applications, which could be the subject of a distinct book. In what follows we describe only those MEMS and NEMS devices with direct relevance to nanoelectronics and in particular, to the instrumentation used at the nanoscale. The specific instrumentation of nanoelectronics based on MEMS and NEMS, such as the above-mentioned AFM, will be subsequently described, with an emphasis on its capabilities and performances.

MEMS

Microelectromechanical systems (MEMS) are small integrated devices or systems that combine electrical and mechanical components. They range in size from the sub micrometer (or sub micron) level to the millimetre level, and there can be any number, from a few to millions, in a particular system.

MEMS extend the fabrication techniques developed for the integrated circuit industry to add mechanical elements such as beams, gears, diaphragms, and springs to devices.

MEMS are separate and distinct from the hypothetical vision of molecular nanotechnology or molecular electronics. MEMS are made up of components between 1 to 100 micrometres in size (*i.e.* 0.001 to 0.1 mm), and MEMS devices generally range in size from 20 micrometres (20 millionths of a metre) to a millimetre (*i.e.* 0.02 to 1.0 mm). They usually consist of a central unit that processes data (the microprocessor) and several components that interact with the outside such as microsensors. At these size scales, the standard constructs of classical physics are not always useful. Because of the large surface area to volume ratio of MEMS, surface effects such as electrostatics and wetting dominate volume effects such as inertia or thermal mass.

The potential of very small machines was appreciated before the technology existed that could make them, for example, Richard Feynman's famous 1959 lecture There's Plenty of Room at the Bottom. MEMS became practical once they could be fabricated using modified semiconductor device fabrication technologies, normally used to make electronics. These include moulding and plating, wet etching (KOH, TMAH) and dry etching (RIE and DRIE), electro discharge machining (EDM), and other technologies capable of manufacturing small devices. An early example of a MEMS device is the resonistor – an electromechanical monolithic resonator.

HISTORICAL BACKGROUND

The creation of the transistor at Bell Telephone Laboratories in 1947 sparked a fast-growing microelectronic technology. Jack Kilby of Texas Instruments built the first integrated circuit (IC) in 1958 using germanium (Ge) devices. It consisted of one transistor, three resistors, and one capacitor. The IC was implemented on a sliver of Ge that was glued on a glass slide. Later that same year Robert Noyce of Fairchild Semiconductor announced the development of a planar double-diffused Si IC. The complete transition from the original Ge transistors with grown and alloyed junctions to silicon (Si) planar double-diffused devices took about 10 years. The success of Si as an electronic material was due partly to its wide availability from silicon dioxide (SiO_2) (sand), resulting in potentially lower material costs relative to other semiconductors.

Since 1970, the complexity of ICs has doubled every two to three years. The minimum dimension of manufactured devices and ICs has decreased from 20 microns to the sub micron levels of today. Current ultra-large-scale-integration (ULSI) technology enables the fabrication of more than 10 million transistors and capacitors on a typical chip.

IC fabrication is needy upon sensors to provide input from the surrounding environment, just as control systems need actuators (also referred to as transducers) in order to carry out their desired functions. Due to the availability of sand as a material, much effort was put into developing Si processing and characterization tools. These tools are now being used to advance transducer technology. Today's IC technology far outstrips the original sensors and actuators in performance, size, and cost.

Attention in this locale was first focused on microsensor (*i.e.*, microfabricated sensor) development. The first microsensor, which has also been the most successful, was the Si pressure sensor. In 1954 it was discovered that the piezoresistive effect in Ge and Si had the potential to produce Ge and Si strain gauges with a gauge factor (*i.e.*, instrument sensitivity) 10 to 20 times greater than those based on metal films. As a result, Si strain gauges began to be developed commercially in 1958. The first high-volume pressure sensor was marketed by National Semiconductor in 1974. This sensor included a temperature controller for constant-temperature operation. Improvements in this technology since then have included the utilization of ion implantation for improved control of the piezoresistor fabrication. Si pressure sensors are now a billion-dollar industry.

Around 1982, the term micromachining came into use to designate the fabrication of micromechanical parts (such as pressure-sensor diaphragms or accelerometer suspension beams) for Si microsensors. The micromechanical parts were fabricated by selectively etching areas of the Si substrate away in order to leave behind the desired geometries. Isotropic etching of Si was developed in the early 1960s for transistor fabrication. Anisotropic etching of Si then came about in 1967. Various etch-stop techniques were subsequently developed to provide further process flexibility.

These techniques also shape the basis of the bulk micromachining processing techniques. Bulk micromachining designates the point at which the bulk of the Si substrate is etched away to leave behind the desired micromechanical elements. Bulk micromachining has remained a powerful technique for the fabrication of micromechanical elements. However, the need for flexibility in device design and performance improvement has motivated the development of new concepts and techniques for micromachining.

Among these is the sacrificial layer technique, first demonstrated in 1965 by Nathanson and Wickstrom, in which a layer of material is deposited between structural layers for mechanical separation and isolation. This layer is removed during the release etch to free the structural layers and to allow mechanical devices to move relative to the substrate. A layer is releasable when a sacrificial layer separates it from the substrate. The application of the sacrificial layer technique to micromachining in 1985 gave rise to surface micromachining, in which the Si substrate is primarily used as a mechanical support upon which the micromechanical elements are fabricated.

Prior to 1987, these micromechanical structures were incomplete in motion. During 1987-1988, a turning point was reached in micromachining when, for the first time, techniques for integrated fabrication of mechanisms (*i.e.* rigid bodies connected by joints for transmitting, controlling, or constraining relative movement) on Si were demonstrated. During a series of three separate workshops on microdynamics held in 1987, the term MEMS was coined. Equivalent terms for MEMS are microsystems (preferred in Europe) and micromachines (preferred in Japan).

MEMS TECHNOLOGY

Micro-Electro-Mechanical Systems, or MEMS, is a technology that in its most general form can be defined as miniaturized mechanical and electro-mechanical elements (*i.e.*, devices and structures) that are made using the techniques of microfabrication. The critical physical dimensions of MEMS devices can vary from well below one micron on the lower end of the dimensional spectrum, all the way to several millimetres. Likewise, the types of MEMS devices can vary from relatively simple structures having no moving elements, to extremely complex electromechanical systems with multiple moving elements under the control of integrated microelectronics. The one main criterion of MEMS is that there are at least some elements having some sort of mechanical functionality whether or not these elements can move. The term used to define MEMS varies in different parts of the world. In the United States they are predominantly called MEMS, while in some other parts of the world they are called "Microsystems Technology" or "micromachined devices".

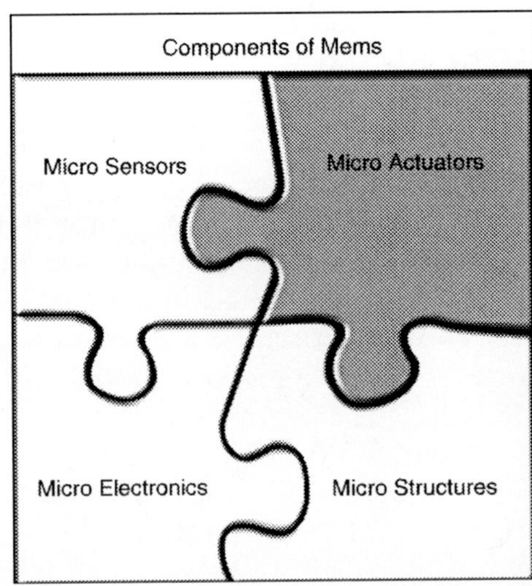

Whereas the functional elements of MEMS are miniaturized structures, sensors, actuators, and microelectronics, the most notable (and perhaps most interesting) elements are the microsensors and microactuators. Microsensors and microactuators are appropriately categorized as "transducers", which are defined as devices that convert energy from one form to another. In the case of microsensors, the device typically converts a measured mechanical signal into an electrical signal.

Over the precedent several decades MEMS researchers and developers have demonstrated an extremely large number of microsensors for almost every possible sensing modality including temperature, pressure, inertial forces, chemical species, magnetic fields, radiation, etc. Remarkably, many of these micromachined sensors have demonstrated performances exceeding those of their macroscale counterparts.

That is, the micromachined version of, for example, a pressure transducer, usually outperforms a pressure sensor made using the most precise macroscale level machining techniques. Not only is the performance of MEMS devices exceptional, but their technique of production leverages the same batch fabrication techniques used in the integrated circuit industry – which can translate into low per-device production costs, as well as many other benefits. Consequently, it is possible to not only achieve stellar device performance, but to do so at a relatively low cost level. Not surprisingly, silicon based discrete microsensors were quickly commercially exploited and the markets for these devices continue to grow at a rapid rate.

More newly, the MEMS research and development community has demonstrated a number of microactuators including: microvalves for control of gas and liquid flows; optical switches and mirrors to redirect or modulate light beams; independently controlled micromirror arrays for displays, microresonators for a number of different applications, micropumps to develop positive fluid pressures, microflaps to modulate airstreams on airfoils, as well as many others. Surprisingly, even though these microactuators are extremely small, they frequently can cause effects at the macroscale level; that is, these tiny actuators can perform mechanical feats far larger than their size would imply. For example, researchers have placed small microactuators on the leading edge of airfoils of an aircraft and have been able to steer the aircraft using only these microminiaturized devices.

The real possible of MEMS starts to become fulfilled when these miniaturized sensors, actuators, and structures can all be merged onto a common silicon substrate along with integrated circuits (*i.e.*, microelectronics). While the electronics are fabricated using integrated circuit (IC) process sequences (*e.g.*, CMOS, Bipolar, or BICMOS processes), the micromechanical components are fabricated using compatible "micromachining" processes that selectively etch away parts of the silicon wafer or add new structural layers to form the mechanical and electromechanical devices.

It is even more interesting if MEMS can be merged not only with microelectronics, but with other technologies such as photonics, nanotechnology, etc. This is sometimes called "heterogeneous integration."

Clearly, these technologies are filled with numerous commercial market opportunities.

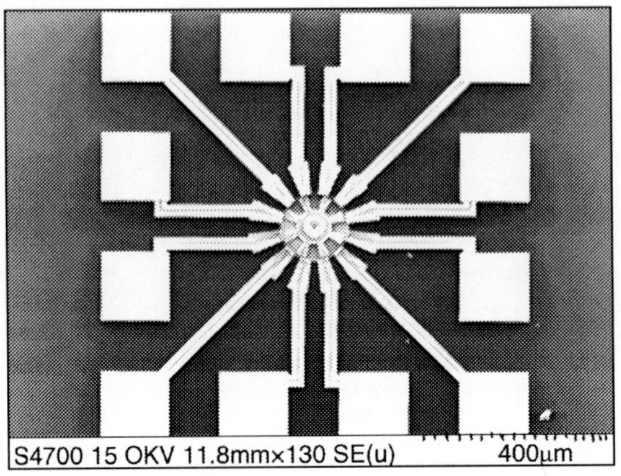

Figure: A surface micromachined electro-statically-actuated micromotor fabricated by the MNX. This device is an example of a MEMS-based microactuator.

While extra complex levels of integration are the future trend of MEMS technology, the present state-of-the-art is more modest and usually involves a single discrete microsensor, a single discrete microactuator, a single microsensor integrated with electronics, a multiplicity of essentially identical microsensors integrated with electronics, a single microactuator integrated with electronics, or a multiplicity of essentially identical microactuators integrated with electronics. Nevertheless, as MEMS fabrication techniques advance, the promise is an enormous design freedom wherein any type of microsensor and any type of microactuator can be merged with microelectronics as well as photonics, nanotechnology, etc., onto a single substrate.

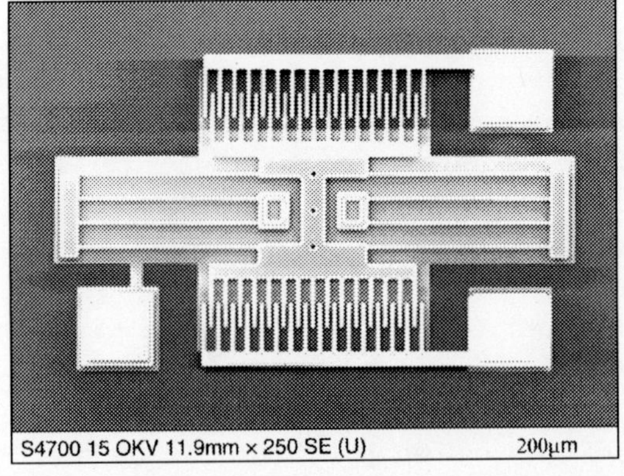

Figure: A surface micromachined resonator fabricated by the MNX. This device can be used as both a microsensor as well as a microactuator.

This apparition of MEMS whereby microsensors, microactuators and microelectronics and other technologies, can be integrated onto a single microchip is expected to be one of the most important technological breakthroughs of the future. This will enable the development of smart products by augmenting the computational ability of microelectronics with the perception and control capabilities of microsensors and microactuators. Microelectronic integrated circuits can be thought of as the "brains" of a system and MEMS augments this decision-making capability with "eyes" and "arms", to allow microsystems to sense and control the environment. Sensors gather information from the environment through measuring mechanical, thermal, biological, chemical, optical, and magnetic phenomena. The electronics then process the information derived from the sensors and through some decision making capability direct the actuators to respond by moving, positioning, regulating, pumping, and filtering, thereby controlling the environment for some desired outcome or purpose. Furthermore, because MEMS devices are manufactured using batch fabrication techniques, similar to ICs, unprecedented levels of functionality, reliability, and sophistication can be placed on a small silicon chip at a relatively low cost. MEMS technology is extremely diverse and fertile, both in its expected application areas, as well as in how the devices are designed and manufactured. Already, MEMS is revolutionizing many product categories by enabling complete systems-on-a-chip to be realised.

Nanotechnology is the aptitude to manipulate matter at the atomic or molecular level to make something useful at the nano-dimensional scale. Basically, there are two approaches in implementation: the top-down and the bottom-up. In the top-down approach, devices and structures are made using many of the same techniques as used in MEMS except they are made smaller in size, usually by employing more advanced photolithography and etching techniques. The bottom-up approach typically involves deposition, growing, or self-assembly technologies. The advantages of nano-dimensional devices over MEMS involve benefits mostly derived from the scaling laws, which can also present some challenges as well.

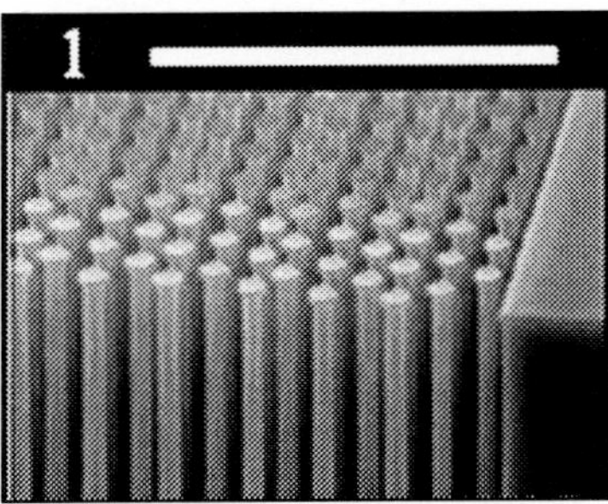

Figure: An array of sub-micron posts made using top-down nanotechnology fabrication techniques.

A number of experts believe that nanotechnology promises to: a). allow us to put essentially every atom or molecule in the place and position desired – that is, exact positional control for assembly, b). allow us to make almost any structure or material consistent with the laws of physics that can be specified at the atomic or molecular level; and c). allow us to have manufacturing costs not greatly exceeding the cost of the required raw materials and energy used in fabrication (*i.e.*, massive parallelism).

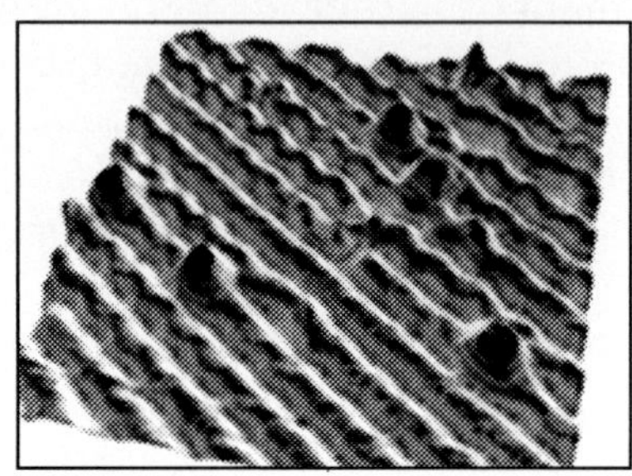

Figure: A colorized image of a scanning-tunnelling microscope image of a surface, which is a common imaging technique used in nanotechnology.

Even though MEMS and Nanotechnology are sometimes cited as separate and distinct technologies, in reality the distinction between the two is not so clear-cut. In fact, these two technologies are highly dependent on one another. The well-known scanning tunnelling-tip microscope (STM) which is used to detect individual atoms and molecules on the nanometer scale is a MEMS device. Similarly the atomic force microscope (AFM) which is used to manipulate the placement and position of individual atoms and molecules on the surface of a substrate is a MEMS device as well. In fact, a variety of MEMS technologies are required in order to interface with the nanoscale domain.

Similarly, many MEMS technologies are becoming dependent on nanotechnologies for successful new products. For example, the crash airbag accelerometers that are manufactured using MEMS technology can have their long-term reliability degraded due to dynamic in-use stiction effects between the proof mass and the substrate. A nanotechnology called Self-Assembled Monolayers (SAM) coatings are now routinely used to treat the surfaces of the moving MEMS elements so as to prevent stiction effects from occurring over the product's life.

A lot of experts have concluded that MEMS and nanotechnology are two different labels for what is essentially a technology encompassing highly miniaturized things that cannot be seen with the human eye. Note that a similar broad definition exists in the integrated circuits domain which is frequently referred to as microelectronics technology even though state-of-the-art IC technologies typically have devices with dimensions of tens of nanometers. Whether or not MEMS and nanotechnology are one in the same, it is unquestioned that there are overwhelming mutual dependencies between these two technologies that will only increase in time. Perhaps what is most important are the common benefits afforded by these technologies, including: increased information capabilities; miniaturization of systems; new materials resulting from new science at miniature dimensional scales; and increased functionality and autonomy for systems.

MATERIALS FOR MEMS MANUFACTURING

The manufacture of MEMS evolved from the process technology in semiconductor device fabrication, *i.e.* the basic techniques are deposition of material layers, patterning by photolithography and etching to produce the required shapes.

Silicon

Silicon is the fabric used to create most integrated circuits used in consumer electronics in the modern world. The economies of scale, ready availability of cheap high-quality materials and ability to incorporate electronic functionality make silicon attractive for a wide variety of MEMS applications. Silicon also has significant advantages engendered through its material properties. In single crystal form, silicon is an almost perfect Hookean material, meaning that when it is flexed there is virtually no hysteresis and hence almost no energy dissipation. As well as making for highly repeatable motion, this also makes silicon very reliable as it suffers very little fatigue and can have service lifetimes in the range of billions to trillions of cycles without breaking.

Polymers

Still though the electronics industry provides an economy of scale for the silicon industry, crystalline silicon is still a complex and relatively expensive material to produce. Polymers on the other hand can be produced in huge volumes, with a great variety of material characteristics. MEMS devices can be made from polymers by processes such as injection moulding, embossing or stereolithography and are especially well suited to microfluidic applications such as disposable blood testing cartridges.

Metals

Metals be able to also be used to create MEMS elements. While metals do not have some of the advantages displayed by silicon in terms of mechanical properties, when used within their limitations, metals can exhibit very high degrees of reliability. Metals can be deposited by electroplating, evaporation, and sputtering processes. Commonly used metals include gold, nickel, aluminium, copper, chromium, titanium, tungsten, platinum, and silver.

Ceramics

The nitrides of silicon, aluminium and titanium as well as silicon carbide and other ceramics are increasingly applied in MEMS fabrication due to advantageous combinations of material properties. AlN crystallizes in the wurtzite structure and thus shows pyroelectric and piezoelectric properties enabling sensors, for instance, with sensitivity to normal and shear forces. TiN, on the other hand, exhibits a high electrical conductivity and large elastic modulus allowing to realise electrostatic MEMS actuation schemes with ultrathin membranes. Moreover, the high resistance of TiN against biocorrosion qualifies the material for applications in biogenic environments and in biosensors.

MEMS MANUFACTURING TECHNOLOGIES

Bulk Micromachining

Bulk micromachining is the oldest example of silicon based MEMS. The whole thickness of a silicon wafer is used for building the micromechanical structures. Silicon is machined using various etching processes. Anodic bonding of glass plates or additional silicon wafers is used for adding features in the third dimension and for hermetic encapsulation. Bulk micromachining has been essential in enabling high performance pressure sensors and accelerometers that have changed the shape of the sensor industry in the 80's and 90's.

Surface Micromachining

Surface micromachining uses layers deposited on the surface of a substrate as the structural materials, rather than using the substrate itself. Surface micromachining was created in the late 1980s to render micromachining of silicon more compatible with planar integrated circuit technology, with the goal of combining MEMS and integrated circuits on the same silicon wafer. The original surface micromachining concept was based on thin polycrystalline silicon layers patterned as movable mechanical structures and released by sacrificial etching of the underlying oxide layer. Interdigital comb electrodes were used to produce in-plane forces and to detect in-plane movement capacitively. This MEMS paradigm has enabled the manufacturing of low cost accelerometers for *e.g.* automotive air-bag systems and other applications where low performance and/or high g-ranges are sufficient. Analog Devices have pioneered the industrialization of surface micromachining and have realised the co-integration of MEMS and integrated circuits.

High Aspect Ratio (HAR) Silicon Micromachining

Both size and surface silicon micromachining are used in the industrial production of sensors, ink-jet nozzles, and other devices. But in many cases the distinction between these two has diminished. A new etching technology, deep reactive-ion etching, has made it possible to combine good performance typical of bulk micromachining with comb structures and in-plane operation typical of surface micromachining. While it is common in surface micromachining to have structural layer thickness in the range of 2 μm, in HAR silicon micromachining the thickness can be from 10 to 100 μm. The materials commonly used in HAR silicon micromachining are thick polycrystalline silicon, known as epi-poly, and bonded silicon-on-insulator (SOI) wafers although processes for bulk silicon wafer also have been created (SCREAM). Bonding a second wafer by glass frit bonding, anodic bonding or alloy bonding is used to protect the MEMS structures. Integrated circuits are typically not combined with HAR silicon micromachining. The consensus of the industry at the moment seems to be that the flexibility and reduced process complexity obtained by having the two functions separated far outweighs the small penalty in packaging.

A forgotten the past regarding surface micromachining revolved around the choice of polysilicon A or B. Fine grained (<300A grain size, US4897360), post strain annealed pure polysilicon was advocated by Prof Henry Guckel (U. Wisconsin); while a larger grain, doped stress controlled polysilicon was advocated by the UC Berkeley group.

MANUFACTURE TECHNOLOGIES

The three characteristic features of MEMS fabrication technologies are miniaturization, multiplicity, and microelectronics. Miniaturization enables the production of compact, quick-response devices. Multiplicity refers to the batch fabrication inherent in semiconductor processing, which allows thousands or millions of components to be easily and concurrently fabricated. Microelectronics provides the intelligence to MEMS and allows the monolithic merger of sensors, actuators, and logic to build closed-loop feedback components and systems. The successful miniaturization and multiplicity of traditional electronics systems would not have been possible without IC fabrication technology. Therefore, IC fabrication technology, or microfabrication, has so far been the primary enabling technology for the development of MEMS. Microfabrication provides a powerful tool for batch processing and miniaturization of mechanical systems into a dimensional domain not accessible by conventional (machining) techniques. Furthermore, microfabrication provides an opportunity for integration of mechanical systems with electronics to develop high-performance closed-loop-controlled MEMS.

Advances in IC technology in the last decade have brought about corresponding progress in MEMS fabrication processes. Manufacturing processes allow for the monolithic integration of microelectromechanical structures with driving, controlling, and signal-processing electronics. This integration promises to improve the performance of micromechanical devices as well as reduce the cost of manufacturing, packaging, and instrumenting these devices.

IC Fabrication

Any conversation of MEMS requires a basic understanding of IC fabrication technology, or microfabrication, the primary enabling technology for the development of MEMS. The major steps in IC fabrication technology are film growth, doping, lithography, etching, dicing, and packaging.

Film Growth: Usually, a polished Si wafer is used as the substrate, on which a thin film is grown. The film, which may be epitaxial Si, SiO_2, silicon nitride (Si_3N_4), polycrystalline Si (polysilicon), or metal, is used to build both active or passive components and interconnections between circuits.

Doping: To modulate the properties of the device layer, a low and controllable level of an atomic impurity may be introduced into the layer by thermal diffusion or ion implantation.

Lithography: A pattern on a mask is then transferred to the film by means of a photosensitive (*i.e.*, light sensitive) chemical known as a photoresist. The process of pattern generation and transfer is called photolithography. A typical mask consists of a glass plate coated with a patterned chromium (Cr) film.

Etching: Next is the selective removal of unwanted regions of a film or substrate for pattern delineation. Wet chemical etching or dry etching may be used. Etch-mask materials are used at various stages in the removal process to selectively prevent those portions of the material from being etched. These materials include SiO2, Si3N4, and hard-baked photoresist.

Dicing: The finished wafer is sawed or machined into small squares, or dice, from which electronic components can be made.

Packaging: The individual sections are then packaged, a process that involves physically locating, connecting, and protecting a device or component. MEMS design is strongly coupled to the packaging requirements, which in turn are dictated by the application environment.

Bulk Micromachining and Wafer Bonding

Bulk micromachining is an addition of IC technology for the fabrication of 3D structures. Bulk micromachining of Si uses wet- and dry-etching techniques in conjunction with etch masks and etch stops to sculpt micromechanical devices from the Si substrate. The two key capabilities that make bulk micromachining a viable technology are:

- Anisotropic etchants of Si, such as ethylenediamine and pyrocatechol (EDP), potassium hydroxide (KOH), and hydrazine (N2H4). These preferentially etch single crystal Si along given crystal planes.

- Etch masks and etch-stop techniques that can be used with Si anisotropic etchants to selectively prevent regions of Si from being etched. Good etch masks are provided by SiO_2 and Si_3N_4, and some metallic thin films such as Cr and Au (gold).

A drawback of wet anisotropic etching is that the microstructure geometry is defined by the internal crystalline structure of the substrate. Consequently, fabricating multiple, interconnected micromechanical structures of free-form geometry is often difficult or impossible.

Two additional processing techniques have extended the range of traditional bulk micromachining technology: deep anisotropic dry etching and wafer bonding.

Reactive gas plasmas can perform deep anisotropic dry etching of Si wafers, up to a depth of a few hundred microns, while maintaining smooth vertical sidewall profiles. The other technology, wafer bonding, permits a Si substrate to be attached to another substrate, typically Si or glass. Used in combination, anisotropic etching and wafer bonding techniques can construct 3D complex microstructures such as microvalves and micropumps.

Surface Micromachining

Surface micromachining enables the fabrication of compound multicomponent integrated micromechanical structures that would not be possible with traditional bulk micromachining. This technique encases specific structural parts of a device in layers of a sacrificial material during the fabrication process.

The substrate wafer is used primarily as a mechanical support on which multiple alternating layers of structural and sacrificial material are deposited and patterned to realise micromechanical structures. The sacrificial material is then dissolved in a chemical etchant that does not attack the structural parts. The most widely used surface micromachining technique, polysilicon surface micromachining, uses SiO_2 as the sacrificial material and polysilicon as the structural material.

The microelectronic fabrication industry typically grows polysilicon, silicon nitride, and silicon dioxide films using recipes that minimize time. Unfortunately, a deposition process that is optimized to speed does not always create a low internal stress film. In fact, most of these films have internal stresses that are highly compressive (tending to contract). A freestanding plate of highly compressive polysilicon that is held at all its edges will buckle (*i.e.*, collapse or give way). This is highly undesirable. The solution is to modify the film deposition process to control the internal stress by making it stress-free or slightly tensile.

One way to do this is to dope the film with boron, phosphorus, or arsenic. However, a doped polysilicon film is conductive, and this property may interfere with the mechanical devices incorporated electronics. Another problem with doped polysilicon is that it is roughened by hydrofluoric acid (HF), which is commonly used to free sections of the final mechanical device from the substrate.

Rough polysilicon has different mechanical properties than smooth polysilicon. Therefore, the amount of roughening must be taken into account when designing the mechanical parts of the micro device.

A better way to manage the stress in polysilicon is through post annealing, which involves the deposition of pure, fine-grained, compressive (*i.e.*, can be compressed) polysilicon. Annealing the polysilicon after deposition at elevated temperatures can change the film to be stress-free or tensile. The annealing temperature sets the film's final stress. After this, electronics can then be incorporated into polysilicon films through selective doping, and hydrofluoric acid will not change the mechanical properties of the material.

Deposition temperature and the film's silicon to nitride ratio can control the stress of a silicon nitride (Si_3N_4) film. The films can be deposited in compression, stress-free, or in tension. Deposition temperature and post annealing can control silicon dioxide (SiO_2) film stress. Because it is difficult to control the stress of SiO_2 accurately, SiO_2 is typically not used as a mechanical material by itself, but as electronic isolation or as a sacrificial layer under polysilicon.

Micromolding Process

In the micromolding process, microstructures are fabricated using molds to define the deposition of the structural layer. The structural material is deposited only in those areas constituting the microdevice structure, in contrast to bulk and surface micromachining, which feature blanket deposition of the structural material followed by etching to realise the final device geometry. After the structural layer deposition, the mold is dissolved in a chemical etchant that does not attack the structural material. One of the most prominent micromolding processes is the LIGA process. LIGA is a German acronym standing for lithographie, galvanoformung, und abformung (lithography, electroplating, and moulding). This process can be used for the manufacture of high-aspect-ratio 3D microstructures in a wide variety of materials, such as metals, polymers, ceramics, and glasses. Photosensitive polyimides are also used for fabricating plating molds. The photolithography process is similar to conventional photolithography, except that polyimide works as a negative resist.

FABRICATING MEMS AND NANOTECHNOLOGY

MEMS fabrication is an extremely exciting endeavour due to the customised nature of process technologies and the diversity of processing capabilities. MEMS fabrication uses many of the same techniques that are used in the integrated circuit domain such as oxidation, diffusion, ion implantation, LPCVD, sputtering, etc., and combines these capabilities with highly specialized micromachining processes.

Bulk Micromachining

The oldest micromachining knowledge is bulk micromachining. This technique involves the selective removal of the substrate material in order to realise miniaturized mechanical components. Bulk micromachining can be accomplished using chemical or physical means, with chemical means being far more widely used in the MEMS industry.

A extensively used bulk micromachining technique is chemical wet etching, which involves the immersion of a substrate into a solution of reactive chemical that will etch exposed regions of the substrate at measurable rates. Chemical wet etching is popular in MEMS because it can provide a very high etch rate and selectivity.

Furthermore, the etch rates and selectivity can be modified by: altering the chemical composition of the etch solution; adjusting the etch solution temperature; modifying the dopant concentration of the substrate; and modifying which crystallographic planes of the substrate are exposed to the etchant solution.

There are two universal types of chemical wet etching in bulk micromachining: isotropic wet etching and anisotropic wet etching. In isotropic wet etching, the etch rate is not dependent on the crystallographic orientation of the substrate and the etching proceeds in all directions at equal rates. In theory, lateral etching under the masking layer etches at the same rate as the etch rate in normal direction. However, in practice lateral etching is usually much slower without stirring, and consequently isotropic wet etching is almost always performed with vigorous stirring of the etchant solution. The profile of the etch using an isotopic wet etchant with and without stirring of the etchant solution.

Any etching process requires a masking material to be used, with preferably a high selectivity relative to the substrate material. Common masking materials for isotropic wet silicon etching include silicon dioxide and silicon nitride. Silicon nitride has a much lower etch rate compared to silicon dioxide and therefore is more frequently used.

The scrape rate of some isotropic wet etchant solution mixtures are dependent on the dopant concentration of the substrate material. For example: the commonly used mixture of $HC_2H_3O_2$:HNO_3:HF in the ratio of 8:3:1 will etch highly doped silicon ($> 5 \times 1018$ atoms/cm^3) at a rate of 50 to 200 microns/hour, but will etch lightly doped silicon material at a rate 150 times less. Nevertheless, the etch rate selectivity with respect to dopant concentration is highly dependent on solution mixture.

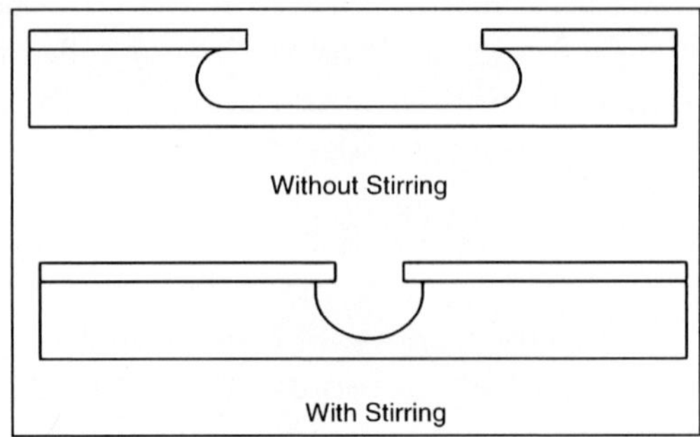

Figure : Illustration of the etch profile, with and without stirring, using an isotropic wet chemical etchant.

The a great deal more widely used wet etchants for silicon micromachining are anisotropic wet etchants. Anisotropic wet etching involves the immersion of the substrate into a chemical solution wherein the etch rate is dependent on crystallographic orientation of the substrate. The mechanism by which the etching varies according to silicon crystal planes is attributed to the different bond configurations and atomic density that the different planes exposed to the etchant solution. Wet anisotropic chemical etching is typically described in terms of etch rates according to the different normal crystallographic places, usually <100>, <110>, and <111>. In general, silicon anisotropic etching etches more slowly along the <111> planes than all the other planes in the lattice and the difference in etch rate between the different lattice directions can be as high as 1000 to 1. It is thought that the reason for the slower etch rate of the <111> planes is that these planes have the highest density of exposed silicon atoms in the etchant solution, as well as 3 silicon bonds below the plane, thereby leading to some amount of chemical shielding of the surface.

The aptitude to delineate the different crystal planes of the silicon lattice in anisotropic wet chemical etching provides a high-resolution etch capability with reasonably tight dimensional control. It also provides the ability for two-sided processing to embody self-isolated structures wherein only one side is exposed to the environment. This assists in packaging of the device and is very useful for MEMS devices exposed to harsh environments, such as pressure sensors. Anisotropic etching techniques have been around for over 25 years and are commonly used in the manufacturing of silicon pressure sensors as well as bulk micromachined accelerometers.

An illustration of some of the shapes that are possible using anisotropic wet etching of <100> oriented silicon substrates including an inverted pyramidal and a flat bottomed trapezoidal etch pit. Note that the shape of the etch pattern is primarily determined by the slower etching <111> planes. SEM photographs of a silicon substrate after an anisotropic wet etching. A trapezoidal etch pit that has been subsequently diced across the etch pit, the backside of a thin membrane that could be used to make a pressure sensor. It is important to note that the etch profiles are only for a <100> oriented silicon wafer; substrates with other crystallographic orientations will exhibit different shapes. Occasionally, substrates with other orientations are used in MEMS fabrication, but given the cost, lead times and availability, the vast majority of substrates used in bulk micromachining have <100> orientation.

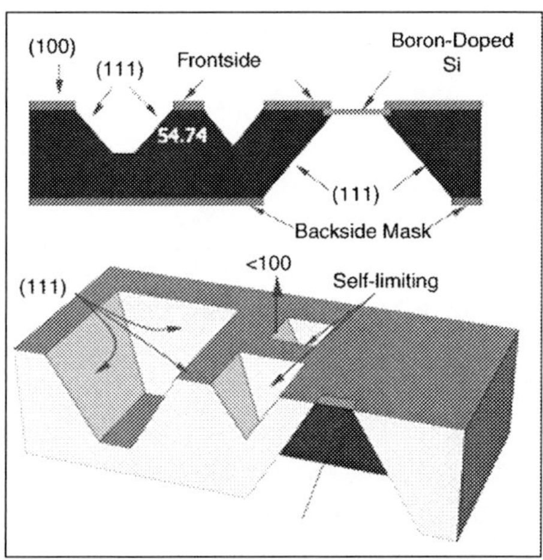

Figure : Illustration of shape of the etch profiles of a <100> oriented silicon substrate after immersion in an anisotropic wet etchant solution.

Figures : a and b: SEMS of a <100> oriented silicon substrate after immersion in an anisotropic wet etchant.

Functional anisotropic wet etching requires the ability to successfully mask certain areas of the substrate and consequently an important criterion for selecting an etchant is the availability of good masking materials. Silicon nitride is a commonly used masking material for anisotropic wet etchants since it has a very low etch rate in most etchant solutions. Some care must be exercised in the type of silicon nitride used, since any pin hole defects will result in the attack of the underlying silicon. Also, some low-stress silicon-rich nitrides can etch at much higher rates compared to stoichiometric silicon nitride formulations. Thermally grown SiO_2 is frequently used as a masking material, but some care must be exercised to ensure a sufficiently thick masking layer when using KOH etchants, since the etch rates of oxide can be high. Photoresists are unusable in any anisotropic etchant. Many metals including Ta, Au, Cr, Ag, and Cu hold up well in EDP and Al holds up in TMAH under certain conditions.

In common, the etch rate, etch rate ratios (100)/(111), and etch selectivities of anisotropic etchants are strongly dependent on chemical composition and temperature of the etchant solution regularly, when using bulk micromachining it is desirable to make thin membranes of silicon or control the etch depths very precisely. As with any chemical process, the uniformity of the etching can vary across the substrate, making this difficult. Timed etches whereby the etch depth is determined by multiplying the etch rate by the etch time are difficult to control and etch depth is very dependent on sample thickness uniformity, etchant species diffusion effects, loading effects, etchant aging, surface preparation, etc. To allow a higher level of precision in anisotropic etching the MEMS field has developed solutions to this problem, namely etch stops. Etch stops are very useful to control the etching process and provide uniform etch depths across the wafer, from wafer to wafer, and from wafer lot to wafer lot. There are two basic types of etch stop techniques that are used in micromachining: dopant etch stops and electrochemical etch stops.

SURFACE MICROMACHINING TECHNOLOGY

Surface micromachining is one more very popular technology used for the fabrication of MEMS devices. There are a very large number of variations of how surface micromachining is performed, depending on the materials and etchant combinations that are used. However, the common theme involves a sequence of steps starting with the deposition of some thin-film material to act as a temporary mechanical layer onto which the actual device layers are built; followed by the deposition and patterning of the thin-film device layer of material which is referred to as the structural layer; then followed by the removal of the temporary layer to release the mechanical structure layer from the constraint of the underlying layer, thereby allowing the structural layer to move. An illustration of a surface micromachining process, wherein an oxide layer is deposited and patterned. This oxide layer is temporary and is commonly referred to as the sacrificial layer. Subsequently, a thin film layer of polysilicon is deposited and patterned and this layer is the structural mechanical layer. Lastly, the temporary sacrificial layer is removed and the polysilicon layer is now free to move as a cantilever.

A number of the reasons surface micromachining is so popular is that it provides for precise dimensional control in the vertical direction. This is due to the fact that the structural and sacrificial layer thicknesses are defined by deposited film thicknesses which can be accurately controlled. Also, surface micromachining provides for precise dimensional control in the horizontal direction, since the structural layer tolerance is defined by the fidelity of the photolithography and etch processes used. Other benefits of surface micromachining are that a large variety of structure, sacrificial and etchant combinations can be used; some are compatible with microelectronics devices to enable integrated MEMS devices. Surface micromachining frequently exploits the deposition characteristics of thin-films such as conformal coverage using LPCVD. Lastly, surface micromachining uses single-sided wafer processing and is relatively simple. This allows higher integration density and lower resultant per die cost compared to bulk micromachining.

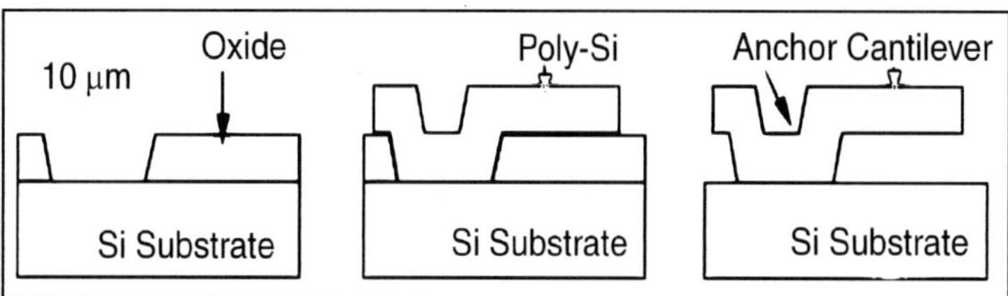

Figure : Illustration of a surface micromachining process.

One of the disadvantages of outside micromachining is that the mechanical properties of most deposited thin-films are usually unknown and must be measured. Also it is common for these types of films to have a high state of residual stress, frequently necessitating a high temperature anneal to reduce residual stress in the structural layer. Also, the reproducibility of the mechanical properties in these films can be difficult to achieve. Additionally, the release of the structural layer can be difficult due to a stiction effect whereby the structural layer is pulled down and stuck to the underlying substrate due to capillary forces during release. Stiction can also occur in use and an anti-stiction coating material may be needed.

The majority commonly used surface micromachining process and material combination is a PSG sacrificial layer, a doped polysilicon structural layer, and the use of Hydrofluoric acid as the etchant to remove the PSG sacrificial layer and release the device. This type of surface micromachining process is used to fabricate the Analog Devices integrated MEMS accelerometer device used for crash airbag deployment. SEMs of two surface micromachined polysilicon MEMS devices.

Another variation of the surface micromachining process is to use a metal structural layer, a polymer layer as the sacrificial layer, and an O_2 plasma as the etchant. The advantage of this process is that the temperature of the sacrificial and structural layer depositions are sufficiently low so as not to degrade any microelectronics in the underlying silicon substrate, thereby integrating MEMS with electronics. Also, since the sacrificial layer is removed without immersion in a liquid, problems associated with stiction during release are avoided. A process similar to this is used to produce the Texas Instruments Digital Light Processor (DLP) device used in projection systems.

Micromachining Process and Wafer Bonding

Wafer bonding is a micromachining process that is analogous to welding in the macroscale world and involves the joining of two (or more) wafers together to create a multi-wafer stack. There are three basic types of wafer bonding including: direct or fusion bonding; field-assisted or anodic bonding; and bonding using an intermediate layer. In general, all bonding techniques require substrates that are very flat, smooth, and clean, in order for the wafer bonding to be successful and free of voids.

Direct or synthesis bonding is typically used to mate two silicon wafers together or alternatively to mate one silicon wafer to another silicon wafer that has been oxidized. Direct wafer bonding can be performed on other combinations, such as bare silicon to a silicon wafer with a thin-film of silicon nitride on the surface as well.

The wafer bonding is analogous to welding in the macroscale world. Wafer bonding is used to attach a thick layer of single crystal silicon onto another wafer. This can be extremely useful when it is desired to have a thick layer of material for applications requiring appreciable mass or in applications where the material properties of single crystal silicon are advantageous over those of thin-film LPCVD materials. Direct wafer bonding is also used to fabricate Silicon-On-Insulator (SOI) wafers having device layers several microns or more in thickness.

One more popular wafer bonding technique is anodic bonding. In anodic bonding a silicon wafer is bonded to a Pyrex 7740 wafer using an electric field and elevated temperature. The two wafers can be pre-processed prior to bonding and can be aligned during the bonding procedure. The mechanism by which anodic bonding works is based on the fact that Pyrex 7740 has a high concentration of Na+ ions; a positive voltage applied to the silicon wafer drives the Na+ ions from the Pyrex glass surface, thereby creating a negative charge at glass surface. The elevated temperature during the bonding process allows the Na^+ ions to migrate in the glass with relative ease. When the Na+ ions reach the interface, a high field results between silicon and glass, and this combined with the elevated temperatures fuses the two wafers together. As with direct wafer bonding, it is imperative that the wafers are flat, smooth, and clean and that the anodic bonding process is performed in a very clean environment. An advantage of this process is that Pyrex 7740 has a thermal expansion coefficient nearly equal to silicon and therefore there is a low value of residual stress in the layers. Anodic bonding is a widely used technique for MEMS packaging. In addition to direct and anodic bonding there are other wafer bonding techniques that are used in MEMS fabrication. One technique is eutectic bonding and involves the bonding of a silicon substrate to another silicon substrate at an elevated temperature using an intermediate layer of gold on the surface of one of the wafers. Eutectic bonding works because the diffusion of gold into silicon is extremely rapid at elevated temperatures. In fact this is a preferred technique of wafer bonding at relatively low temperatures. One more wafer bonding technique used in MEMS is glass frit bonding. In this process a glass is spun or screen-printed onto a substrate surface. Subsequently this wafer is physically contacted to another wafer and the composite is annealed to flow the glass intermediate layer and bond the two wafers. Finally, various polymers can be used as intermediate layers to bond wafers including epoxy resins, photoresists, polyimides, silicones, etc. This technique is commonly used during various fabrication steps in MEMS such as when the device wafer becomes too fragile to handle without mechanical support.

HIGH-ASPECT RATIO MEMS FABRICATION TECHNOLOGIES

Deep Reactive Ion Etching of Silicon

Deep reactive ion etching or DRIE is a relatively new fabrication technology that has been extensively adopted by the MEMS community. This technology enables very high aspect ratio etches to be performed into silicon substrates. The sidewalls of the etched holes are nearly vertical and the depth of the etch can be hundreds or even thousands of microns into the silicon substrate.

The etch is a dry plasma etch and uses a high density plasma to alternately etch the silicon and deposit an etch resistant polymer layer on the sidewalls. The etching of the silicon is performed using a SF6 chemistry whereas the deposition of the etch resistant polymer layer on the sidewalls uses a C4F8 chemistry. Mass flow controllers alternate back and forth between these two chemistries during the etch. The protective polymer layer is deposited on the sidewalls as well as on the bottom of the etch pit, but the anisotropy of the etch removes the polymer at the bottom of the etch pit faster than the polymer is removed from the sidewalls. The sidewalls are not perfected or optically smooth and if the sidewall is magnified under SEM inspection, a characteristic washboard or scalloping pattern is seen in the sidewalls. The etch rates on most commercial DRIE systems varies from 1 to 4 microns per minute. DRIE systems are single wafer tools. Photoresist can be used as a masking layer for DRIE etching. The selectivity with photoresist and oxide is about 75 to 1 and 150 to 1, respectively. For a through wafer etch a relatively thick photoresist masking layer will be required. The aspect ratio of the etch can be as high as 30 to 1, but in practice tends to be 15 to 1. The process recipe depends on the amount of exposed silicon due to loading effects in the system, with larger exposed areas etching at a much faster rate compared to smaller exposed areas. Consequently, the etch must frequently be characterized for the exact mask feature and depth to obtain desirable results. This is a SEM of a MEMS component fabricated using DRIE and wafer bonding. This device was made using an SOI wafer wherein a backside etch was performed through the handle wafer, stopping on the buried oxide layer, and a frontside DRIE was performed on the SOI device layer. Then the buried oxide was removed to release the microstructure, allowing it to freely move. This is a cross section SEM of a silicon microstructure fabricated using DRIE technology. As can be seen, the etch is very deep into the silicon substrate and the sidewalls are nearly vertical.

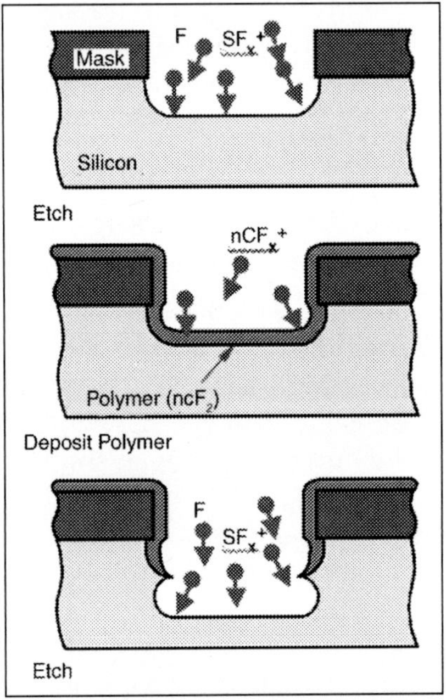

Figure : Illustration of how deep reactive ion etching works.

Figure : SEM of a MEMS device fabricated using two sided DRIE etching technology on an SOI wafer.

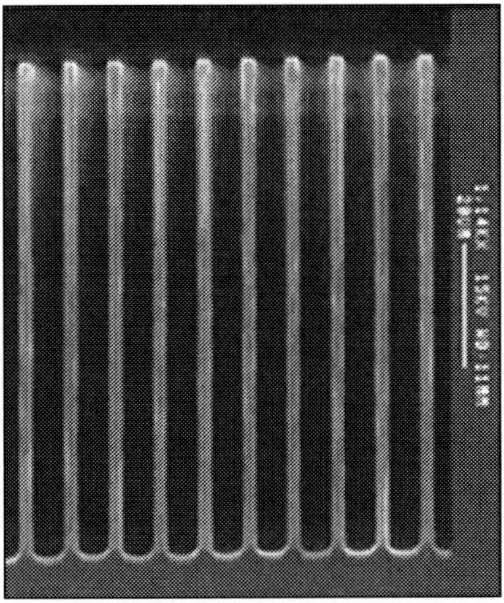

Figure : SEM of the cross section of a silicon wafer demonstrating high-aspect ratio and deep trenches that can be fabricated using DRIE technology.

Deep Reactive Ion Etching of Glass

Glass substrates can also be etched deep into the material with high aspect ratios and this technology has been gaining in popularity in MEMS fabrication. A structure fabricated into glass using this technology. The typical etch rates for high aspect ratio glass etching range between 250 and 500 nm per minute. Depending on the depth of the photoresist, metal or a polysilicon can be used as a mask.

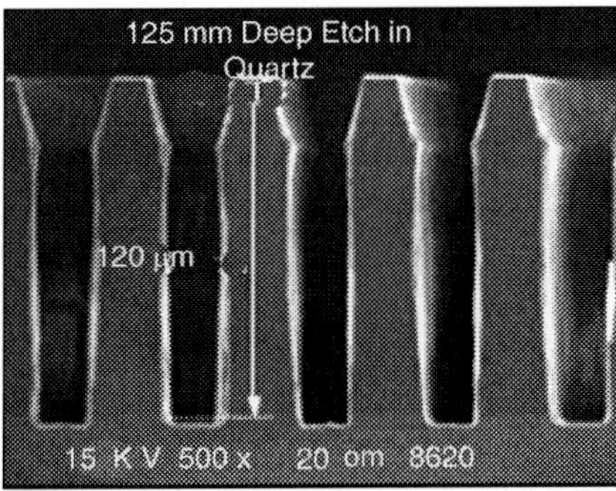

Figure : SEM of high aspect ratio structures etched into a glass substrate fabricated by MNX.

LIGA

One more popular high aspect ratio micromachining technology is called LIGA, which is a German acronym for "Lithographie Galvanoformung

Adformung." This is primarily a non-silicon based technology and requires the use of synchrotron generated x-ray radiation. The basic process is outline and starts with the cast of an x-ray radiation sensitive PMMA onto a suitable substrate. A special x-ray mask is used for the selective exposure of the PMMA layer using x-rays. The PMMA is then developed and will be defined with extremely smooth and nearly perfectly vertical sidewalls. Also, the penetration depth of the x-ray radiation into the PMMA layer is quite deep and allows exposure through very thick PMMA layers, up to and exceeding 1 mm. After the development, the patterned PMMA acts as a polymer mold and is placed into an electroplating bath and Nickel is plated into the open areas of the PMMA. The PMMA is then removed, thereby leaving the metallic microstructure.

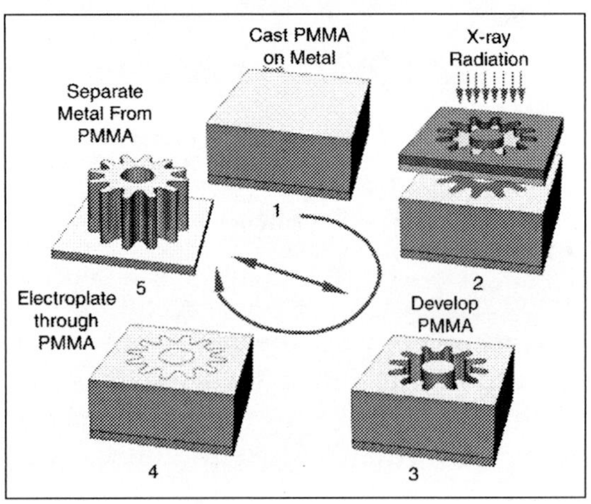

Figure : An illustration of the steps involved in the LIGA process to fabricate high aspect ratio MEMS devices.

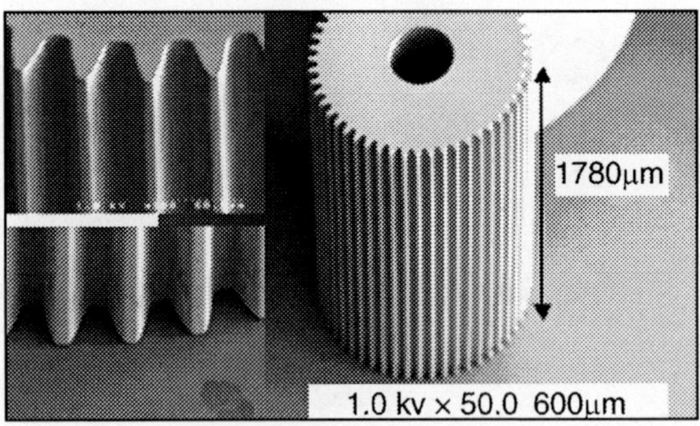

Figure : A tall, high aspect ratio gear made using LIGA technology.

Since LIGA requires a special mask and a synchrotron (X-ray) radiation source for the exposure, the cost of this process is relatively expensive. A variation of the process which reduces the cost of the micromachined parts made with this process is to reuse the fabricated metal part (step 5) as a tool insert to imprint the shape of the tool into a polymer layer (step 3), followed by electroplating of metal into the polymer mold (step 4) and removal of the polymer mold (step 5). Obviously this sequence of steps eliminates the need for a synchrotron radiation source each time a part is made and thereby significantly lowers the cost of the process. The dimensional control of this process is quite good and the tool insert can be used many times before it is worn out.

Procedure of Hot Embossing

A procedure to replicate deep high-aspect ratio structures in polymer materials is to fabricate the metal tool insert using LIGA or a comparable technology and then to emboss the tool insert pattern into a polymer substrate, which is then used as the part. A mold insert is made using an appropriate fabrication technique having the inverse pattern made into it. The mold insert is placed into a hot embossing system that includes a chamber in which a vacuum can be drawn. The substrate and polymer are heated to above the glass transition temperature, Tg, of the polymer material and the mold insert is pressed into the polymer substrate. The vacuum is critical for the polymer to faithfully replicate the features in the mold insert since otherwise air would be trapped between the two surfaces resulting in distorted features. Subsequently, the substrates are cooled to below the glass transition temperature of the polymer material and force is applied to de-emboss the substrates.

The hot embossing can successfully replicate complicated, deep and high-aspect ratio features. This process can make imprints into a polymer hundreds of microns deep with very good dimensional control. The advantage of this process is that the cost of the individual polymer parts can be very low compared to the same structures made using other technologies. Because of the overwhelming cost advantages combined with very good performance, this polymer moulding process is very popular for producing microfluidic components for medical applications.

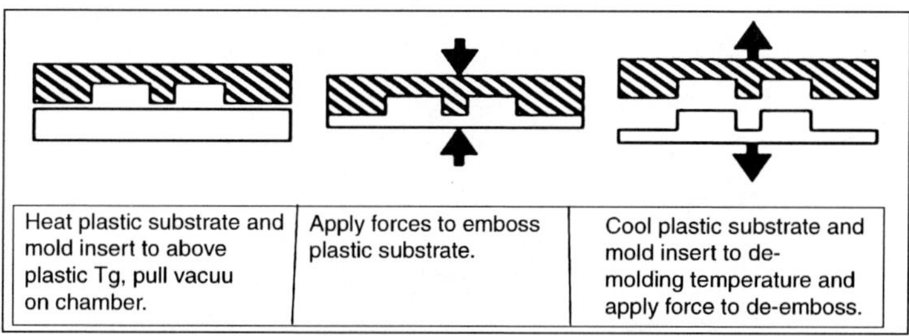

Figure : Illustration of the hot embossing process to create microdevices.

Heat plastic substrate and mold put in to above plastic Tg, pull vacuum on chamber. Apply forces to emboss plastic substrate. Cool plastic substrate and mold insert to de-moulding temperature and apply force to de-emboss.

Figure : Photograph of a hot embossing platform during use.

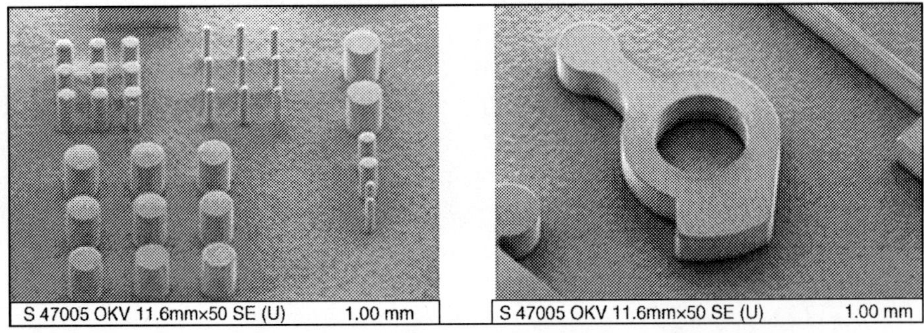

Figure : SEM of a variety of small test structures made in a plastic substrate using hot embossing technology at the MNX. The height of the plastic microstructures is nearly 300 um and the smallest features have a diameter of about 25 um.

OTHER MICROMACHINING TECHNOLOGIES

In adding to bulk micromachining, surface micromachining, wafer bonding, and high aspect ratio micromachining technologies, there are a number of other techniques used to fabricate MEMS devices.

XeF2 Dry Phase Etching

Xenon Difluoride (XeF2) in a vapour state is an isotropic etchant for silicon. This etchant is highly selective with respect to other materials commonly used in microelectronics fabrication including: LPCVD silicon nitride, thermal SiO2, aluminum, titanium and others. Since this etchant is a completely dry release process, it does not suffer from the stiction problems of wet release processes. This etchant is popular with micromachining of microstructures in pre-processed CMOS wafers where openings in the passivation layers on the surface of the substrate are made to expose the silicon for etching.

Electro-Discharge Micromachining

Electro-discharge micromachining or micro-EDM is a procedure used to machine a conductive material using electrical breakdown discharges to remove material. A working electrode is made from a metal material onto which high voltage pulses are applied. The working electrode is brought into close proximity to the material to be machined, which is immersed in a dielectric fluid. The minimum sized features that can be made with micro-EDM are dependent on the size of the working electrode and how it is fixtured, but holes as small as tens of microns have been made using this technique. One issue with micro-EDM is that it is a slow serial process.

Laser Micromachining

Lasers can generate an intense quantity of energy in very short pulses of light and direct that energy onto a selected region of material for micromachining. Among the many types of lasers now in use for micromachining include: CO_2, YAG, excimer, etc. Each has its own unique properties and capabilities suited to particular applications. Factors that determine the type of laser to use for a particular application include laser wavelength, energy, power, and temporal and spatial modes; material type; feature sizes and tolerances; processing speed; and cost. The action of CO_2 and Nd:YAG lasers is essentially a thermal process, whereby focusing optics are used to direct a predetermined energy/power density to a well-defined location on the work piece to melt or vaporize the material. Another mechanism, which is nonthermal and referred to as photoablation, occurs when organic materials are exposed to ultraviolet radiation generated from excimer, harmonic YAG, or other UV source. Similar to microEDM, laser micromachining can produce features on the order of tens of microns, but it is a serial process and therefore slower.

Focused Ion Beam Micromachining

Another adaptable tool for performing micromachining is the focused ion beam (FIB). The accelerating voltages are adjustable from a few keV to several hundred keV. The spot sizes can be focused down to below 25 nm, making it capable of producing extremely small structures. The user can input a 3-D CAD solid model of desired etching topology; the computer-controlled stage with sub-micron positional accuracy allows very precise registration of sample. In addition to material removal, the FIB can also be used to perform ion induced deposition, lithography, implantation doping, mask repair, device repair, and device diagnostics. Many of these tools can also be outfitted with a secondary column for mass analysis of particles removed from the substrate using uSIMS.

MEMS PROCESS INTEGRATION

Process integration for our purpose is defined as understanding, characterizing and optimizing to the extent possible, the interrelationship of the individual processing steps in a process sequence. Given the customization of MEMS process sequences, it should not be surprising that process integration is of critical importance. Also, MEMS developers must have good data on the electrical materials properties as well as data on the mechanical material properties. Coupled with the enormous diversity of materials and processing techniques used in MEMS fabrication, means that process integration can be a major part of a product development effort. Fortunately, the MNX has been working on this problem and has developed project-tested solutions so as to simultaneously provide the flexibility in the process sequence that is needed in MEMS while also affording a good level of reproducibility.

Advantages of MEMS and Nano Manufacturing There are many beneficial attributes of MEMS and Nanotechnology. Firstly, these devices are made using integrated circuit-like processes, which enables the ability to integrate multiple functionalities onto a single microchip. The ability to integrate miniaturized sensors, miniaturized actuators and miniaturized structures along with microelectronics has far-reaching implications in countless products and applications.

Secondly, MEMS and Nanotechnology borrow many of the production techniques of batch fabrication from the integrated circuit industry and therefore, the per-unit device or microchip cost of complex miniaturized electromechanical systems can be radically reduced - similar to the per die cost reductions we have experienced in the IC industry. Although the cost of the production equipment and each wafer can be relatively high, the fact that this cost can be spread over many die in batch fabrication production can drastically lower the per part cost.

Thirdly, integrated circuit fabrication techniques coupled with the tremendous advantages of silicon and many other thin-film materials in mechanical applications allows the reliability of miniaturized electromechanical systems to be radically improved. It is well known that the most expensive and most unreliable components of a conventional macroscale control system are the sensors and actuators. It is expected that as these miniaturized sensors and actuators are integrated onto a single microchip with electronics, we will see similar improvements in system reliability such as we have experienced in the transition from discrete electronics on a printed circuit board to integrated circuits. The lower cost of these miniaturized electromechanical systems also allows them to be easily and massively deployed and more easily maintained and replaced as needed.

Fourthly, miniaturization of micro- and nanosystems enables many benefits including: increased portability, lower power consumption, and the ability to place radically more functionality in a smaller amount of space and without any increase in weight.

Fifthly, the ability to make the signal paths smaller and place radically more functionality in a small amount of space allows the overall performance of electromechanical systems to be enormously improved.

In short, MEMS and Nanotechnology translates into products that have lower cost, higher functionality, improved reliability and increased performance.

MEMS AND NANOTECHNOLOGY APPLICATIONS

There are numerous possible applications for MEMS and Nanotechnology. As a breakthrough technology, allowing unparalleled synergy between previously unrelated fields such as biology and microelectronics, many new MEMS and Nanotechnology applications will emerge, expanding beyond that which is currently identified or known. Here are a few applications of current interest:

MEMS and Biotechnology

MEMS and Nanotechnology is enabling new discoveries in science and engineering such as the Polymerase Chain Reaction (PCR) microsystems for DNA amplification and identification, enzyme linked immunosorbent assay (ELISA), capillary electrophoresis, electroporation, micromachined Scanning Tunnelling Microscopes (STMs), biochips for detection of hazardous chemical and biological agents, and microsystems for high-throughput drug screening and selection.

MEMS and Medicine

There are a wide variety of applications for MEMS in medicine. The first and by far the most successful application of MEMS in medicine (at least in terms of number of devices and market size) are MEMS pressure sensors, which have been in use for several decades. The market for these pressure sensors is extremely diverse and highly fragmented, with a few high-volume markets and many lower volume ones. Some of the applications of MEMS pressure sensors in medicine include:

- The major market for MEMS pressure sensors in the medical sector is the disposable sensor used to monitor blood pressure in IV lines of patients in intensive care. These devices were first introduced in the early 1980's. They replaced other technologies that cost over

$500 and which had a substantial recurring cost since they had to be sterilized and recalibrated after each use. MEMS disposable pressure sensors are delivered pre-calibrated in a sterilized package from the factory at a cost of around $10.

- MEMS pressure sensors are used to measure intrauterine pressure during birth. The device is housed in a catheter that is placed between the baby's head and the uterine wall. During delivery, the baby's blood pressure is monitored for problems during the mother's contractions.
- MEMS pressure sensors are used in hospitals and ambulances as monitors of a patient's vital signs, specifically the patient's blood pressure and respiration.
- The MEMS pressure sensors in respiratory monitoring are used in ventilators to monitor the patient's breathing.
- MEMS pressure sensors are used for eye surgery to measure and control the vacuum level used to remove fluid from the eye, which is cleaned of debris and replaced back into the eye during surgery
- Special hospital beds for burn victims that employ inflatable mattresses use MEMS pressure sensors to regulate the pressure inside a series of individual inflatable chambers in the mattress. Sections of the mattress can be inflated as needed to reduce pain as well as improve patient healing.
- Physician's office and hospital blood analysers employ MEMS pressure sensors as barometric pressure correction for the analysis of concentrations of O_2, CO_2, calcium, potassium, and glucose in a patient's blood.
- MEMS pressure sensors are used in inhalers to monitor the patient's breathing cycle and release the medication at the proper time in the breathing cycle for optimal effect.
- MEMS pressure sensors are used in kidney dialysis to monitor the inlet and outlet pressures of blood and the dialysis solution and to regulate the flow rates during the procedure.
- MEMS pressure sensors are used in drug infusion pumps of many types to monitor the flow rate and detect for obstructions and blockages that indicate that the drug is not being properly delivered to the patient.

The donation to patient care for all of these applications has been enormous. More recently, MEMS pressure sensors have been developed and are being marketed that have wireless interrogation capability. These sensors can be implanted into a human body and the pressure can be measured using a remotely scanned wand. Another application is MEMS inertial sensors, specifically accelerometers and rate sensors which are being used as activity sensors. Perhaps the foremost application of inertial sensors in medicine is in cardiac pacemakers wherein they are used to help determine the optimum pacing rate for the patient based on their activity level. MEMS devices are also starting to be employed in drug delivery devices, for both ambulatory and implantable applications. MEMS electrodes are also being used in neuro-signal detection and neuro-stimulation applications. A variety of biological and chemical MEMS sensors for invasive and non-invasive uses are beginning to be marketed. Lab-on-a-chip and miniaturized biochemical analytical instruments are being marketed as well.

MEMS and Communications

High frequency circuits are benefiting considerably from the advent of RF-MEMS technology. Electrical components such as inductors and tunable capacitors can be improved significantly compared to their integrated counterparts if they are made using MEMS and Nanotechnology. With the integration of such components, the performance of communication circuits will improve, while the total circuit area, power consumption and cost will be reduced. In addition, the mechanical switch, as developed by several research groups, is a key component with huge potential in various RF and microwave circuits. The demonstrated samples of mechanical switches have quality factors much higher than anything previously available. Another successful application of RF-MEMS is in resonators as mechanical filters for communication circuits.

MEMS and Inertial Sensing

MEMS inertial sensors, specifically accelerometers and gyroscopes, are quickly gaining market acceptance. For example, MEMS accelerometers have displaced conventional accelerometers for crash air-bag deployment systems in automobiles. The previous technology approach used several bulky accelerometers made of discrete components mounted in the front of the car with separate electronics near the air-bag and cost more than $50 per device. MEMS technology has made it possible to integrate the accelerometer and electronics onto a single silicon chip at a cost of only a few dollars. These MEMS accelerometers are much smaller, more functional, lighter, more reliable, and are produced for a fraction of the cost of the conventional macroscale accelerometer elements. More recently, MEMS gyroscopes (*i.e.*, rate sensors) have been developed for both automobile and consumer electronics applications. MEMS inertial sensors are now being used in every car sold as well as notable customer electronic handhelds such as Apple iPhones and the Nintendo Wii.

MEMS EMERGING TECHNOLOGY

MEMS is an emerging technology which uses the tools and techniques that were developed for the Integrated Circuit industry to build microscopic machines. These machines are built on standard silicon wafers. The genuine power of this technology is that many machines can be built at the same time across the surface of the wafer, with no assembly required. Since it is a photographic-like process, it is just as easy to build a million machines on the wafer as it would be to build just one.

These tiny machines are becoming ubiquitous, and are quickly finding their way into a variety of commercial and defence applications. There are several different broad categories of MEMS technologies.

Bulk Micromachining

Bulk micromachining is a manufacture technique which builds mechanical elements by starting with a silicon wafer, and then etching away unwanted parts, and being left with useful mechanical devices. Typically, the wafer is photo patterned, leaving a protective layer on the parts of the wafer that you want to keep. The wafer is then submersed into a liquid etchant, like potassium hydroxide, which eats away any exposed silicon. This is a relatively simple and inexpensive fabrication technology, and is well suited for applications which do not require much complexity, and which are price sensitive.

Nowadays, almost all pressure sensors are built with Bulk Micromachining. Bulk Micromachined pressure sensors offer several advantages over traditional pressure sensors. They cost less, are highly reliable, manufacturable, and there is very good repeatability between devices.

All new cars on the market today have several micromachined pressure sensors, typically used to measure manifold pressure in the engine. The small size and high reliability of micromachined pressure sensors make them ideal for a variety of medical applications as well.

Surface Micromachining

While Bulk micromachining creates devices by etching into a wafer, Surface Micromachining builds devices up from the wafer layer-by-layer.

A typical Surface Micromachining process is a repetitive sequence of depositing thin films on a wafer, photopatterning the films, and then etching the patterns into the films. In order to create moving, functioning machines, these layers are alternating thin films of a structural material (typically silicon) and a sacrificial material (typically silicon dioxide). The structural material will form the mechanical elements, and the sacrificial material creates the gaps and spaces between the mechanical elements. At the end of the process, the sacrificial material is removed, and the structural elements are left free to move and function.

For the container of the structural level being silicon, and the sacrificial material being silicon dioxide, the final "release" process is performed by placing the wafer in Hydrofluoric Acid. The Hydrofluoric Acid quickly etches away the silicon dioxide, while leaving the silicon undisturbed. The wafers are typically then sawn into individual chips, and the chips packaged in an appropriate manner for the given application. Surface Micromachining requires more fabrication steps than Bulk Micromachining, and hence is more expensive. It is able to create much more complicated devices, capable of sophisticated functionality. Surface Micromachining is suitable for applications requiring more sophisticated mechanical elements.

LIGA

LIGA is a technology which creates small, but relatively high aspect ratio devices using x-ray lithography. The process typically starts with a sheet of PMMA. The PMMA is covered with a photomask, and then exposed to high energy x-rays. The mask allows parts of the PMMA to be exposed to the x-rays, while protecting other parts. The PMMA is then placed in a suitable etchant to remove the exposed areas, resulting in extremely precise, microscopic mechanical elements. LIGA is a relatively inexpensive fabrication technology, and suitable for applications requiring higher aspect ratio devices than what is achievable in Surface Micromachining.

Deep Reactive Ion Etching

Deep reactive ion etching is a type of Bulk Micromachining which etches mechanical elements into a silicon wafer. Unlike traditional Bulk Micromachining, which uses a wet chemical etch, Deep Reactive Ion Etching micromachining uses a plasma etch to create features. This allows greater flexibility in the etch profiles, enabling a wider array of mechanical elements. The fabrication tools needed to perform Deep Reactive Ion etching are somewhat expensive, to this technology is typically more expensive than traditional Bulk Micromachining based on wet etching.

Integrated MEMS Technologies

Since MEMS devices are created with the same tools used to create integrated circuits, in some cases it is actually possible to fabricate Micromachines and Microelectronics on the same piece of silicon. Fabricating machines and transistors side by side enables machines that can have intelligence. A number of exciting products are already taking advantage of this capability.

APPLICATIONS OF MEMS

At this time are some examples of MEMS technology:

MEMS Pressure Sensors

MEMS pressure microsensors typically have a flexible diaphragm that deforms in the presence of a pressure difference. The deformation is converted to an electrical signal appearing at the sensor output. A pressure sensor can be used to sense the absolute air pressure within the intake manifold of an automobile engine, so that the amount of fuel required for each engine cylinder can be computed. In this example, piezoresistors are patterned across the edges of a region where a silicon diaphragm will be micromachined. The substrate is etched to create the diaphragm. The sensor die is then bonded to a glass substrate, creating a sealed vacuum cavity under the diaphragm. The die is mounted on a package, where the topside of the diaphragm is exposed to the environment. The change in ambient pressure forces the downward deformation of the diaphragm, resulting in a change of resistance of the piezoresistors. On-chip electronics measure the resistance change, which causes a corresponding voltage signal to appear at the output pin of the sensor package.

Acceleration Sensors of Accelerometers

Accelerometers are acceleration sensors. An inertial mass suspended by springs is acted upon by acceleration forces that cause the mass to be deflected from its initial position. This deflection is converted to an electrical signal, which appears at the sensor output. The application of MEMS technology to accelerometers is a relatively new development.

It is a surface micromachined piezoelectric accelerometer employing a zinc oxide (ZnO) active piezoelectric film. The design is a simple cantilever structure, in which the cantilever beam serves simultaneously as proof mass and sensing element. One of the fabrication approaches developed is a sacrificial oxide process based on polysilicon surface micromachining, with the addition of a piezoelectric layer atop the polysilicon film. In the sacrificial oxide process, a passivation layer of silicon dioxide and low-stress silicon nitride is deposited on a bare silicon wafer, followed by 0.5 micron of liquid phase chemical vapour deposited (LPCVD) phosphorous-doped polysilicon. Then, a 2.0-micron layer of phosphosilicate glass (PSG) is deposited by LPCVD and patterned to define regions where the accelerometer structure will be anchored to the substrate. The PSG film acts as a sacrificial layer that is selectively etched at the end to free the mechanical structures. A second layer of 2.0-micron-thick phosphorus-doped polysilicon is deposited via LPCVD on top of the PSG, and patterned by plasma etching to define the mechanical accelerometer structure. This layer also acts as the lower electrode for the sensing film. A thin layer of silicon nitride is next deposited by LPCVD, and acts as a stress-compensation layer for balancing the highly compressive residual stresses in the ZnO film. By varying the thickness of the Si3N4 layer, the accelerometer structure may be tuned to control bending effects resulting from the stress gradient through the device thickness. A ZnO layer is then deposited on the order of 0.5 micron, followed by sputtering of a 0.2-micron layer of platinum (Pt) deposited to form the upper electrode. A rapid thermal anneal is performed to reduce residual stresses in the sensing film. Afterwards, the Pt, Si3N4, and ZnO layers are patterned in a single ion milling etch step, and the devices are then released by passivating the ZnO film with photoresist, and immersing the wafer in buffered hydrofluoric acid, which removes the sacrificial PSG layer.

Types of Accelerometer and Inertial Sensors

Inertial sensors are a type of accelerometer and are one of the major commercial products that utilize surface micromachining. They are used as airbag-deployment sensors in automobiles, and as tilt or shock sensors. The application of these accelerometers to inertial measurement units (IMUs) is limited by the need to manually align and assemble them into three-axis systems, and by the resulting alignment tolerances, their lack of in-chip analog-to-digital conversion circuitry, and their lower limit of sensitivity. A three-axis force-balanced accelerometer has been designed at the University of California, Berkeley, to overcome some of these limitations. The accelerometer was designed for the integrated MEMS/CMOS technology. This technology involves a manufacturing technique where a single-level (plus a second electrical interconnect level) polysilicon micromachining process is integrated with 1.25-micron CMOS.

Microengines : Fabrication of Devices

A three-level polysilicon micromachining process has enabled the fabrication of devices with increased degrees of complexity. The process includes three movable levels of polysilicon, each separated by a sacrificial oxide layer, plus a stationary level. Operation of the small gears at rotational speeds greater than 300,000 rpm has been demonstrated. Microengines can be used to drive the wheels of microcombination locks. They can also be used in combination with a microtransmission to drive a pop-up mirror out of a plane. This device is known as a micromirror.

Some other Applications

MEMS IC fabrication technologies contain also allowed the manufacture of microtransmissions using sets of small and large gears interlocking with other sets of gears to transfer power.

A newly developed MicroStar cross-connect fabric developed by Bell Labs, a micro-optoelectromechanical system device, is based on MEMS technology. The most pervasive bottlenecks for communications carriers are the switching and cross-connect fabrics that switch, route, multiplex, demultiplex, and restore traffic in optical networks. The optical transmission systems move information as photons, but switching and cross-connect fabrics until now have been largely electronic, requiring costly and time-consuming bandwidth-limiting optical-to-electronic-to-optical conversions at every network connection and cross point. MicroStar is composed of 256 mirrors, each one

0.5 mm in diameter, spaced 1 mm apart, and covering less than 1 square inch of silicon. The mirrors sit within the router so that only one wavelength can illuminate any one mirror. Each mirror can tilt independently to pass its wavelength to any of 256 input and output fibres. The mirror arrays are made using a self-assembly process that causes a frame around each mirror to lift from the silicon surface and lock in place, positioning the mirrors high enough to allow a range of movement. MicroStar is part of Lucent Technology's Lambda Router cross-connect system aimed at helping carriers deliver vast amounts of data unimpeded by conventional bottlenecks.

As a final example, MEMS technology has been used in fabricating vapourization microchambers for vaporizing liquid microthrusters for nanosatellites. The chamber is part of a microchannel with a height of 2-10 microns, made using silicon and glass substrates. The nozzle is fabricated in the silicon substrate just above a thin-film indium tin oxide heater deposited on glass.

THE FUTURE OF MICROSYSTEMS TECHNOLOGY PROCESSES

Each of the three basic microsystems technology processes we have seen, bulk micromachining, sacrificial surface micromachining, and micromolding/LIGA, employs a different set of capital and intellectual resources. MEMS manufacturing firms must choose which specific microsystems manufacturing techniques to invest in. MEMS knowledge has the potential to change our daily lives as much as the computer has. However, the material needs of the MEMS field are at a preliminary stage. A thorough understanding of the properties of existing MEMS materials is just as important as the development of new MEMS materials.

Future MEMS applications will be driven by processes enabling greater functionality through higher levels of electronic-mechanical integration and greater numbers of mechanical components working alone or together to enable a complex action. Future MEMS products will demand higher levels of electrical-mechanical integration and more intimate interaction with the physical world. The high up-front investment costs for large-volume commercialization of MEMS will likely limit the initial involvement to larger companies in the IC industry. Advancing from their success as sensors, MEMS products will be embedded in larger non-MEMS systems, such as printers, automobiles, and biomedical diagnostic equipment, and will enable new and improved systems.

Micro-nano-bio Systems

The different steps of device design, characterization, encapsulation and packaging, as well as customised electronic instrumentation are approached from the initial conception to the final biodevice in order to generate knowledge, micro-nano devices and complete systems with high added value.

Activities include the development of new technologies and tools for the detection, identification, quantification, and monitoring of molecules, cells and tissues of clinical and biomedical relevance. Research focuses in:

- Micro-Nano systems for diagnosis.
- On-chip environmental health monitoring.
- Nano-Bio-Electronic Interfaces.
- NanoBioFuel cells.
- Nanobioelectrochemistry.

NANOMECHANICAL CANTILEVER SENSORS

The creation of the atomic force microscope (AFM) in 1986 introduced the need for small cantilever structures to be used as AFM probes, and these were almost immediately fabricated by using silicon micromachining. AFM cantilevers have since then been continuously improved and are standard commercial products. The availability of such small cantilever structures sensitive to very small forces, in which the deflection can be accurately monitored by the AFM instrument, triggered their use as transducers for physical and chemical measurements. In particular, in the past decade they have been used for biochemical sensing by various techniques.

In our group we have worked in the development of micro and nano cantilevers since 2002. The activity is concentrated in the development of cantilever structures and devices for specific applications:

- Piezoresistice cantilevers optimized for force measurements
- AFM probes for electrical measurements in a liquid environment

Piezoresistive Micro and Nanocantilevers

Micro and nanoelectromechanical devices can be second-hand for the detection of small forces and masses. In particular, the invention of the atomic force microscope (AFM) in 1986 lead to the development of microcantilevers to be used as AFM probes, which have later been used as transducers for physical and chemical measurements, or for biosensing.

Starting during the development of the EU-funded project BioFinger (2002-2006) which we coordinated, we are working in piezoresistive cantilevers for the detection of small forces and biomolecules.

Automatic Structures as Biomolecular Detectors

The forces involved in chemical bonding and intermolecular interactions have values ranging from about 1 nN for a covalent bond to about 10 pN for Van der Waals interactions, with ligand-receptor forces somewhere in between. Typically, [bio]molecular detection requires the ability to perceive forces of about 100 pN. A cantilever ray is one of the simplest mechanical structures.

If a force F is applied at the free end perpendicularly to the beam, the deflection z of the cantilever is, for a material with Young's modulus E:

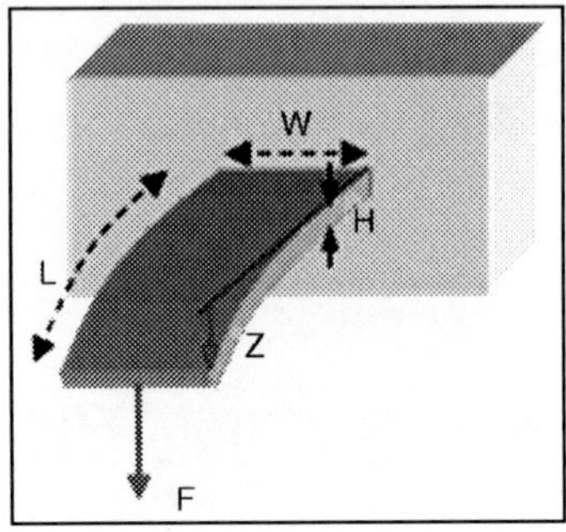

Figure:

$$Z = \frac{4L^3}{Ewh^3} F$$

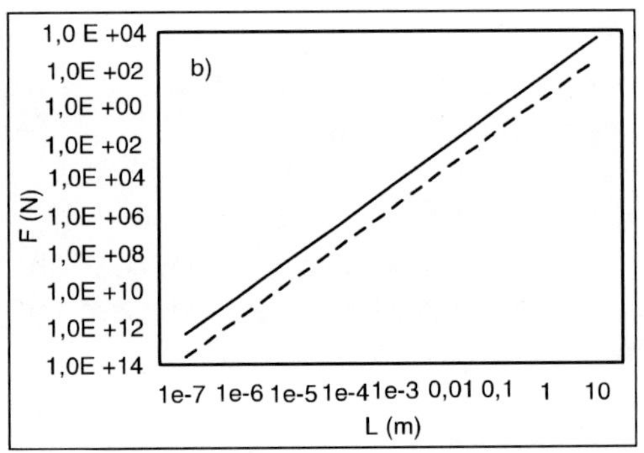

Figure: The force required to produce a deflection equal to the cantilever thickness as a function of the cantilever length, for silicon cantilevers with a fixed length and width to thickness ratio (L/h = 100, w/h = 10).

[Bio]molecular sensing is achieved by detecting chemical interactions between molecules on the cantilever surface and molecules in the chemical solution. Two basic detection techniques are used: static and dynamic.

The static technique is based on the fact that a cantilever structure will bend when its mechanical stress is not uniform along its thickness. The stress at one of the cantilever surfaces may change by physical adsorption or chemical bonding of the analyte molecules.

A connected technique is based on detecting intermolecular forces by approaching and retracting a functionalised cantilever tip to a functionalised surface. If specific molecules are present on the surface so that a ligand-receptor interaction between the tip and surface exists, the cantilever will bend during the retraction. This technique is called molecular force spectroscopy.

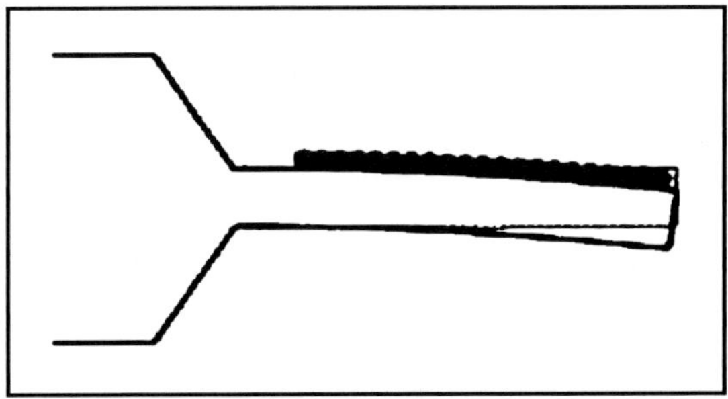

Figure:

The dynamic technique is based on modifying the resonance properties of a vibrating cantilever. A change in the mass of the cantilever will result in a change in the resonant frequency. This makes a cantilever operated in the dynamic mode an extremely sensitive mass detector.

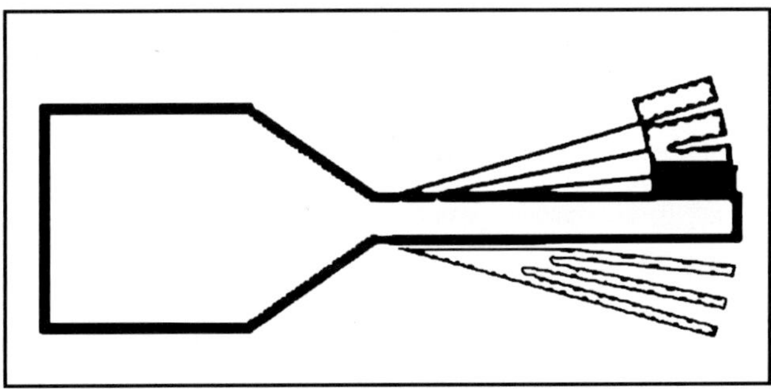

Figure:

In any of the detection modes, a [bio]chemical sensor will only be sensitive if the mechanical deformation of the cantilever can be transformed into an electrical signal. The typical deflections that can be expected for forces in the 100 pN range would be of the order of 1-10 nm. A challenge in cantilever-based [bio]chemical sensors is therefore the development of transducer elements able to obtain useful electrical signals from these displacement values.

Electrical Readout of Mechanical Signals

A digit of transducing principles have been used over the years to convert mechanical displacements into electrical signals. AFM instruments use an optical principle. The cantilever deflection is monitored by measuring the position of a laser beam deflected by the cantilever. This technique has a very high sensitivity, but it losses efficiency for small cantilever and it is difficult to miniaturize.

The capacitive principle has been extensively used to detect the movement of micromechanical structures in MEMS sensors such as accelerometers. It is based in detecting the variation of the capacitance on a two-electrode capacitor, where one electrode is fixed and the other is in the mobile structure. For micro- and nano-cantilevers, however, the small dimensions involved result in extremely small capacitance values and in even smaller capacitance variations due to deflection.

The piezoresistive principle has also been widely used in MEMS such as pressure sensors and accelerometers. The electrical resistance of a conductor changes with the mechanical stress. This effect is especially strong in semiconductors such as silicon, where the change in resistivity is much higher than purely geometrical effects. To detect the deflection of a cantilever, a resistor must be located on one of its surfaces, where the mechanical stress has a maximum value.

We have been working in piezoresistive cantilevers for the detection of biomolecules, based on the static measurement of the deflection when a functionalized cantilever is bonded to a functionalized surface through specific intermolecular forces.

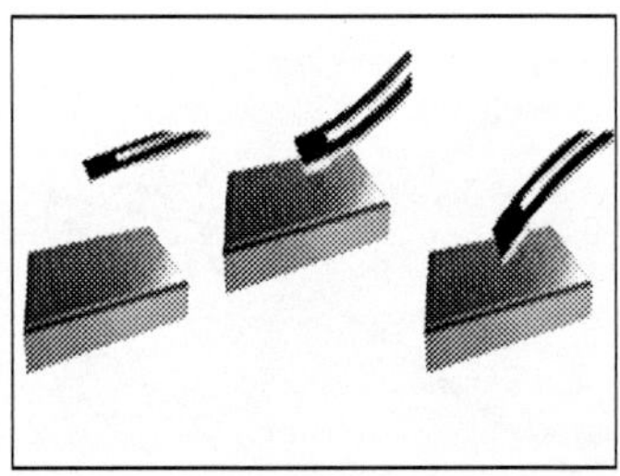

Figure:

A u-shaped structure has been used for the cantilevers, which minimizes the leg thickness. Within the BioFinger project we fabricated cantilevers in polycrystalline silicon. The reason was the compatibility of this structure with the integration of the cantilevers in a CMOS process. The cantilever structure and fabricated cantilevers.

Figure:

The force sensitivity can be calculated as:

$$\frac{\Delta V_0}{F} = \frac{\Pi_l L}{2wt^2} V_{bias}$$

where Π_l is the longitudinal piezoresistive coefficient, and it is assumed that the resistor has uniform doping and a thickness equal to 1/3 of the cantilever thickness. For other doping conditions the factor 1/2 changes.

The smallest amount force that can be measured is given by the noise equivalent force (NEF), which is the noise voltage divided by the force sensitivity. For piezoresistive cantilevers there are three main noise sources. The thermal or Johnson–Nyquist noise is due to the thermal energy of the carriers on a resistor and is independent of the frequency. The flicker ("1/f") noise is proportional to the inverse of the frequency. In semiconductors it is related to charge traps and it has been empirically found to be inversely proportional to the number of carriers on the conductor. The thermomechanical noise results from vibrational movement of the cantilever due to the thermal energy transferred from the surrounding medium.

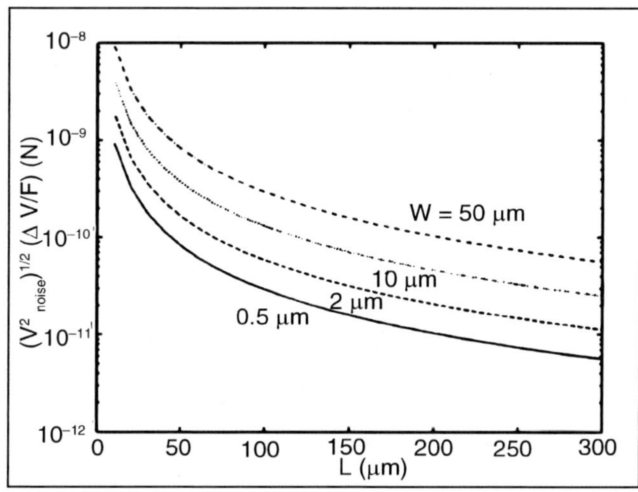

Figure:

To add to the sensitivity and resolution of the measurements, we integrated the piezoresistive cantilevers with on-chip amplifiers on a commercial CMOS technology. A post-processing was used to release the cantilevers.

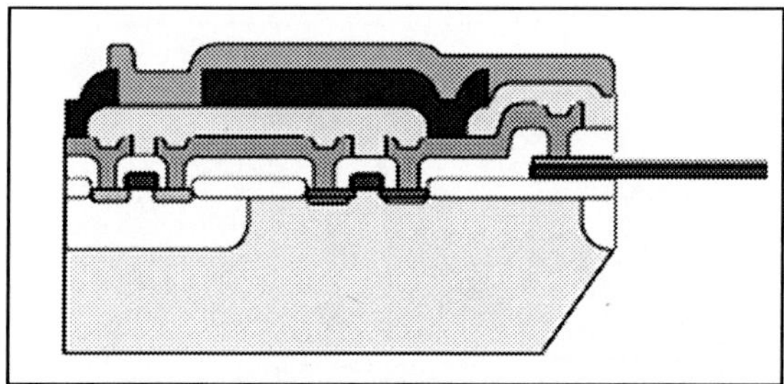

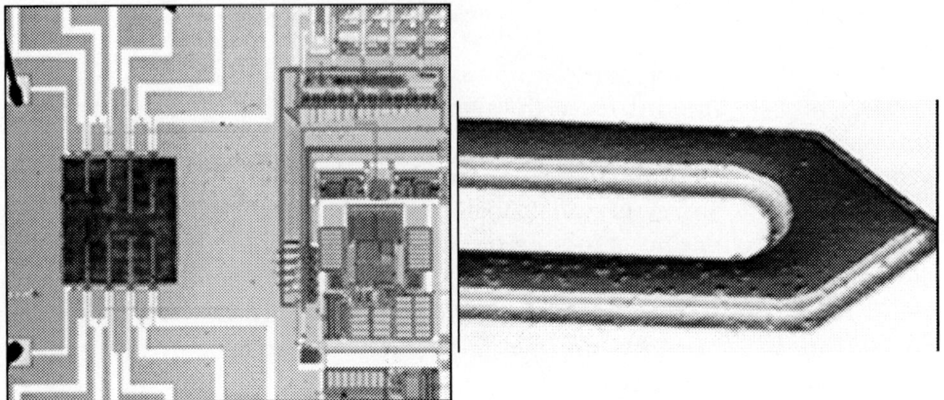

The electromechanical characterization of the cantilevers has been performed by applying a known displacement with an AFM. The AFM is operated in dynamic mode, *i.e.* oscillating until it makes contact with the cantilever. The electrical output of the cantilever and the amplitude of the AFM oscillations.

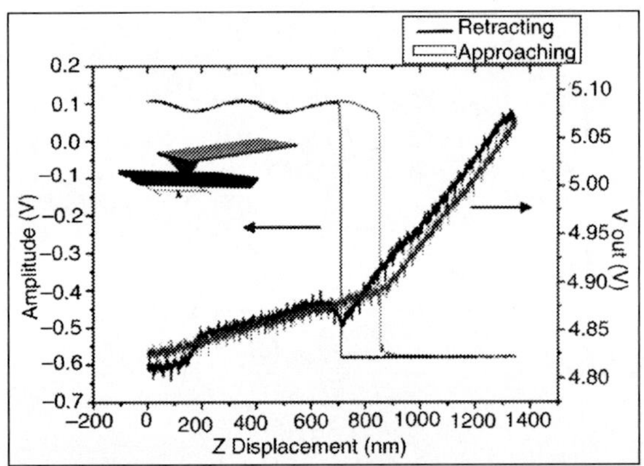

The force sensitivity is obtained from the measured displacement sensitivity and the cantilever spring constant. An enhanced version of the polysilicon cantilevers was fabricated with the micromachining done by deep reactive ion etching (DRIE).

As the force sensitivity depends inversely on the cantilever width, narrow cantilevers would be more sensitive (but the resistance will increase, thus increasing the noise; a tradeoff would be required). We also fabricated submicron width cantilevers by using electron beam lithography.

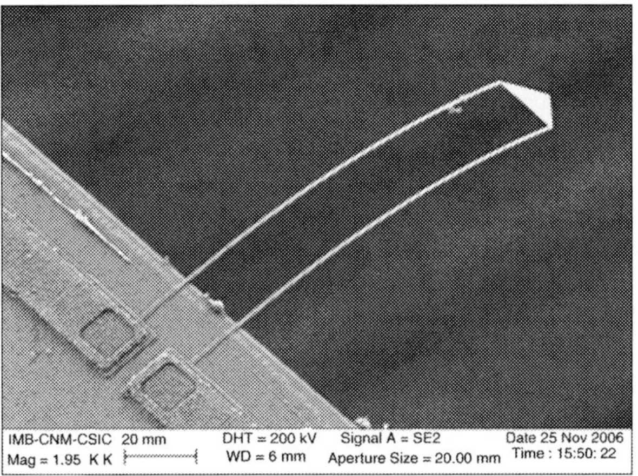

If a hybrid configuration is used, there is no need to use polysilicon to fabricate the cantilevers. Crystalline silicon has a much higher piezoresistive coefficient. We have fabricated such cantilevers using SOI substrates.

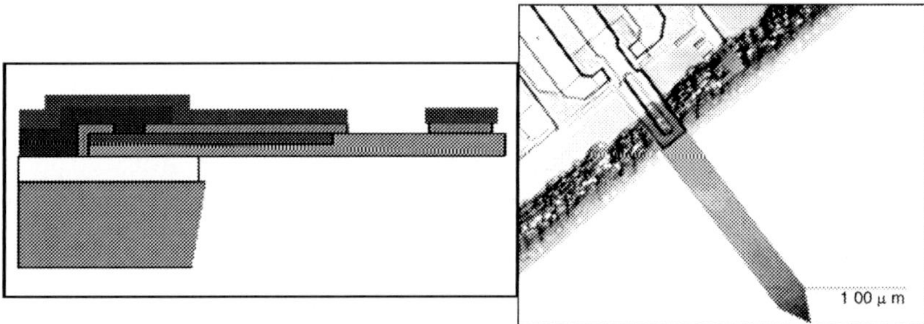

A summary of the main results for the force sensitivity and resolution. The resolution (noise voltage divided by sensitivity) depends on the electronic noise in the measurement systems used.

The results are for a $V_{bias} = 5$ V in a half bridge configuration (measurement cantilever plus reference cantilever). The amplifier gain is set to 640.

	F Sensitivity *V/F (μV/pN)*	*F resolution* *F_{min} (pN)*
CMOS monolithic	11	28
Polysilicon No amplifier	0.011	28000
Polysilicon hybrid amplif.	9.2	34
EBL "narrow" cantilevers	96	7.3
Crystalline Si	60	65

TYPES OF SENSORS BASE ON MICRO AND NANOCANTILEVERS

HUMIDITY SENSORS

The humidity in the surroundings can be measured if one side of microcantilever is coated with gelatin. Gelatin binds to the water vapours present in the atmosphere, thereby causing the bending of the cantilever. Researchers at Oak Ridge National Laboratory (ORNL), USA showed that cantilevers coated with hygroscopic materials such as phosphoric acid can be used as a sensor for detecting water vapour with picogram mass resolution. When water vapours are adsorbed on the coated surface of the cantilever, there is change in the resonance frequency of microcantilevers and cantilever deflection. Sensitivity of microcantilevers can be increased by coating its surface with materials having a high affinity for the analyte.

HERBICIDE SENSORS OF MICROCANTILEVERS

Microcantilevers have been used to detect the concentration of herbicides in the liquid environment by Roberto Raiteri and co-workers. The herbicide 2,4-dichlorophenoxyacetic acid (2,4-D) was coated on the upper surface of the cantilever. The monoclonal antibody against 2,4-D was then provided to the cantilever. The specific interaction between the monoclonal antibody and the herbicide caused the bending of the cantilever. A lot of research is going on to develop antibody coated cantilever immunobiosensors for the detection of organochlorine and organophosphorous pesticides and herbicides present at ng/l concentration in aqueous media. Alvarez and Co-workers demonstrated the use of microcantilevers for the detection of pesticide dichloro dipheny trichloroethane (DDT).

Metal Ion Sensors of Microcantilever

Microcantilever sensors have been employed to detect a concentration of 10^{-9} M CrO_4^{2-} in a flow cell. In this device, a self-assembled layer of triethyl-12-mercaptododecyl ammonium bromide on the gold-coated microcantilever surface was used. Microcantilevers could be used for the chemical detection of a number of gaseous analytes. A multielement sensor array device employing microcantilevers can be made to detect various ions simultaneously.

Temperature Sensors/Heat Sensors

Changes in temperature and temperature bend a cantilever composed of materials with different thermal expansion coefficients by the bimetallic effect. Microcantilever based sensors can measure changes in temperature as small as 10^{-5} K and can be used for photo thermal measurement. They can be used as microcalorimeters to study the heat evolution in catalytic chemical reactions and enthalpy changes at phase transitions. Bimetallic microcantilevers can perform photothermal spectroscopy with a sensitivity of 150 fJ and a sub-millisecond time resolution. They can detect heat changes with attojoule sensitivity.

Viscosity Sensors

Changes in the medium viscoelasticity shift the cantilever resonance frequency. A highly viscous medium surrounding the cantilever as well as an added mass will damp the cantilever oscillation lowering its fundamental resonance frequency. Cantilevers can therefore be vibrated by piezoelectric actuators to resonate and used as viscosity meters.

Calorimetry Sensors

In these sensors, only the temperature changes are to be measured. Most of the chemical reactions are associated with a change in heat. So, calorimetry has got tremendous potential to identify a wide range of compounds. Enzymes like glucose oxidase can be immobilized and coated on the surface of the microcantilever, which will react specifically with glucose in the solution producing a recognizable calorimetric signal. Due to the tiny thermal mass and sensitivity of the cantilever, calorimetry sensors employing cantilevers will be next generation of sensors for detecting temperature changes.

Sensor Detecting Magnetic Beads

Baselt and co-workers explain the possibility of using microcantilevers as force transducers to detect the presence of receptor-coated magnetic beads. It is possible to detect the presence of single μm size magnetic bead sticking onto the functionalized cantilever surface by applying an external magnetic field and measuring the deflection of the microcantilever. An extremely sensitive sensor can be made by labelling the analyte with magnetic beads.

Cantilever Based Telemetry Sensors

Cantilever based telemetry sensors will deploy fieldable devices to relay pertinent data to central collection stations. They will enable the use of mobile units worn or carried by personnel and will replace wired sensors in some applications. Researchers at ORNL are building a microfabricated chip with built-in electronic processing and telemetry. They are also working on a technique to detect different species.

Microsensors to Monitor Missile Storage and Maintenance Needs

Miniaturized microcantilever based sensors with remote wireless monitoring capability have been employed to gain insight into stockpile condition. This technology will evaluate ammunition lifetime based on environmental parameters like humidity, temperature, pressure, shock and corrosion as well as number of other indicators of propellant degradation including NOx. Single chip detectors with electronics and telemetry could be developed with several hundred cantilevers as an array to simultaneously monitor, identify and quantify many important parameters. Corrosion sensors have limited life in moderate to severe environments. Systems have to be build to collect environmental data for better knowledge of environmental conditions. There is a need to develop materials like zeolites for use as sensitizing coatings for specific detection. Zeolites are thermally stable aluminosilicate framework structures used commercially as molecular sieves, catalysts, ion-exchangers and chemical absorbers. They show excellent selectivity and selective thermal desorption properties.

Remote Infrared Radiation Detection Sensors

A remote infrared (IR) radiation detection sensor has been developed by Oden and co-workers. The sensor is made up of a piezoresistive cantilever coated with a heat absorbing layer. Piezoresistive microcantilevers represent an important development in uncooled IR detection technology. The cantilever undergoes bending due to the differential stress between the coating and the substrate. The cantilever bending causes a change in the piezoresistance, which is proportional to the amount of the heat absorbed. Temperature variations can be detected by coating the cantilever with a different material, which causes the bimetallic effect resulting in the bending of the cantilever. Thus, calorimetric detection of chemical reactions can be done. Gold-black would serve as the IR absorbing material. High thermal expansion bimaterial coatings such as Al, Pb and Zn could be used to increase the thermally induced bending of the microcantilever. Two dimensional cantilever arrays can be used for IR imaging as they are simple, highly sensitive and fast responding.

Explosives Detection Devices

It is believed that dogs contain got amazing smelling power, the reason they are widely employed in the detection of explosives. Dogs can detect explosives by sniffing easily vapourized organic chemicals present at concentration as low as parts-per-billion. Many groups are conducting active research with the intention of making a 'nose-on-a-chip' device having the smelling power exactly similar to the dog's nose. In this 'nose-on-a-chip' device, a microcantilever array could be used in which each cantilever will be coated differently to pick up a specific organic compound. It can be incorporated in our everyday use item like shoes, walking cane, purse etc. to detect the explosives without letting the culprits know about the search operation. The device would be a great achievement from the security point of view and would prevent large accidents. A microcantilever coated with platinum or a transition metal can react with trinitrotoluene (TNT) if it is heated to 570°C and held at that temperature for 0.1 second. The reaction of TNT with the cantilever coating will cause a mini-explosion. Thundat and his group are developing a matchbox-size device to detect explosives in airport luggage and landmines based on this technique.

CALIBRATION OF AFM CANTILEVERS

A significant application of the atomic force microscope is in quantitative force measurements, a prerequisite of which is accurate knowledge of the cantilever spring constant.

Calibration against a Known Standard

In this technique, a lever of known spring constant is attached to a sample substrate and the AFM cantilever is deflected by performing a force curve at the end of the lever. By noting the slope of the force curve relative to that obtained on a rigid substrate (or, better yet, relative to a series of different known levers), the spring constant of the cantilever can be deduced.

In a force curve, the *A–B* (deflection signal) is plotted against piezo extension *z*. In general, the *A–B* signal is uncalibrated and will be different for every cantilever used. Therefore, we usually measure the deflection in arbitrary units (*i.e.*, Volts), requiring that a detector calibration factor *s* be found such that the actual cantilever deflection from equilibrium, $y = sV$, where V is the deflection signal relative to the equilibrium cantilever deflection.

It is easy to show that the slope dV/dz is related to the spring constants of the cantilever k_0 and reference lever K according to

$$\left(s\frac{dV}{dz}\right)^{-1} = 1 + \frac{k}{K\cos^2\phi}$$

Note that the angle ø — the tilt angle of the cantilever — has until recently been neglected. The corresponding factor arises because only a cosø component of the vertical force due to the reference lever acts to deflect the cantilever, and only a cosø component of that deflection is along the measured vertical direction. In practice, this correction is typically a few per cent, which is less than the uncertainty in the technique.

Consequently, if a series of reference levers is employed, the cantilever spring constant is determined simply from the slope *m* and intercept *b* of a plot of force-curve slopes d*V*/dz arising from each reference-lever spring constant *K* in the series. We find

$$k = \frac{m}{b}\cos^2\phi$$

Difficulties with this approach include the fact that reference levers with spring constants near that of the cantilever to be calibrated must be available. Care must be taken to position the AFM cantilever so that it contacts the end of the reference lever (the fractional error in position leads to a *cubic* correction). It is obviously also essential that the spring constant of the reference levers be known. In one case (unpublished), we found that a series of calibrations performed using this technique gave a result that disagreed by a factor of nearly two with both the Cleveland and thermal noise techniques (which agreed well with each other), suggesting that the "accurately known" standards were in error. Nevertheless, this is a very direct technique if care is taken.

The Cleveland Added-Mass Technique

The Cleveland technique determines the cantilever spring steady from the decrease in resonant frequency resulting from the attachment of known masses. The theory is very simple: one expects the resonant frequency *f* to depend on added mass *M* according to

$$\omega = 2\pi f = \sqrt{\frac{k}{M^* + M}}$$

where M^* is the effective mass of the cantilever. Thus, if a series of different masses are attached, a plot of *M* vs. $1/f^2$ yields a straight line, the slope *m* of which can be used to determine the spring constant:

$$k = 4\pi^2 m$$

This technique can be employed even without a micromanipulator. We use the coarse positioning screws on the AFM head to move the cantilever into contact with tungsten spheres (used because of their high density) deposited on a clean, dry surface. After picking up a sphere (a somewhat tedious task), we measure the new resonant frequency (which can typically be done with built-in hardware for AFMs capable of intermittent contact mode or other AC imaging techniques). We then remove the tip holder, invert it, and carry it to an optical microscope, which is used to take a high-magnification picture of the cantilever and attached mass. We estimate the mass of the tungsten sphere from its diameter and known density. This process is repeated until either a sufficient number of different resonant frequencies have been recorded, the cantilever is lost or broken, or the operator gets bored.

The further the mass *M* is from the base of the cantilever, the more it will lower the resonant frequency, its mass must be corrected according to its position relative to the cantilever tip :

$$M_{corr} = M\left(\frac{L + \Delta L}{L}\right)^3$$

The major disadvantages of this technique are that it is somewhat time intensive (even with the apparatus and a substrate of tungsten spheres in place, spheres must be picked up by the cantilever and imaged optically) and destroys the cantilever (even if excessive force is avoided during the procedure, one is generally left with a cantilever littered with tungsten spheres at the end of the day). The significant advantage that resonant frequencies are accurately measured is offset, at least partially, by uncertainties in the measurement of position and size of the added masses, both of which enter the calculations cubically.

The Thermal Noise Technique

The thermal noise technique appeals to the *equipartition theorem,* which states that the thermal energy present in all terms in the Hamiltonian of a system that are quadratically dependent on a generalized coordinate is equal to $k_B T/2$, where k_B is Boltzmann's constant and T is the absolute temperature (in Kelvin). If one can treat the cantilever as an ideal spring of constant k, a measurement of the thermal noise $\langle x^2 \rangle$ in its position allows the spring constant to be determined as

$$k = \frac{k_B T}{\langle x^2 \rangle}$$

Unfortunately, the cantilever *cannot* be treated as an ideal spring. A series of improvements, lead to the correct calibration formula

$$k = 0.8174 \frac{k_B T}{s^2 P} \left[\frac{1 - \left(\frac{3D}{2L} \right) \tan\phi}{1 - \left(\frac{2D}{L} \right) \tan\phi} \cos\phi \right]^2$$

where

- the numerical factor 0.8174 is a result of the geometry of the cantilever spring
- s is again the sensitivity calibration factor (in units of m/V)
- P is the positional noise power (in units of V^2) isolated in the fundamental resonant mode only
- D is the height of the tip
- L is the cantilever length (more correctly, the distance from the base of the cantilever to the tip

and the term in square brackets accounts for the cantilever tilt relative to the substrate and the geometry of the optical-lever detection scheme commonly used in commercial AFMs.

The thermal noise technique is imperfect by the sensitivity of the device used to measure the noise in the deflection signal: stiffer cantilevers suffer less thermal vibration. One source of error is the calibration factor s, which can be obtained from a force curve against a rigid substrate. In many cases an oscillatory baseline due to interference between the reflected laser beam and laser light scattered from the substrate can lead to an incorrect estimate of *s* (the calibration against a known standard is also succeptible to such errors).

The primary advantage of the thermal noise technique is ease of use. Once the apparatus and procedures are in place, it takes approximately 5 minutes to calibrate a cantilever, including mounting the cantilever, measuring data, and analysing the results.

THERMAL NOISE TECHNIQUE: THEORY AND BACKGROUND

A guided tour through the history of the thermal noise technique for those who feel the need to work.

The thermal noise technique for calibrating AFM cantilevers dates back to 1993. As presented, it used the equipartition theorem to relate the measured thermal noise in the AFM cantilever deflection signal to the spring constant in the cantilever, which was assumed to behave like an ideal spring. In order to exclude other sources of noise, such as electronic noise, 60 Hz noise, table vibrations, etc., the authors recommended that the noise be measured in the frequency domain.

Butt and Jaschke recognized that because a cantilever spring is not an ideal Hookean spring, its potential energy is not simply $kx^2/2$. In order to relate the cantilever spring constant to the noise in its end position, one must express the Hamiltonian of the cantilever spring in terms of its normal modes. Moreover, the same authors pointed out that AFMs that use the optical-lever technique to sense the cantilever deflection actually measure the end *slope* of the cantilever as that is the quantity that determines the position of the reflected laster spot. The end slope is related to the end deflection, but the relationship is different for a cantilever who's end is supported by a surface (the configuration used to calibrate the deflection sensitivity) and one who's end is free to oscillate (the configuration in which the thermal noise is measured). The authors calculated a correction factor which also included the thermal noise in *all* cantilever vibrational modes, despite the technical difficulty (there is no way to discriminate against instrumental noise) and, indeed, the impossibility of obtaining such data (no instrument has infinite band width).

Walters, *et al.* were it seems that the first to note the need to limit the result of Butt and Jaschke to include the contribution of the fundamental vibrational mode only. When this is done, the original result needs to be modified by a numerical factor of 0.8174. This allowed them to calibrate cantilevers from the area of the fundamental resonant peak in power spectra of the vibrational noise. Since then, occasional claims have been made (by others) that overlap of this peak with the second vibrational mode can be significant; however, we have not seen this and believe that, except perhaps in cases with *very* broad resonances, this will not be important.

The effects of cantilever tilt on its response to forces, this author revisited the issue of calibration by thermal noise and obtained the correction for cantilever tilt and aspect ratio *D/L* included in the "final" thermal calibration formula presented here, as well as for the other two techniques. Physically, cantilever tilt (which is necessary in all mounting systems) results is a cosine-squared correction: only a component of vertical forces exerted by the surface acts to deflect the cantilever, and only a component of that deflection is resolved along the vertical axis. The cantilever aspect ratio also results in a small correction. Although these corrections are small in typical cases, they become more important for the high-aspect ratio cantilevers favoured for high-resolution studies, and similar considerations can be important for colloidal probes.

It is worth note that both the reference lever and added noise techniques calibrate the deflection sensitivity by bending the cantilever with vertical forces exerted by a flat substrate. Therefore, the tilt effect was entirely compensated in previous measurements that employed either of those calibration schemes, leaving a much weaker aspect-ratio correction. This author recommends that, in future work, the intrinsic cantilever spring constant (*i.e.*, tilt-independent response of the cantilever to perpendicular forces) be quoted, and measured forces be corrected for the cantilever tilt (and for the aspect-ratio in cases where it makes a significant difference).

Is this the last word on the thermal noise technique? Perhaps not — implicit in the numerical corrections to the thermal noise result (as well as to force measurements in general, irrespective of the calibration technique employed) is the assumption of a rectangular cantilever. Although many more rectangular cantilevers are commercially available now than were a decade ago, force data is often acquired with triangular cantilevers, which have different bending modes. Fortunately, numerical simulations of such cantilevers predict only minor corrections so this is unlikely to be a concern until cantilevers can be calibrated to the 1% level rather than the current state-of-the-art of only 5-10%.

MEMS SWITCH USES MAGNETIC ACTUATION

MEMS RF switches contain distinct advantages over solid-state switches. They provide much lower insertion loss, higher isolation, better linearity, use lower power and are potentially cheaper. They also outdo conventional electromechanical switches and relays in terms of switching speed, integration capabilities, power and cost.

However, most MEMS RF work has focused on electrostatic actuation, presumably because other actuation mechanisms, such as magnetic and thermal actuation, consume much power and thus are not suitable for many applications.

Microlab and Arizona State University have demonstrated a magnetic latching and switching mechanism that eliminates the need for the static power supply and that can latch on or off with zero power required to hold this state. The switch is nonvolatile and bistable. Because of the long-range magnetic forces, the switch-it has been named MagLatch-requires low operation voltage (less than 5 V).

The design consists of a cantilever, an embedded planar coil, a permanent magnet and the necessary electrical contacts. As an electrical switch, the cantilever is a two-layer composite consisting of a soft magnetic material NiFe permalloy on its topside and a highly conductive material, such as gold, at the bottom surface. The cantilever is supported by torsion flexures from the two sides. The contact end to the right of the cantilever can be deflected up or down by applying a current through the coil. When down, the cantilever makes electrical contact with the bottom conductors and the switch is on, or closed; when the contact end is up, or opened, the switch is off. The permanent magnet holds the cantilever up or down after switching, making the device a latching relay. Single-pole-double-throw and RF switches can be designed. Latching optical switches can be made based on similar principles.

The principle behind the latching characteristics is the preferential magnetization of a cantile ver made of soft magnetic material (for example, permalloy). In a constant, nearly perpendicular magnetic field, a cantilever can have a clockwise or a counter-clockwise torque depending on the angle between the cantilever and the field, which leads to the bistability. To switch the relay, a second magnetic field, in this case generated by a short current pulse through a coil, realigns the magnetization of the cantilever causing it to flip. A static external magnetic field instantly latches the switch in the closed or open position, respectively. The switch maintains this state until the next switching signal realigns the cantilever. The relay consumes no power to maintain the latched state.

NEMS

NEMS is an acronym for "Nano Electro Mechanical Systems". The company NEMS AB develops devices based on a proprietary technology platform under US patent number 6,818,959. These can have a broad variety of applications such as bio sensors, electrical switches and filters, and optical devices. Despite the variety of uses, they share some common characteristics. Due to nanometer dimensions, and the use of light materials, the devices achieve performance in terms of speed, sensitivity and versatility that challenges or surpasses the limitations of technological achievement to date. Another common denominator is the possibility of simple, low cost fabrication, made possible by developments in Nano Imprint Lithography (NIL).

Nanoelectromechanical systems (NEMS) are devices integrating electrical and mechanical functionality on the nanoscale. NEMS form the logical next miniaturization step from so-called microelectromechanical systems, or MEMS devices. NEMS typically integrate transistor-like nanoelectronics with mechanical actuators, pumps, or motors, and may thereby form physical, biological, and chemical sensors. The name derives from typical device dimensions in the nanometer range, leading to low mass, high mechanical resonance frequencies, potentially large quantum mechanical effects such as zero point motion, and a high surface-to-volume ratio useful for surface-based sensing mechanisms. Uses include accelerometers, or detectors of chemical substances in the air.

FUTURE OF NEMS

Previous to NEMS devices can actually be implemented, reasonable integrations of carbon based products must be created. The focus is currently shifting from experimental work towards practical applications and device structures that will implement and profit from the use of carbon nanotubes. At this point in NEMS research, there is a general understanding of the properties of carbon nanotubes and graphene. The next challenge to overcome involves understanding all of the properties of these carbon based tools, and using the properties to make efficient and durable NEMS with low failure rates.

NEMS devices, if implemented into everyday technologies, could further reduce the size of modern devices and allow for better performing sensors. Carbon based materials have served as prime materials for NEMS use, because of their highlighted mechanical and electrical properties. Once NEMS interactions with outside environments are integrated with effective designs, they will likely become useful products to everyday technologies.

BASIC PRINCIPLES OF ATOMIC FORCE MICROSCOPY (AFM)

Figure: Electron micrograph of a used AFM cantilever image width ~100 micrometers...

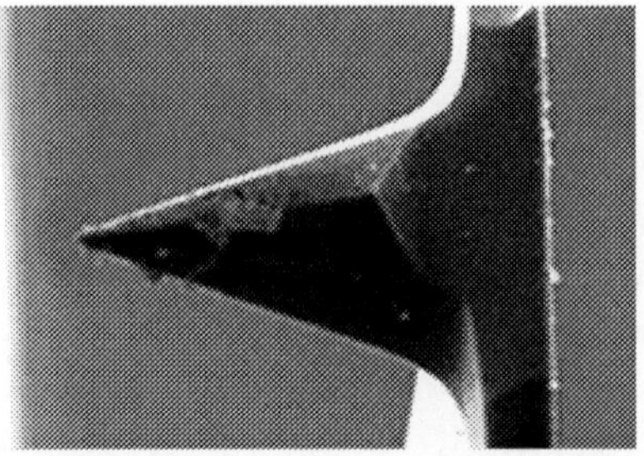

Figure: and ~30 micrometers

The AFM consists of a cantilever with a sharp tip (probe) at its end that is used to scan the specimen surface. The cantilever is typically silicon or silicon nitride with a tip radius of curvature on the order of nanometers. When the tip is brought into proximity of a sample surface, forces between the tip and the sample lead to a deflection of the cantilever according to Hooke's law. Depending on the situation, forces that are measured in AFM include mechanical contact force, van der Waals forces, capillary forces, chemical bonding, electrostatic forces, magnetic forces, Casimir forces, solvation forces, etc. Along with force, additional quantities may simultaneously be measured through the use of specialized types of probe. Typically, the deflection is measured using a laser spot reflected from the top surface of the cantilever into an array of photodiodes. Other techniques that are used include optical interferometry, capacitive sensing or piezoresistive AFM cantilevers. These cantilevers are fabricated with piezoresistive elements that act as a strain gauge. Using a Wheatstone bridge, strain in the AFM cantilever due to deflection can be measured, but this technique is not as sensitive as laser deflection or interferometry.

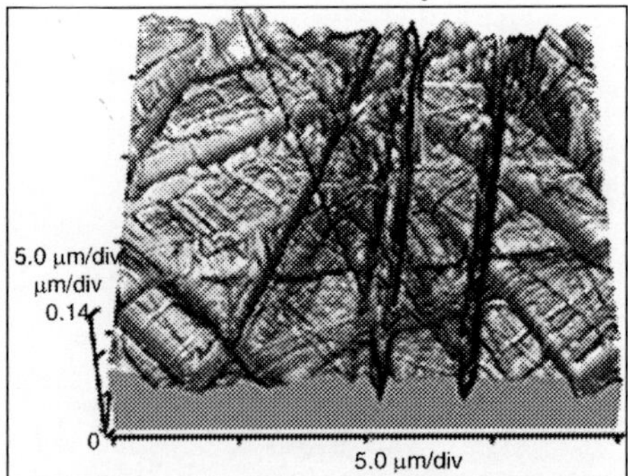

Figure: Atomic force microscope topographical scan of a glass surface. The micro and nano-scale features of the glass can be observed, portraying the roughness of the material. The image space is (x,y,z) = (20 µm × 20 µm × 420 nm).

If the tip be scanned at a constant height, a risk would exist that the tip collides with the surface, causing damage. Hence, in most cases a feedback mechanism is employed to adjust the tip-to-sample distance to maintain a constant force between the tip and the sample. Traditionally, the sample is mounted on a piezoelectric tube, that can move the sample in the z direction for maintaining a constant force, and the x and y directions for scanning the sample. Alternatively a 'tripod' configuration of three piezo crystals may be employed, with each responsible for scanning in the x,y and z directions. This eliminates some of the distortion effects seen with a tube scanner. In newer designs, the tip is mounted on a vertical piezo scanner while the sample is being scanned in X and Y using another piezo block. The resulting map of the area $z = f(x,y)$ represents the topography of the sample.

The AFM can be operated in a number of modes, depending on the application. In general, possible imaging modes are divided into static (also called *contact*) modes and a variety of dynamic (non-contact or "tapping") modes where the cantilever is vibrated.

Imaging Modes

The primary modes of operation for an AFM are static mode and dynamic mode. In static mode, the cantilever is "dragged" across the surface of the sample and the contours of the surface are measured directly using the deflection of the cantilever. In the dynamic mode, the cantilever is externally oscillated at or close to its fundamental resonance frequency or a harmonic. The oscillation amplitude, phase and resonance frequency are modified by tip-sample interaction forces. These changes in oscillation with respect to the external reference oscillation provide information about the sample's characteristics.

Contact Mode

In the static mode operation, the static tip deflection is used as a feedback signal. Because the measurement of a static signal is prone to noise and drift, low stiffness cantilevers are used to boost the deflection signal. However, close to the surface of the sample, attractive forces can be quite strong, causing the tip to "snap-in" to the surface. Thus static mode AFM is almost always done in contact where the overall force is repulsive. Consequently, this technique is typically called "contact mode". In contact mode, the force between the tip and the surface is kept constant during scanning by maintaining a constant deflection.

Non-contact Mode

In this form, the tip of the cantilever does not contact the sample surface. The cantilever is instead oscillated at a frequency slightly above its resonant frequency where the amplitude of oscillation is typically a few nanometers (<10 nm). The van der Waals forces, which are strongest from 1 nm to 10 nm above the surface, or any other long range force which extends above the surface acts to decrease the resonance frequency of the cantilever. This decrease in resonant frequency combined with the feedback loop system maintains a constant oscillation amplitude or frequency by adjusting the average tip-to-sample distance. Measuring the tip-to-sample distance at each (x,y) data point allows the scanning software to construct a topographic image of the sample surface.

Non-contact mode AFM does not suffer from tip or sample degradation effects that are sometimes observed after taking numerous scans with contact AFM. This makes non-contact AFM preferable to contact AFM for measuring soft samples. In the case of rigid samples, contact and non-contact images may look the same. However, if a few monolayers of adsorbed fluid are lying on the surface of a rigid sample, the images may look quite different. An AFM operating in contact mode will penetrate the liquid layer to image the underlying surface, whereas in non-contact mode an AFM will oscillate above the adsorbed fluid layer to image both the liquid and surface.

Schemes for dynamic mode operation include frequency modulation and the more common amplitude modulation. In frequency modulation, changes in the oscillation frequency provide information about tip-sample interactions. Frequency can be measured with very high sensitivity and thus the frequency modulation mode allows for the use of very stiff cantilevers. Stiff cantilevers provide stability very close to the surface and, as a result, this technique was the first AFM technique to provide true atomic resolution in ultra-high vacuum conditions.

In amplitude modulation, changes in the oscillation amplitude or phase provide the feedback signal for imaging. In amplitude modulation, changes in the phase of oscillation can be used to discriminate between different types of materials on the surface. Amplitude modulation can be operated either in the non-contact or in the intermittent contact regime. In dynamic contact mode, the cantilever is oscillated such that the separation distance between the cantilever tip and the sample surface is modulated.

Amplitude modulation has also been used in the non-contact regime to image with atomic resolution by using very stiff cantilevers and small amplitudes in an ultra-high vacuum environment.

Tapping Mode

In ambient conditions, most samples develop a liquid meniscus layer. Because of this, keeping the probe tip close enough to the sample for short-range forces to become detectable while preventing the tip from sticking to the surface presents a major problem for non-contact dynamic mode in ambient conditions. Dynamic contact mode (also called intermittent contact or tapping mode) was developed to bypass this problem.

In *tapping mode,* as well called "AC Mode" or "intermittent contact mode", the cantilever is driven to oscillate up and down at near its resonance frequency by a small piezoelectric element mounted in the AFM tip holder similar to non-contact mode. However, the amplitude of this oscillation is greater than 10 nm, typically 100 to 200 nm. The interaction of forces acting on the cantilever when the tip comes close to the surface, van der Waals force, dipole-dipole interaction, electrostatic forces, etc. cause the amplitude of this oscillation to decrease as the tip gets closer to the sample. An electronic servo uses the piezoelectric actuator to control the height of the cantilever above the sample. The servo adjusts the height to maintain a set cantilever oscillation amplitude as the cantilever is scanned over the sample. A *tapping AFM* image is therefore produced by imaging the force of the intermittent contacts of the tip with the sample surface.

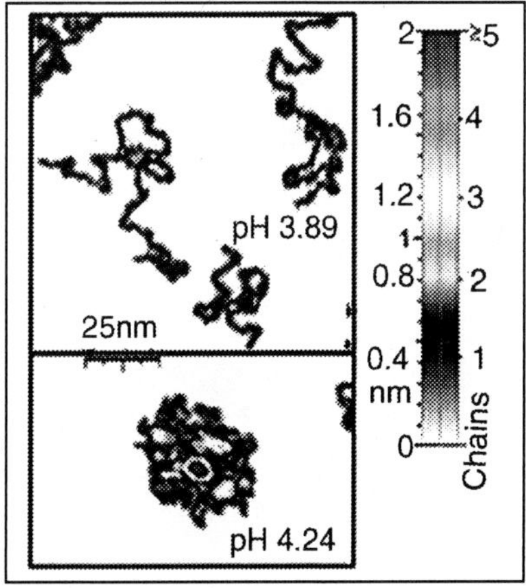

Figure: Single polymer chains (0.4 nm thick) recorded in a tapping mode under aqueous media with different pH.

This method of "tapping" lessens the damage done to the surface and the tip compared to the amount done in contact mode. Tapping mode is gentle enough even for the visualization of supported lipid bilayers or adsorbed single polymer molecules (for instance, 0.4 nm thick chains of synthetic polyelectrolytes) under liquid medium. With proper scanning parameters, the conformation of single molecules can remain unchanged for hours.

AFM Resolution

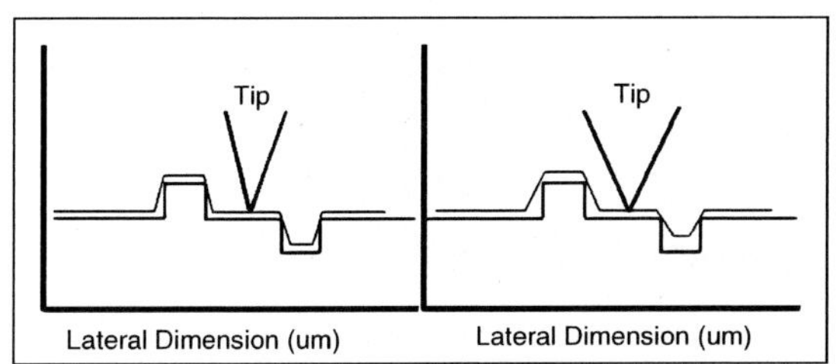

The concept of resolution in AFM is different from radiation based microscopies because AFM imaging is a three dimensional imaging technique. The ability to distinguish two separate points on an image is the standard by which lateral resolution is usually defined. There is clearly an important distinction between images resolved by wave optics and scanning probe techniques.

The former is limited by diffraction, and later primarily by apical probe geometry and sample geometry. Usually the width of a DNA molecule is loosely used as a measure of resolution, because it has a known diameter of 2.0 nm in the B form. Some of the best values for AFM imaging are 3.0 nm quoted form DNA in propanol. Unfortunately, this definition of resolution can be misleading because the sample height clearly effects this value.

Indeed, numerous authors have seen that it is the radius of curvature that significantly influences the resolving ability of the AFM. Images of DNA made by the sharper tip have shown dramatic improvements in resolution widths. Even greater improvements in resolution have been attained with Tappingmode but contact imaging still is capable of high resolution imaging.

Vibration Isolation

In arrange to obtain good AFM results, the vibration isolation platform is needed. The vibration isolation consists of a large mass attached to bungy cords firmly anchored to the building. Damping of the oscillation is believed to result from rubbing of the rubber fibres inside of the bungy cord against the outside lining material. Between the low resonance frequency of the bungy cord system and the high resonance frequency of the microscope hardware itself (> 10 kHz), the AFM effectively comprises a band pass filter. This allows the microscopists to safely image their samples in the intermediate range of about 1 - 100 Hz and obtain atomic resolution.

Comparison of AFM and other Imaging Techniques

AFM versus STM

It's interesting to compare AFM and its precursor — Scanning Tunnelling Microscope. In some cases, the resolution of STM is better than AFM because of the exponential dependence of the tunnelling current on distance. The force-distance dependence in AFM is much more complex when characteristics such as tip shape and contact force are considered. STM is generally applicable only to conducting samples while AFM is applied to both conductors and insulators.

In terms of versatility, needless to say, the AFM wins. Furthermore, the AFM offers the advantage that the writing voltage and tip-to-substrate spacing can be controlled independently, whereas with STM the two parameters are integrally linked.

AFM versus SEM

Compared with Scanning Electron Microscope, AFM provides extraordinary topographic contrast direct height measurements and unobscured views of surface features (no coating is necessary).

AFM versus TEM

Compared with Transmission Electron Microscopes, three dimensional AFM images are obtained without expensive sample preparation and yield far more complete information than the two dimensional profiles available from cross-sectioned samples.

AFM versus Optical Microscope

Compared with Optical Interferometric Microscope (optical profiles), the AFM provides unambiguous measurement of step heights, independent of reflectivity differences between materials.

The Common AFM Modes

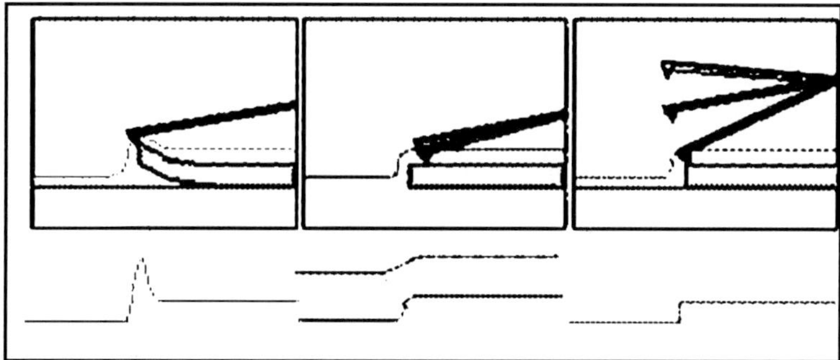

Many AFM modes have appeared for special purpose while the technique of AFM is becoming mature. Here I only specify the three commonly used techniques: Contact Mode(left), Non-contact Mode(middle) and Tapping Mode(right). Look here for a summary.

Contact Mode

The contact mode where the tip scans the sample in close contact with the surface is the common mode used in the force microscope. The force on the tip is repulsive with a mean value of 10^{-9} N. This force is set by pushing the cantilever against the sample surface with a piezoelectric positioning element. In contact mode AFM the deflection of the cantilever is sensed and compared in a DC feedback amplifier to some desired value of deflection. If the measured deflection is different from the desired value the feedback amplifier applies a voltage to the piezo to raise or lower the sample relative to the cantilever to restore the desired value of deflection.

The voltage that the feedback amplifier applies to the piezo is a measure of the height of features on the sample surface. It is displayed as a function of the lateral position of the sample. A few instruments operate in UHV but the majority operate in ambient atmosphere, or in liquids. Problems with contact mode are caused by excessive tracking forces applied by the probe to the sample. The effects can be reduced by minimizing tracking force of the probe on the sample, but there are practical limits to the magnitude of the force that can be controlled by the user during operation in ambient environments. Under ambient conditions, sample surfaces are covered by a layer of adsorbed gases consisting primarily of water vapour and nitrogen which is 10-30 monolayers thick.

When the probe touches this contaminant layer, a meniscus forms and the cantilever is pulled by surface tension toward the sample surface. The magnitude of the force depends on the details of the probe geometry, but is typically on the order of 100 nanoNewtons. This meniscus force and other attractive forces may be neutralized by operating with the probe and part or all of the sample totally immersed in liquid. There are many advantages to operate AFM with the sample and cantilever immersed in a fluid. These advantages include the elimination of capillary forces, the reduction of Van der Waals' forces and the ability to study technologically or biologically important processes at liquid solid interfaces. However there are also some disadvantages involved in working in liquids. These range from nuisances such as leaks to more fundamental problems such as sample damage on hydrated and vulnerable biological samples.

In adding, a large class of samples, including semiconductors and insulators, can trap electrostatic charge (partially dissipated and screened in liquid). This charge can contribute to additional substantial attractive forces between the probe and sample. All of these forces combine to define a minimum normal force that can be controllably applied by the probe to the sample. This normal force creates a substantial frictional force as the probe scans over the sample. In practice, it appears that these frictional forces are far more destructive than the normal force and can damage the sample, dull the cantilever probe and distort the resulting data. Also many samples such as semiconductor wafers can not practically be immersed in liquid. An attempt to avoid these problem is the Non-contact Mode.

Noncontact Mode

A novel era in imaging was opened when microscopists introduced a system for implementing the non-contact mode which is used in situations where tip contact might alter the sample in subtle ways. In this mode the tip hovers 50 - 150 Angstrom above the sample surface. Attractive Van der Waals forces acting between the tip and the sample are detected, and topographic images are constructed by scanning the tip above the surface. Unfortunately the attractive forces from the sample are substantially weaker than the forces used by contact mode. Therefore the tip must be given a small oscillation so that AC detection techniques can be used to detect the small forces between the tip and the sample by measuring the change in amplitude, phase, or frequency of the oscillating cantilever in response to force gradients from the sample. For highest resolution, it is necessary to measure force gradients from Van der Waals forces which may extend only a nanometer from the sample surface. In general, the fluid contaminant layer is substantially thicker than the range of the Van der Waals force gradient and therefore, attempts to image the true surface with non-contact AFM fail as the oscillating probe becomes trapped in the fluid layer or hovers beyond the effective range of the forces it attempts to measure.

Tapping Mode

Tapping form is a key advance in AFM. This potent technique allows high resolution topographic imaging of sample surfaces that are easily damaged, loosely hold to their substrate, or difficult to image by other AFM techniques. Tapping mode overcomes problems associated with friction, adhesion, electrostatic forces, and other difficulties that an plague conventional AFM scanning techniques by alternately placing the tip in contact with the surface to provide high resolution and then lifting the tip off the surface to avoid dragging the tip across the surface. Tapping mode imaging is implemented in ambient air by oscillating the cantilever assembly at or near the cantilever's resonant frequency using a piezoelectric crystal. The piezo motion causes the cantilever to oscillate with a high amplitude (typically greater than 20nm) when the tip is not in contact with the surface. The oscillating tip is then moved toward the surface until it begins to lightly touch, or tap the surface. During scanning, the vertically oscillating tip alternately contacts the surface and lifts off, generally at a frequency of 50,000 to 500,000 cycles per second. As the oscillating cantilever begins to intermittently contact the surface, the cantilever oscillation is necessarily reduced due to energy loss caused by the tip contacting the surface. The reduction in oscillation amplitude is used to identify and measure surface features.

During tapping mode operation, the cantilever oscillation amplitude is maintained constant by a feedback loop. Selection of the optimal oscillation frequency is software-assisted and the force on the sample is automatically set and maintained at the lowest possible level. When the tip passes over a bump in the surface, the cantilever has less room to oscillate and the amplitude of oscillation decreases. Conversely, when the tip passes over a depression, the cantilever has more room to oscillate and the amplitude increases (approaching the maximum free air amplitude). The oscillation amplitude of the tip is measured by the detector and input to the NanoScope III controller electronics. The digital feedback loop then adjusts the tip-sample separation to maintain a constant amplitude and force on the sample.

At what time the tip contacts the surface, the high frequency (50k - 500k Hz) makes the surfaces stiff (viscoelastic), and the tip-sample adhesion forces is greatly reduced. TappingMode inherently prevents the tip from sticking to the surface and causing damage during scanning. Unlike contact and non-contact modes, when the tip contacts the surface, it has sufficient oscillation amplitude to overcome the tip-sample adhesion forces. Also, the surface material is not pulled sideways by shear forces since the applied force is always vertical. Another advantage of the TappingMode technique is its large, linear operating range. This makes the vertical feedback system highly stable, allowing routine reproducible sample measurements.

Tapping mode operation in fluid has the same advantages as in the air or vacuum. However imaging in a fluid medium tends to damp the cantilever's normal resonant frequency. In this case, the entire fluid cell can be oscillated to drive the cantilever into oscillation. This is different from the tapping or non-contact operation in air or vacuum where the cantilever itself is oscillating. When an appropriate frequency is selected (usually in the range of 5,000 to 40,000 cycles per second), the amplitude of the cantilever will decrease when the tip begins to tap the sample, similar to TappingMode operation in air. Alternatively, the very soft cantilevers can be used to get the good results in fluid. The spring constant is typically 0.1 N/m compared to the tapping mode in air where the cantilever may be in the range of 1-100 N/m.

Biological Applications of AFM

Single of the advantages of AFM is that it can image the non-conducting surfaces. So it was immediately extended to the biological systems, such as analysing the crystals of amino acids and organic monolayers. Applications of AFM in the biosciences include: DNA and RNA analysis; Protein-nucleic acid complexes; Chromosomes; Cellular membranes; Proteins and peptides; Molecular crystals; Polymers and biomaterials; Ligand-receptor binding. Bio-samples have been investigated on lysine-coated glass and mica substrate, and in buffer solution. By using phase imaging technique one can distinguish the different components of the cell membranes. Little sample preparation is required for bioimaging with the AFM. In most cases it is as simple as spotting a few microliters of solution on mica or glass. Of course contaminations that cover surface features have to be avoided or removed. First the substrate-adsorbate should be rinsed with a large excess of buffer. The following procedures are dialysis, centrifugation and homogenization. In order to get good contrast and to reduce mechanical damage of the soft biological materials, the samples can be stabilized by adding covalent cross-linking agents or certain cations that are able to link the constituents of the sample to each other or to the substrate. Cooling can also stiffen the sample. Nevertheless, all these techniques have significant influence on the properties of biomolecules. The sharper tip now is available commercially and really does help a lot.

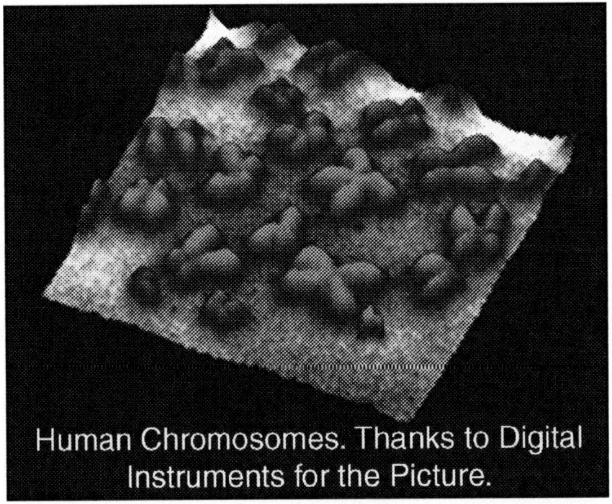

One area of significant progress is the imaging of nucleic acids. The ability to generate nanometer-resolved images of unmodified nucleic acids has broad biological applications. Chromosome mapping, transcription, translation and small molecule-DNA interactions such as intercalating mutagens, provide exciting topics for high-resolution studies. The first highly reproducible AFM images of DNA were obtained only in 1991. Four major advances that have enabled clear resolution of nucleic acids are: Control of the local imaging environment including sample modification; TappingMode scanning techniques; Improved AFM probes (such as standard silicon nitride probes modified by electron beam deposition and Oxide Sharpened NanoProbes) and Compatible substrates (such as salinized mica and carbon coated mica).

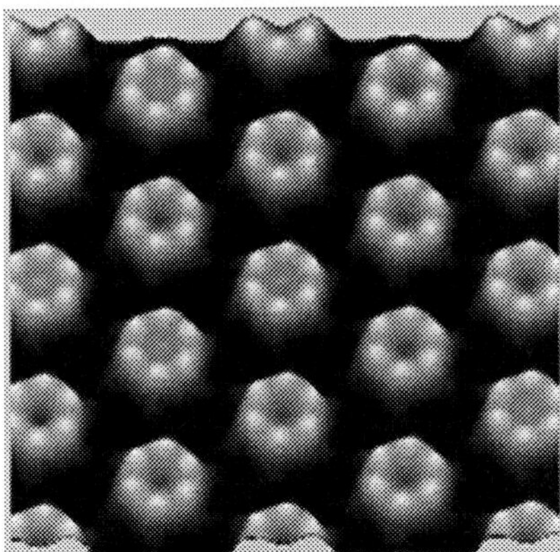

Figure: Protein surface layer of D. Radiodurans.

Cell biologists have applied the AFM's unique capabilities to study the dynamic behaviour of living and fixed cells such as red and white blood cells, bacteria, platelets, cardiac myocytes, living renal epithelial cells, and glial cells. For example, plasma membrane in migrating epithelial cells has been imaged in real time. The dynamic membrane invagination process was observed in the presence of calcium and when calcium levels were reduced the process was prevented. 30nm lipidic pore formation could also be resolved during calcium reduction. AFM imaging of cells usually achieves a resolution of only 20-50 nm, not sufficient for resolving membrane proteins but still suitable for imaging other surface features, such as rearrangements of plasma membrane or movement of submembrane filament bundles. The requirement for the imaging buffer is not restrictive, as long as the buffer does not severely affect the integrity of the cells. EDTA should be avoided because cells will detach from the substrate in the absence of divalent cations. Cultured cells normally adhere well to the substrate and are not displaced by modest probe forces.

The recent success imaging individual proteins and other small molecules with the AFM such as collogen. Smaller molecules that do not have a high affinity for common AFM substrates have been successfully imaged by employing selective affinity binding procedures. Thiol incorporation at both the 5' and 3' ends of short PCR products has been shown to confer a high affinity for ultraflat gold substrates. A similar approach was used to immobilize antibodies (IgG1) on treated mica. In this case, the low affinity that IgG molecules have for mica was overcome by cloning a metal-chelating peptide into the carboxy terminus sequence of the IgG's heavy chain. The recombinant sequence was transformed into cells that expressed the complementary light chain. The purified IgG containing the metal-chelating peptide was shown to bind in a regiospecific manner to nickel-treated mica. Covalent binding of biological structures to derivatized glass substrates has also enabled high resolution imaging of some samples that are not stable on untreated glass substrates. New approaches in AFM have provided a solid foundation from which research is expanding into more complex analyses. Higher resolution imaging of a variety of small molecules is improving at a rapid pace.

The recent innovation, such as Digital Instruments BioScope system, which combines the high resolution of AFM with the ease of use and familiarity of inverted optical microscopes, has further added to the attractiveness of AFM for biological imaging. Bright-field, fluorescence and other optical techniques can be used to identify structures of interest while the AFM simultaneously generates nanometer-resolved images of the sample surface.

Countless biological processes - DNA replication, protein synthesis, drug interaction, and many others - are largely governed by intermolecular forces. AFM has the ability to measure forces in the nanonewton range. This makes it possible to quantify the molecular interaction in biological systems such as a variety of important ligand-receptor interactions. Another application of AFM force measurements is to image or quantify electrical surface charge.

The dynamics of many biological systems depends on the electrical properties of the sample surface. In addition to measuring binding forces and electrostatic forces, the AFM can also probe the micromechanical properties of biological samples. Specifically, the AFM can observe the elasticity and, in fact, the viscosity of samples ranging from live cells and membranes to bone and cartilage.

AFM gives scientists a key tool to investigate the real world.

THE SCANNING TUNNELING MICROSCOPE (STM)

The scanning tunnelling microscope (STM) is a kind of electron microscope that shows three-dimensional images of a sample. In the STM, the structure of a surface is studied using a stylus that scans the surface at a fixed distance from it.

Currents Control the Surface

An extremely fine conducting probe is held close to the sample. Electrons tunnel between the surface and the stylus, producing an electrical signal. The stylus is extremely sharp, the tip being formed by one single atom. It slowly scans across the surface at a distance of only an atom's diameter. The stylus is raised and lowered in order to keep the signal constant and maintain the distance. This enables it to follow even the smallest details of the surface it is scanning. Recording the vertical movement of the stylus makes it possible to study the structure of the surface atom by atom. A profile of the surface is created, and from that a computer-generated contour map of the surface is produced.

Important in Many Sciences

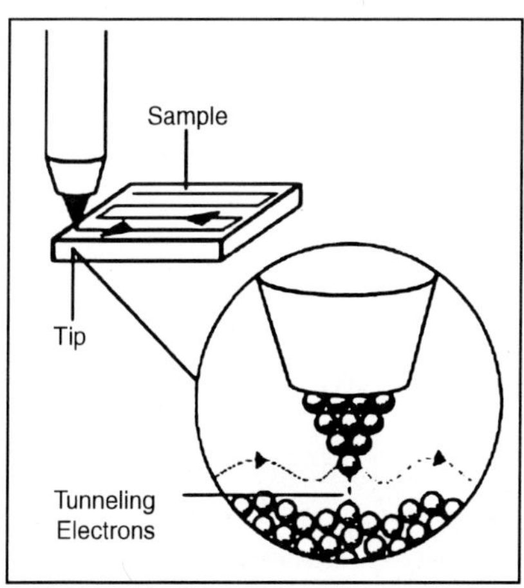

The study of surfaces is an important part of physics, with particular applications in semiconductor physics and microelectronics. In chemistry, surface reactions also play an important part, for example in catalysis. The STM works best with conducting materials, but it is also possible to fix organic molecules on a surface and study their structures. For example, this technique has been used in the study of DNA molecules.

PRINCIPLE OF SCANNING PROBE MICROSCOPY

Universal Microscopes

Modern microscopes, such as the so-called Scanning Tunnelling Microscope (STM), can image the surfaces of materials with unparalleled magnification. The magnification is so extreme, that *individual atoms* become visible. With its ultimate resolution, this remarkable instrument forms the basis of an enormous development within physics. But also in the fields of chemistry and biology, the STM and derived microscopes have conquered an important position within a very short time. With an STM one can only image surfaces of materials that conduct electrical currents. But the principle of the STM is so flexible, that with relatively modest changes in the technology also non-conductive materials can be imaged. In this way, also oxides and even biological materials, such as DNA, can be investigated on the sub-nanometer scale. By now, there is a family of some twenty different types of microscopes, derived from the STM, which are referred to as Scanning Probe Microscopes (SPM).

In 1986, very soon after their first publications about the STM in 1981, the inventors of this marvelous instrument, Gert Binnig and Heinrich Rohrer from the IBM Research Laboratory in Rüschlikon (Switzerland), were awarded the Nobel Prize in Physics. Nowadays, SPMs can be found in many academic and industrial physics, chemistry and biology laboratories. They are used both as standard analysis tools and as high-level research instruments. There is an impressive literature about SPM-technology and application. The following books provide a good introduction.

The STM Works like a Record Player

The standard of the STM is remarkably simple, and can be compared best with that of an old-fashioned record player. Just like in a record player, the instrument uses a sharp needle, referred to as the 'tip', to interrogate the shape of the surface. But in contrast with a normal record player, the STM tip does not touch the surface. This is done by the technique, indicated in the schematic picture. A voltage is applied between the metallic tip and the specimen, typically between a few milliVolts (mV) and a few Volts (V). When the tip touches the surface of the specimen, the voltage will, of course, result in a current. When the tip is far away from the surface, the current is zero.

The STM operates in the regime of extremely small distances between the tip and the surface of only 0.5 to 1.0 nm, *i.e.* 2 to 4 atomic diameters. At these distances, the electrons can 'jump' from the tip to the surface or *vice-versa*. This 'jumping' is a quantum mechanical process, known as 'tunnelling'. Hence the name of this microscope: tunnelling microscope. The tunnelling process is very difficult, which implies that the tunnelling current is always very low. STMs usually operate at tunnelling currents between a few picoAmperes (pA) and a few nanoAmperes (nA).

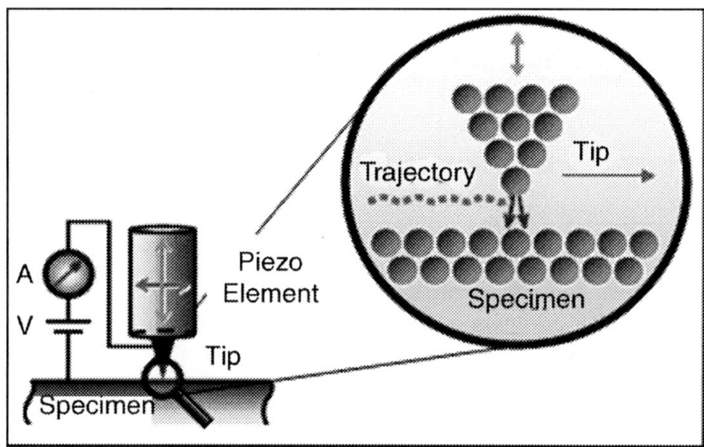

The tunnelling current depends very critically on the precise distance between the last atom of the tip and the nearest atom or atoms of the underlying specimen. When this distance is increased only a little bit, the tunnelling current decreases strongly. As a rule of thumb, for every extra atom diameter that is added to the distance, the current becomes a factor 1000 lower! This means that the tunnelling current provides a highly sensitive measure of the distance between the tip and the surface.

Feedback

The STM tip is attached to a piezo-electric element. This is a piece of material with the useful property that it changes its length a little bit, when it is put under an electrical voltage. By adjusting the voltage on the piezo element, the distance between the tip and the surface can be regulated. In most STMs, the voltage on the piezo element is adjusted such, that the tunnelling current always has the same value, for example 1 nA. In this way, the distance between the last atom on the tip and the nearest atoms in the surface is being kept constant. This distance regulation is performed automatically, by so-called feedback electronics, which continually measure the deviation of the tunnelling current from the desired value – *e.g.* 1 nA – and retract the tip when the current is too high or advance it when the current is too low.

Scanning

While this criticism system is active, two other parts of the piezo element are used to move the tip in the X- and Y-directions, parallel to the surface, to scan over the surface, line by line, similar to the way a television or computer screen image is built up. The combination of small displacements in the X-, Y-, and Z-directions, parallel and perpendicular to the surface, is obtained with suitable combinations or geometries of piezo-electric elements.

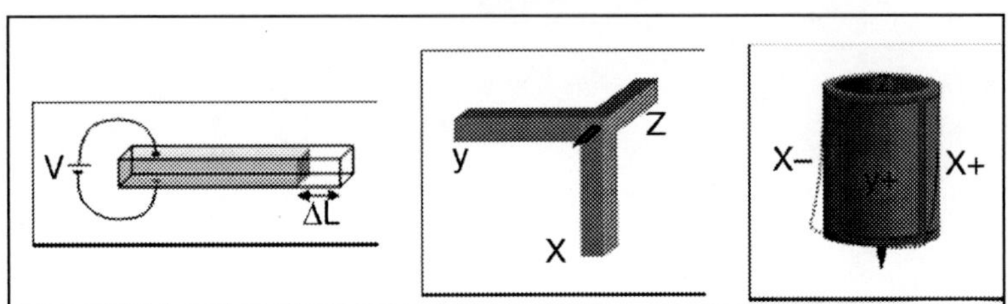

Figure: (1) Principle of piezo element. The applied voltage makes the element longer or shorter. (2) The combination of three piezo elements makes it possible to move the STM tip in the X-, Y-, and Z-directions. (3) In most modern scanning probe microscopes, one uses a tube geometry.

In most modern scanning probe microscopes, one uses a tube geometry. Each of the four indicated sections can be made longer or shorter individually. If all four sections are made longer or shorter by the same amount, the tip moves in the z-direction. If the X+ side is made longer, and at the same time the X- side is made shorter by the same amount, the tube deforms a little bit, as indicated. For small deformations, this makes the tip move primarily in the X-direction. The same can be done in the Y-direction.

Making STM-images

In the XY-scan, every time that the last atom of the tip is precisely above a surface atom, the tip needs to be retracted a little bit, while it has to be brought slightly closer when the tip atom is between the surface atoms. Thus, the tip automatically follows a bumpy trajectory, which mimics the atomic corrugation of the surface. The information about this trajectory is available in the form of the voltages that have been applied by the feedback electronics to the piezo element during the XY-scan. The last step is to visualize the tip trajectories. There are several ways to do this, for example in the form of a collection of individual height lines, or in the form of a gray-scale or colour-scale representation, or in the form of some three-dimensional perspective view. It is important to remember that the STM is 'colour blind', and that all gray scales and colour scales represent nothing else than heights. Also note that the height scale in most perspective views is grossly exaggerated, in order to bring out the height variations more clearly.

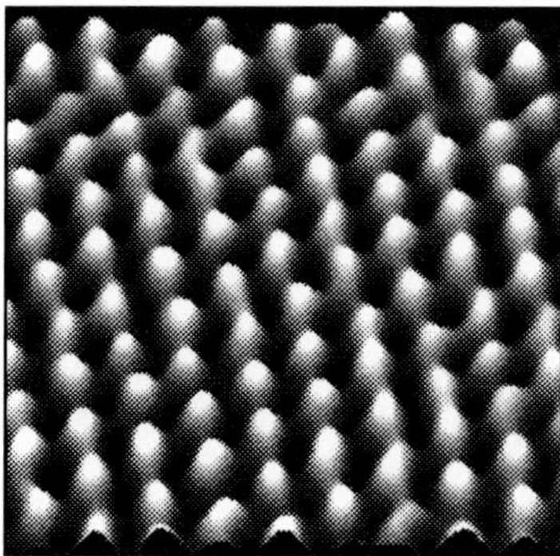

Figure: Perspective colour view of graphite surface.

Electronics

The circuit that is second-hand for the feedback electronics schematically looks as follows. The tunnelling current I_t, which is typically 1 nA, is converted by a so-called pre-amplifier to an ordinary voltage V_t. Often the conversion is such that a current I_t of 1 nA at the input of the preamplifier results in a voltage V_t of 1 V at the output. In the next step, this voltage V_t is amplified logarithmically. Because the tunnelling current – and therefore also the voltage V_t – is an exponential function of the distance d between the tip and the surface, the logarithm of the voltage V_t is a measure for the distance between the tip and the surface. We refer to the result of the logarithmic conversion as V_{log}.

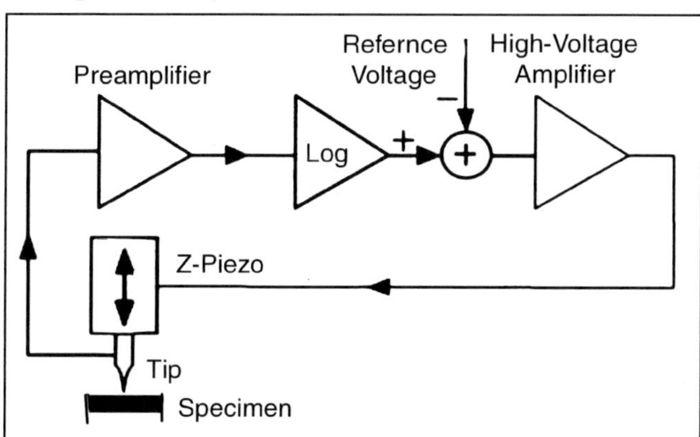

The relation between V_{log} and the distance d is then $V_{log}=a + b{\cdot}d$, where a and b are constants. From this voltage V_{log} we subtract a reference voltage V_{ref}. This reference value can be chosen at will. It forms a measure for the tip-surface distance that we want to work at. When the difference voltage between V_{log} and V_{ref} is equal to zero, the tip-surface distance is precisely right. When the difference is positive, the distance is too small; and when it is negative, the distance is too large. This difference voltage forms the input for a high-voltage amplifier, which amplifies it very strongly and passes it on to the piezo element, with which we control the height of the tip. This closes the feedback loop and makes the system complete. Many extra elements are necessary to make the feedback circuit work in practice. Examples are an absolute value amplifier, which allows the STM to work at both positive and negative tunnelling voltages, appropriate filters, to avoid spontaneous ringing of the STM, and combinations of linear, integrating and differentiating amplifiers, in order to obtain a more ideal response.

Why STM works...?

How it is likely to obtain atomic resolution with a simple instrument as the STM can be understood on the basis of the following simple arguments. First, consider the electronic structure of a metal, as symbolized in the picture on the left. The electrons of the metal occupy all available energy levels up to the energy E_F, at which they precisely compensate the positive charge of the metal ions.

For an electron to leave the metal, it needs to acquire an extra amount of energy of F above the Fermi energy. This brings it up to the vacuum level, at which point it is free to move away from the metal. The energy F is known as the work function of the metal. The middle picture shows the situation of a tip and a specimen in close proximity. There is only a narrow region of space between the two, but there is no conductive connection. Classically, electrons still need to have an extra energy F above the Fermi energy to move from specimen to tip or *vice-versa*.

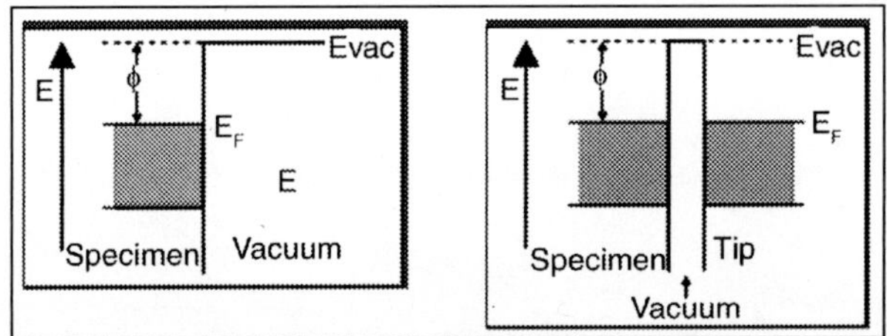

Figure: (1) In a metal, the energy levels of the electrons are filled up to a particular energy, known as the 'Fermi energy' E_F. In order for an electron to leave the metal, it needs an additional amount of energy F, the so-called 'work function'. (2) When the specimen and the tip are brought close to each other, there is only a narrow region of empty space left between them. On either side, the electrons are present up to the Fermi energy. They need to overcome a barrier F to travel from tip to specimen or *vice-versa.*

However, quantum mechanics allows the electrons to go 'right through' the barrier, a process known as *tunnelling*. When an electrical voltage *V* is applied between specimen and tip, this tunnelling phenomenon results in a net electrical current, the 'tunnelling current'. This current depends on the tip-surface distance *d*, on the voltage *V*, and on the height of the barrier F:

$$I(d) = \text{constant} \times eV\, ex\left(-2\frac{\sqrt{2m\Phi}}{\hbar}d\right)$$

This (approximate) equation shows that the tunnelling current obeys Ohm's law, *i.e.* the current *I* is proportional to the voltage *V*. It depends exponentially on the distance *d*. The other quantities in the equation are the work function F, the electron charge and mass *e* and *m*, and Planck's constant. For a typical value of the work function F of 4 electronVolt (eV), the tunnelling current reduces by a factor 10 for every 0.1 nm increase in *d*. This means that over a typical atomic diameter of *e.g.* 0.3 nm, the tunnelling current changes by a factor 1000! This is what makes the STM so sensitive. The tunnelling current depends so strongly on the distance that it is dominated by the contribution flowing between the last atom of the tip and the nearest atom in the specimen.

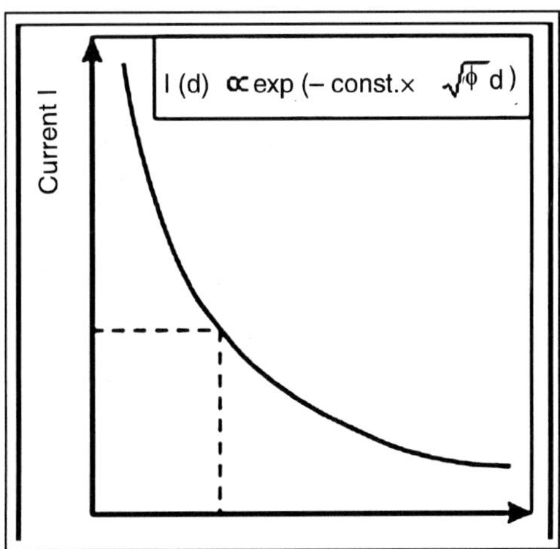

STM in the Interface Physics Group

In the Interface Physics Group, Scanning Tunnelling Microscopy is used to investigate the structure and dynamic behaviour of metal surfaces. We investigate surfaces both in the highly idealized model environment of ultrahigh vacuum (UHV), and in the more realistic context of high gas pressures. The UHV experiments are used to study various types of surface diffusion, growth phenomena, and surface phase transitions. In the high-pressure STM experiments we investigate the influence of high pressures of reactive gas mixtures on the structure of metal surfaces, and the resulting effects on the catalytic activity of these metals. For both types of STM experiments, the group has developed special STM equipment. For the UHV-experiments, this is a *high-speed, variable-temperature STM*. The imaging rate can be as high as 10 images per second (we are presently working *on video-rate STM*).

The STM can keep a single region on the surface 'in view' over an appreciable temperature range of up to 300 K. For the high-pressure experiments, we have constructed a special, so-called *'Reactor-STM'*, in which only the STM tip is allowed inside a tiny reactor chamber, together with the metal specimen, while the rest of the STM is kept outside. More about these special instruments and their applications can be found on our research pages about *Thermo-STM, surface dynamics*, and *Reactor-STM*.

Atomic Force Microscopy

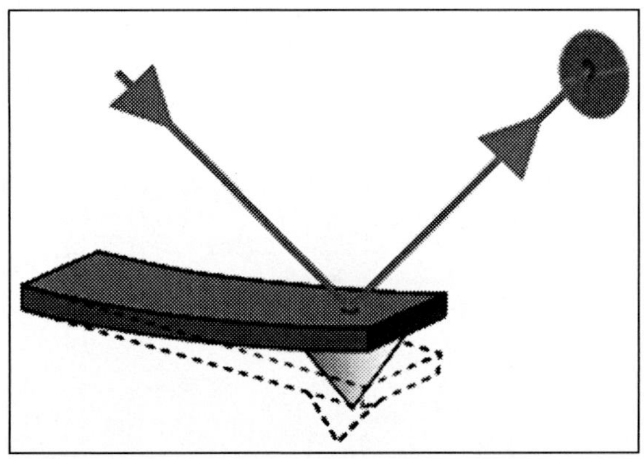

The second scanning probe microscope, which was developed soon after the STM, is the Atomic Force Microscope (AFM). The significant difference between the AFM and the STM is that in the AFM, the tip is not kept at a short distance from the surface. Instead, the AFM tip gently touches the surface.

The AFM does not record the tunnelling current but the small force between the tip and the surface. To this end, the AFM tip is attached to a tiny leaf spring, the cantilever, which has a low spring constant. The bending of this cantilever is detected, often with the use of a laser beam, which is reflected from the cantilever. Thus, rather than to measure contours of constant tunnelling current, the AFM measures contours of constant attractive or repulsive force. The detection is made so sensitive that the forces that can be detected can be as small as a few picoNewton. NanoNewton are usually sufficiently low to avoid damage to either the surface or the tip. Because the AFM does not rely on the presence of a tunnelling current, it can also be used on non-conductive materials.

In the Interface Physics Group, Atomic Force Microscopy is used mainly to investigate the structure and behaviour of *biological model systems*. We are modifying existing microscopes and developing new ones, in order to improve imaging speed, force sensitivity, and increase functionality. More about these special AFM applications can be found on our research pages about *bio-AFM*.

Friction Force Microscopy

Soon after the introduction of the AFM, it was realised that the same instrument could be used to also measure forces in the direction(s) parallel to the surface, *i.e.* the *friction* forces.

For this application, the AFM usually detects not only the deflection of the cantilever perpendicular to the surface, but also the torsion of the cantilever, resulting from one component of the lateral force. Again, the sensitivity can go down to the atomic level, so that this instrument allows us to investigate the atomic-scale origin of the friction force.

This field of research is known as *nanoTribology*. Unfortunately, the geometry of a traditional AFM cantilever is not very sensitive to friction, unless the cantilever is made very thin, in which case it becomes 'over-sensitive' to the forces normal to the surface.

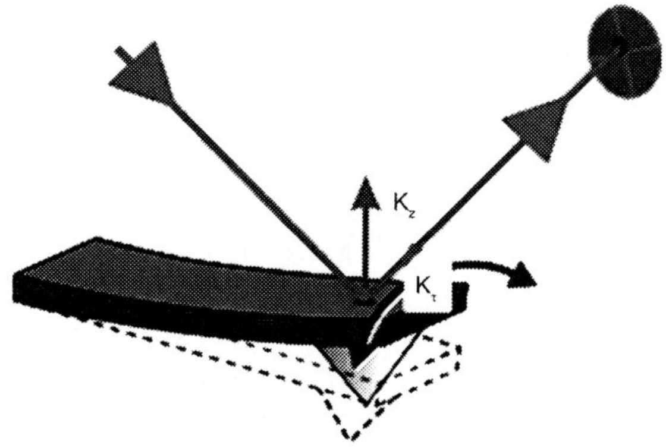

In order to overcome this problem, the Interface Physics Group has designed and constructed a special Friction Force Microscope (FFM), which is highly sensitive to both components of the force parallel to the surface, while it has the regular sensitivity, typical for a traditional AFM, to the perpendicular force. More about atomic-scale friction and our research on friction can be found on the research pages of our group about *nanoTribology*.

SPM on the Market

Although several academic research groups, such as the Interface Physics Group in Leiden, develop new types of SPMs, or improve existing types of SPM, most SPM investigations in academic and industrial research groups are carried out with commercial SPMs nowadays. There is a wide collection of general-purpose or more specialized scanning probe microscopes, which can be obtained from a growing number of companies.

SCANNING NEAR-FIELD OPTICAL MICROSCOPY

Farfield vs Near-Field Imaging

The wave-like scenery of light causes it to diffract, which limits the spatial resolution of a microscope. Under certain assumptions, the minimum detectable separation of two light scatterers for a given optical system is the Rayleigh Criterion. This limits traditional light-microscopy to a resolution of 200-300 nm, at best, with the exception of cutting edge photolithographic systems whose 100 nm resolution is achieved by using vacuum ultraviolet light. There are other techniques available to extend resolution, but by and large, this is what is possible.

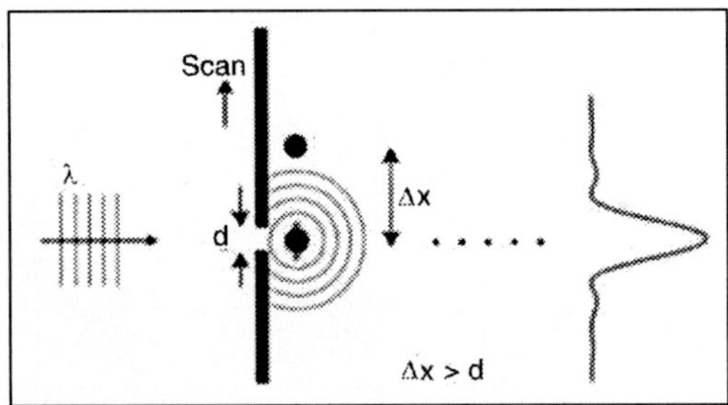

Aperture scanning near-field microscopy is a technique that allows for arbitrarily small details to be resolved. It works by scanning a small aperture over the object. Light can only pass through the apperture, and so this size determines the resolution of the system. This technique is typically implemented by tapering a fibre optic to a narrow point and coating all but the tip with metal. By this technique, images with resolution far beyond what is possible with traditional microscopy can be recorded.

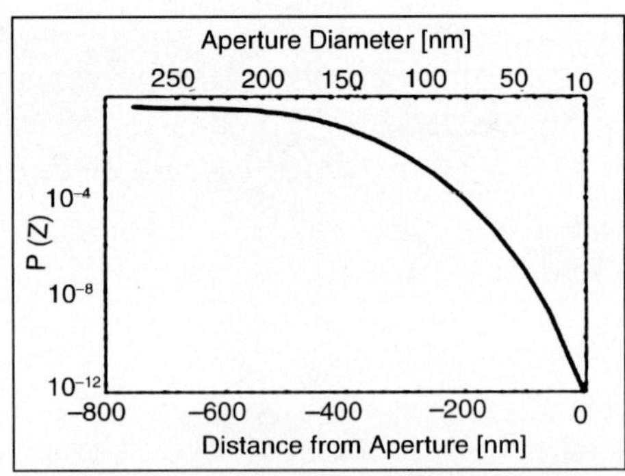

However, the amount of light that can be transmitted by a small aperture poses a limit on how small it can be made before nothing gets though. To a degree this can be lived with, as more optical power can be generated, but the cutoff is so severe that it cannot be made smaller. As the figures illustrate, this is not a subtle extinction.

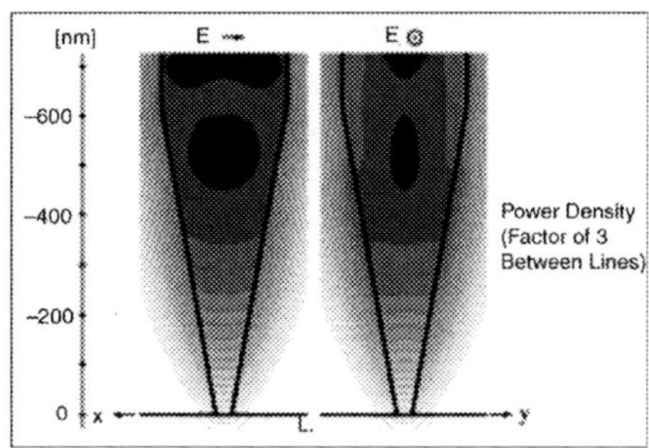

When the aperture is 100 nm, the cutoff is down four orders of magnitude, and when it reaches 50 nm, only one part in 10^8 makes it through. Furthermore, the input power cannot be increased arbitrarily because 1/3 of the power is absorbed in the coating. Increasing the input power above approximately 10mW will destroy the coating. This severely limits the signal-to-noise ratio of small apertures, and is the reason our group uses another approach.

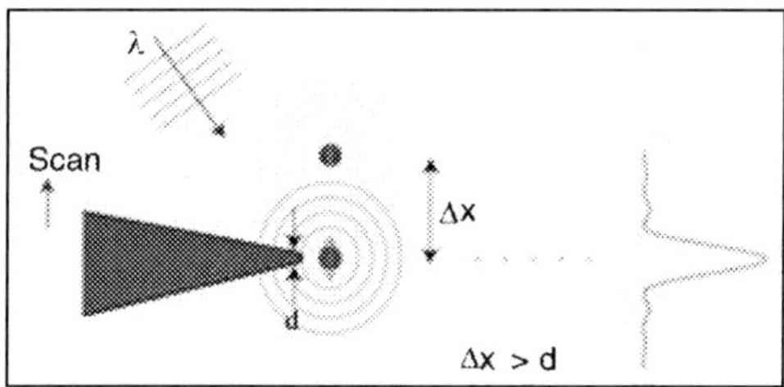

Instead of using a small aperture, we use a metal tip to provide a local excitation. If a sharp metal tip is placed in the focus of a laser beam, an effect called local field enhancement will cause the electric field to become roughly 1000 times stronger. This enhancement is localized to the tip, which has a typical diameter of 10 nm. As this tip is scanned over the surface, an image can be formed with a resolution as fine as the tip.

History of NSOM

The idea of an ultra-high resolution optical microscope. Synge's proposal suggested a new type of optical microscope that would bypass the diffraction limit, but required fabrication of a 10-nanometer aperture (much smaller than the light wavelength) in an opaque screen. A stained and embedded specimen would be ground optically flat and scanned in close proximity to the aperture. While scanning, light illuminating one side of the screen and passing through the aperture would be confined by the dimensions of the aperture, and could be used to illuminate the specimen before undergoing diffraction. As long as the specimen remained within a distance less than the aperture diameter, an image with a resolution of 10 nanometers could be generated. In addition, Synge accurately outlined a number of the technical difficulties that building a near-field microscope would present. Included in these were the challenges of fabricating the minute aperture, achieving a sufficiently intense light source, specimen positioning at the nanometer scale, and maintaining the aperture in close proximity to the specimen. The proposal, although visionary and simple in concept, was far beyond the technical capabilities of the time.

New verification of the feasibility of Synge's proposals had to wait until 1972 when E. A. Ash and G. Nicholls demonstrated the near-field resolution of a sub-wavelength aperture scanning microscope operating in the microwave region of the electromagnetic spectrum. Utilizing microwaves, with a wavelength of 3 centimetres, passing through a probe-forming aperture of 1.5 millimetres, the probe was scanned over a metal grating having periodic line features. Both the 0.5-millimeter lines and 0.5-millimeter gaps in the grating were easily resolvable, demonstrating sub-wavelength resolution having approximately one-sixtieth (0.017) the period of the imaging wavelength.

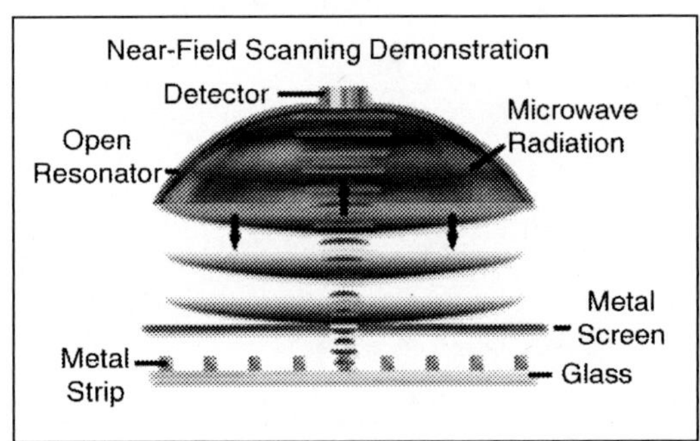

Extension of Synge's concepts to the shorter wavelengths in the visible spectrum presented significantly greater technological challenges (in aperture fabrication and positioning), which were not overcome until 1984 when a research group at IBM Corporation's Zurich laboratory reported optical measurements at a subdiffraction resolution level. An independent group working at Cornell University took a somewhat different approach to overcome the technological barriers of near-field imaging at visible wavelengths, and the two groups' results began the development that has led to the current NSOM instruments. The IBM researchers employed a metal-coated quartz crystal probe on which an aperture was fabricated at the tip, and designated the technique scanning near-field optical microscopy (SNOM). The Cornell group used electron-beam lithography to create apertures, smaller than 50 nanometers, in silicon and metal. The IBM team was able to claim the highest optical resolution (to date) of 25 nanometers, or one-twentieth of the 488-nanometer radiation wavelength, utilizing a test specimen consisting of a fine metal line grating.

Though the achievement of non-diffraction-limited imaging at visible light wavelengths had demonstrated the technical feasibility of the near-field aperture scanning approach, it was not until after 1992 that the NSOM began to evolve as a scientifically useful instrument. This advance in utility can be primarily attributed to the development of shear-force feedback systems and to the employment of a single-mode optical fibre as the NSOM probe, both of which were adapted for the near-field technique by Eric Betzig while working at AT&T Bell Laboratories.

Oscillatory Feedback Techniques

In order to improve signal-to-noise ratios for the feedback signal, the NSOM tip is almost always oscillated at the resonance frequency of the probe. This allows lock-in detection techniques (basically a bandpass filter with the centre frequency set at the reference oscillation frequency) to be utilized, which eliminates positional detection problems associated with low-frequency noise and drift. As the oscillating tip approaches the specimen, forces between the tip and specimen damp the amplitude of the tip oscillation.

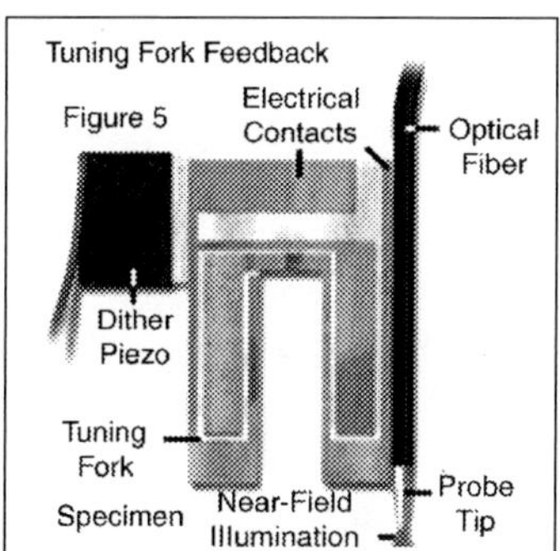

A measure of the mechanical (or electrical) oscillator quality is given by a dimensionless parameter called the quality factor, or Q-factor, or simply Q. The quality factor is defined as the oscillator's resonance frequency divided by its resonance width. It is generally beneficial to maximize the Q of the probe oscillation in order to achieve greater stability and more sensitive tip height regulation. The lower the Q of the oscillating probe, the lower the signal-to-noise ratio, which results in correspondingly lower quality topographic information being obtained from the oscillatory feedback mechanism. Historically, the letter Q has been used to represent the ratio of reactance to resistance of an electrical circuit element. With regard to oscillator characteristics, the term "quality factor" was introduced after the symbol Q was arbitrarily chosen.

Typically, both the peak resonance and the Q-factor are found to change upon approach of the probe tip to the specimen surface. The tip oscillation amplitude and frequency can be monitored by several different techniques, which generally can be categorized within two groups. The shear-force mode utilizes lateral oscillation shear forces generated between the tip and specimen (parallel to the surface) to control the tip-specimen gap during imaging. In contrast, the tapping mode relies on atomic forces occurring during oscillation of the tip perpendicular to the specimen surface (as in AFM) to generate the feedback signal for tip control. Each oscillatory mode has several advantages and disadvantages.

Shear-Force Feedback

The shear-force feedback technique laterally dithers the probe tip at a mechanical resonance frequency in proximity to the specimen surface. The dither amplitude is usually kept low (less than 10 nanometers) to prevent adversely affecting the optical resolution. For optimum image quality, shear-force feedback techniques are usually restricted to use with specimens that have relatively low surface relief, and longer scan times are required compared to operation in tapping mode. However, the straight probes typically employed in shear-force feedback techniques are easier to fabricate and have a lower cost per probe than their bent probe counterparts.

By means of respect to light throughput, the straight probe has a decided advantage over the bent probe, exhibiting much lower loss in propagation intensity. Shear-force imaging with a straight probe, however, is usually very difficult to perform in a liquid medium because the additional viscous damping of the fluid causes a dramatic decrease of the probe oscillation amplitude. In typical operation, as the oscillating probe approaches the specimen surface, the amplitude, phase, and frequency of oscillation each change, due to dissipative and adiabatic forces present at the tip of the probe. The probe oscillation damping due to tip-specimen interaction increases nonlinearly with decreasing tip-specimen separation.

The nature of the shear forces that are responsible for damping the probe tip oscillations during near-field specimen approach is the subject of much research interest. One group of investigators used electron tunnelling current measurements between a metallized NSOM probe and specimen, in shear-force feedback mode, to conclude that the probe actually contacts the surface during the approach cycle of the oscillation. Measurements of the tunnelling current, made as the tip approaches the specimen, indicate that the tip touches the specimen initially as the probe goes into feedback and continues to lightly touch the surface once per oscillatory cycle. From this information it is clear that the most beneficial approach is to make the feedback set-point as high as possible (for example, approximately 99.9 per cent of the original undamped signal) in order to reduce the physical interactions between the probe and the specimen. In practice, the upper limit on the feedback set-point is determined by the signal-to-noise ratio of the feedback signal.

Optical feedback techniques of monitoring the tip vibration amplitude were the most commonly employed during early development of shear-force techniques in NSOM, and can also be applied in the tapping mode. In this approach, for either the straight or bent probe types, a laser is tightly focused as close to the end of the NSOM probe as possible. With the straight probe variation, when under laser illumination, a shadow is cast by the probe onto a split photodiode. In the case of the bent probe technique, the laser is reflected from the top surface of the probe to the split photodiode (similar to the optical feedback techniques employed in the AFM). With the laser feedback established, the probe is then vibrated in either tapping mode or shear-force mode, at a known frequency, utilizing a dither piezo. The split photodiode collects the laser light, and the difference between the signals from each side of the detector is determined. A higher signal-to-noise ratio can be obtained by using a lock-in amplifier to select a portion of the signal that is at the same frequency as the dither piezo drive signal.

The main problem associated with this type of feedback mechanism is that the light source (for example, a laser), which is used to detect the tip vibration frequency, phase, and amplitude, becomes a potential source of stray photons that can interfere with the detection of the NSOM signal. One mechanism for dealing with this effective increase in background signal is to provide a feedback light source that has a different wavelength (usually longer) than the near-field source. This scheme requires additional filtration in front of the detector to selectively block the unwanted photons originating within the feedback system. In most cases, the added filters also block a small percentage of the near-field photons, resulting in reduced signal levels. A non-optical feedback technique is not subject to problems of this nature, and is a primary reason that techniques such as the tuning-fork technique have become increasingly popular.

Piezoelectric quartz tuning forks were first introduced into scanning probe microscopy for use in scanning near-field acoustic microscopy. Later, tuning forks were incorporated into the NSOM to serve as inexpensive and simple, non-optical excitation and detection devices in distance control functions. Quartz crystals have the property of generating an electric field when placed under pressure and, conversely, of changing dimensions when an electric field is applied. This property is termed piezoelectric and occurs when the crystal is composed of molecules that lack both centers and planes of symmetry. Quartz crystals suitable for use in precision oscillators (digital clocks) and highly selective wave filters are mass-produced in huge quantities, making them relatively inexpensive. When quartz tuning forks are utilized for regulation in a feedback loop, their very high mechanical quality factor, Q (as high as approximately 10000), and corresponding high gain, provides the system with high sensitivity to small forces, typically on the order of a piconewton.

The essential configuration of the tuning-fork technique used for shear-force tip feedback consists of a single mode optical fibre attached to one arm of a quartz crystal tuning fork, which is oscillated at the tuning fork's resonance frequency. The equivalent circuit for the tuning fork is a series RLC resonator in parallel with package capacitance. The most common tuning fork resonance frequency is 32,768 hertz (Hz), but the devices are available with resonances ranging from 10 kilohertz to several tens of megahertz.

The single mode optical fibre, routed to the NSOM head, is physically coupled to the crystal tuning fork, which in turn can be driven internally (electrically) or externally by a dither piezo to which the fork is rigidly attached. The mode of oscillation of the tuning fork depends upon the means of excitation. If the fork is (directly) driven electrically, the arms vibrate in opposite directions, whereas external mechanical excitation produces an oscillation in which both arms of the tuning fork move in the same direction. A schematic of a quartz tuning fork configured with an attached fibre for shear-force detection. The piezoelectric potential is acquired from electrodes on the fork and then amplified with a gain of approximately 100 (using an instrumentation amplifier) to produce a signal on the order of a few tens of millivolts. The signal is then fed into a lock-in amplifier and referenced to the drive signal of the oscillating tuning fork. The output from the lock-in amplifier (amplitude, phase, or a combination of amplitude and phase such as the x or y signals) is then compared to a user-specified reference signal in the control loop to maintain the probe in feedback above the specimen.

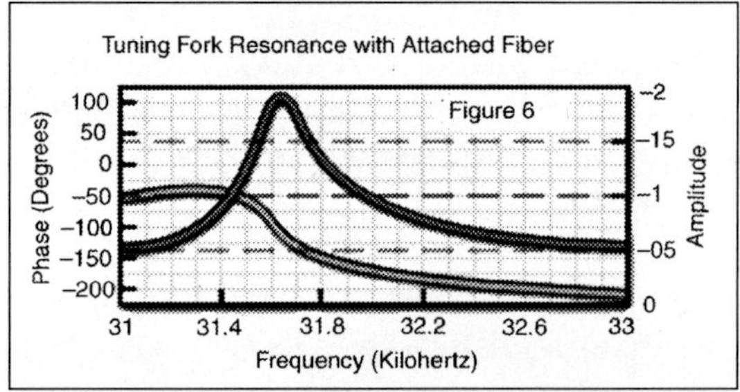

An example resonance curve produced by a 32.7-kHz tuning fork with the attached NSOM fibre. The fork response was measured by sweeping the frequency from 31 kHz to 33 kHz and simultaneously measuring the amplitude and phase of the signal. Upon attachment of the fibre the resonance frequency shifts and the Q-factor of the resonance drops from approximately 20,000 to less than 1000. The Q is defined as:

$$f_r/\Delta f$$

where f(r) is the frequency at the maximum amplitude and Δf is the width of the resonance peak at the points where the amplitude is equivalent to the peak amplitude divided by the square root of 2 (termed the root-mean-square), or approximately 70.7 per cent of the peak amplitude.

There are more than a few advantages of the tuning fork technique that have led to its increased favour over optical techniques of tip regulation. Since the detection of the tip motion is not optical, there is no risk of additional stray light being introduced in the vicinity of the aperture that might interfere with the NSOM signal detection. Additionally, the tuning fork system does not require the tedious alignment procedures of a separate external laser source and associated focusing optical components. Because of the compactness and relative ease of use, the tuning fork technique lends itself to applications requiring remote operation, such as those employed in vacuum systems or environmental control chambers.

Tapping-Mode Feedback

Tapping-mode feedback is another popular technique for tip-to-specimen distance control, and is implemented using several different probe types. A useful design consists of a modified AFM cantilever and transparent tip, usually fabricated from silicon nitride and coated with metal on the bottom of the probe tip. The most commonly employed probe for tapping-mode techniques is the conventional fibre optic probe having a near 90-degree bend close to the tip aperture. The resolution of the tapping-mode near-field image is defined not only by the radius of the tip but also by the amplitude of the oscillation occurring perpendicular to the specimen surface. This is due to the acute sensitivity of the optical signal to the tip-to-specimen separation. In order to maintain high near-field resolution, it is necessary to either maintain a small oscillation amplitude relative to the tip aperture, or to compensate for larger oscillations. A mechanism that has been demonstrated to improve resolution is to synchronize the collected NSOM signal with the cycle of tip oscillation. Modulating the light coupled into the probe, and adjusting the phase such that the specimen is only illuminated when the tip is at its closest approach point, allows maintaining high resolution imaging at fairly large tip oscillation amplitudes.

There are several drawbacks in the application of bent optical probes, each of which can be attributed to the bend itself. A significant problem is the increased difficulty of the probe fabrication, especially when applying a metal coating to the tip. An additional disadvantage is the increased optical loss that occurs due to the bend in the probe. This loss in throughput efficiency is significant, and some published measurements indicate that bent optical fibre probes are at least an order of magnitude less efficient than conventional straight fibre probes. In certain operational modes of NSOM, the intensity loss is not a serious limitation because additional light can be coupled into the fibre to compensate, assuming sufficient laser power is available. The increase in optical coupling is an option because the optical losses, as well as increased heating, occur at the bend in the fibre and not at the aperture of the probe, where local heating would present a major problem. Another potential drawback with bent probes is a change in certain tip properties that occurs due to the presence of the bend, such as a decrease in extinction coefficients when performing polarized light measurements. Extinction ratios of approximately 70:1 have been measured in the far-field utilizing bent tips, as compared to values of greater than 100:1 with conventional straight fibre probes.

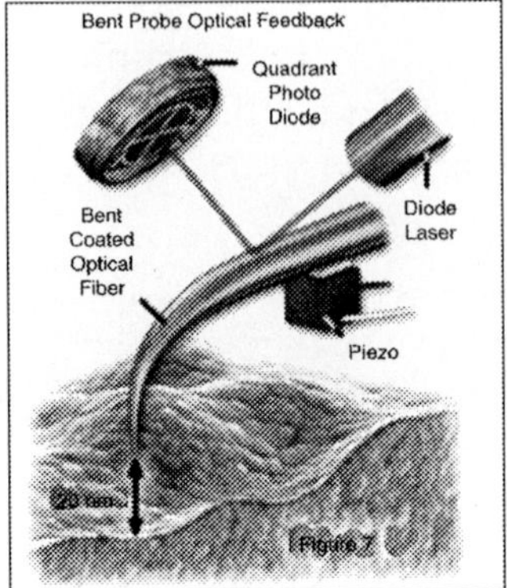

In the twisted probe feedback mode, the probe is oscillated perpendicular to the specimen surface similar to tapping-mode AFM. The amplitude of oscillation can be monitored either mechanically, with a piezo-electric device such as a quartz tuning fork, or optically by reflecting a laser from the top surface of the tip cantilever. The probe is excited to oscillate in one of its eigenmodes and the tip-specimen distance is recorded and used dynamically as the feedback signal. The tip of the probe is prevented from adhering to the specimen due to the oscillation, which provides both a short contact time and a reverse driving force due to the cantilever bending. The success of this feedback technique is dependent on the increased sensitivity of the tip-specimen distance control resulting from the resonance enhancement of the tip vibration. The sharpness and sensitivity of the tip vibration is characterized by the Q of the cantilever (similar to the measured Q in the shear-force oscillation). The Q value of the probe is substantially reduced by the viscosity of a liquid environment, and this is usually accompanied by a large shift in the resonance frequency.

An advantage of the tapping-mode over the shear-force mode is the relative ease with which nanoscale topographic images can be acquired, even when the specimen and probe are immersed in an aqueous or other fluid medium. A variety of research groups have used tapping-mode feedback in single molecule detection, in studies of biological systems, and for imaging in water, among other applications. When the oscillating tip approaches the specimen surface, a decrease in the oscillation amplitude of the tuning fork or in the optical feedback signal is observed. The origin of this damping is still not completely understood; however, several different mechanisms have been proposed, including capillary forces, van der Waals forces, and actual contact between the probe tip and specimen. The action of damping forces on the probe tip can be conceptualized by envisioning a thin layer of water covering the specimen surface (which is actually the case if the specimen is in ambient conditions). When the tip is lowered into the thin layer of surface water, a drag force is imparted to the probe tip as it enters the water. The increased drag force decreases the oscillation amplitude of the end of the fibre probe, and shifts the fibre's resonance frequency. Both of these changes decrease the output signal from the tuning fork (for the non-optical technique). When the reduced signal falls below the threshold of the reference signal, the tip is interpreted as being "engaged" and the feedback control system regulates the height of the tip above the specimen (based on the user specified reference signal). Near-field scanning optical microscopy is continuing to grow in use, especially for the microscopist interested in obtaining the highest possible optical resolution. However, NSOM is not limited to serving solely as an imaging/microscopy instrument; it can also be utilized for specimen manipulation, fabrication, and processing on a nanometric scale. A wide variety of NSOM applications outside of the imaging realm are evolving, including precision laser machining, nanometer-scale optical lithography, and localized release of caged compounds.

Some of the limitations of near-field optical microscopy include:

- Practically zero working distance and an extremely small depth of field.
- Extremely long scan times for high resolution images or large specimen areas.
- Very low transmissivity of apertures smaller than the incident light wavelength.
- Only features at the surface of specimens can be studied.
- Fibre optic probes are somewhat problematic for imaging soft materials due to their high spring constants, especially in shear-force mode.

NSOM is currently still in its infancy, and more research is needed toward developing improved probe fabrication techniques and more sensitive feedback mechanisms. The future of the technique may actually rest in refinement of apertureless near-field techniques (including interferometric), some of which have already achieved resolutions on the order of 1 nanometer. However, typical resolutions for most NSOM instruments range around 50 nanometers, which is only 5 or 6 times better than that achieved by scanning confocal microscopy. This moderate increase in resolution comes at a considerable cost in time required to set up the NSOM instrument for proper imaging, and in the complexity of operation. The greatest advantage of NSOM probably rests in its ability to provide optical and spectroscopic data at high spatial resolution, in combination with simultaneous topographic information. Combining atomic force measurements and near-field scanning optical microscopy has proven to be an extremely powerful approach in certain areas of research, providing new information about a variety of specimen types that is simply not attainable with far-field microscopy.

5

Silicon MOSFETs and Quantum Transport Devices

SILICON MOSFETS

CMOS technology has evolve to a stage where device performance should be assessed against upper limits.

The purpose of this work is to do so. We compare the measured performance of a 0.35mm technology against its upper limit characteristics and show that the measured on-current is ~45% of the upper limit. We show that the channel resistance has a lower limit, no matter how short the channel is or how high its mobility and that the magnitude of this fundamental channel resistance is large enough to influence practical devices. Real devices operate below their fundamental limit because of channel back-scattering. We present a procedure for extracting the channel back-scattering coefficient, r, as a function of gate and drain bias.

We show that the values extracted from measurements are consistent with simple estimates and discuss the differences between n and p-channel MOSFET's. We show that a simple, one parameter scattering model accurately predicts the characteristics of a "well-tempered MOSFET" all the way to their limits. We also show that conventional MOSFET scaling practices maintain a constant r and that the NTRS on current targets near the end of the roadmap are not achievable by conventional device scaling.

Traditional silicon MOSFET scaling has driven the majority of the semiconductor industry for the past four decades. In recent years, new materials and processes have been introduced to maintain pace with Moore's law. High-k gate dielectrics are in volume production and stressor technology is extending the performance of the silicon channel to its limits. III-V materials are widely regarded as leading candidates to fill the nMOS performance gap with both III-V and Ge-based channels being considered for the pMOS device.

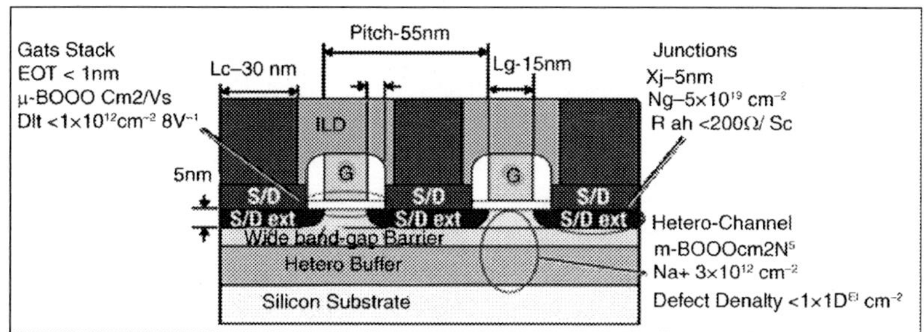

Figure : nFET 15nm technology generation key dimensions and module targets.

A representation of a possible 15nm technology generation nFET. It has a high mobility III-V channel, enabling high performance at low supply voltage. The wide band-gap barrier gives ultra thin body like immunity to short channel effects by improved electrostatic control over the bulk architecture. Heterointegration on a silicon platform is required for III-V integration to be cost competitive.

Gate Stack

As III-V materials lack a good inhabitant oxide interface, a great deal of research is focused on decreasing high-k III-V interface states. Additionally, there are several other concerns associated with high-k dielectrics. These include Coulomb scattering from bulk oxide charges and interface fixed charges, surface roughness scattering, remote phonon scattering, and dielectric charge trapping associated with reliability problems.

These issues need to be understood and addressed concurrently. SEMATECH has investigated:

- The impact of manufacturable ex situ and in situ high-k/higher-k dielectrics;
- Conventional process routes and materials such as HfO_2, ZrO_2, Al_2O_3 by manufacturable atomic-layer-deposition (ALD) to form quality gate stacks;
- Engineering the high-k-channel interface by several means including evaluation of bi-layer schemes to address the dual targets of thin EOT and low D_{it};
- Defect states and their influence on device properties via detailed electrical and physical characterization and helped in the development of schemes to passivate/eliminate these defects.

Additional specifically, we have addressed key gate stack issues including: a) EOT scalability for high performance and electrostatic control with acceptable leakage current, this meets the basic EOT-J_g requirements of a good insulator; b) understanding the source and impact of charge trapping by the insertion of either a $LaAlO_3$ or Al_2O_3 inter-layer, which reduced the mid-gap Dit by ~5x thermal stability on InGaAs. The transmission electron microscopy (TEM) images of various ALD gate stacks on InGaAs and confirm that the interface is atomically sharp and smooth.

Ultrashallow Junctions

In addition to the architecture of the gate region, considerable effort is also ongoing to select the best ohmic contact and junction technology. The physical properties of III-Vs are such that implanted junctions and salicide that have served silicon so well may not be suitable for compound semiconductors. In addition to the challenges that silicon ultrashallow junctions face, implantation of III-Vs suffers from a number of issues including: lower maximum activated doping density, loss of stoichiometry due to preferential group V evaporation, and difficulty in recovering from implant amorphization because of the compound nature of III-Vs.

A number of alternatives are being pursued as a replacement for implantation as the junction technology of choice. One promising approach is selective regrowth. In this technique, either MBE or MOCVD is used to selectively regrow very heavily doped source/drain regions. This has three distinct advantages over implantation: high doping density, abrupt doping profile and reduced leakage current caused by the absence of implant damage. In addition to providing low parasitic series resistance, regrown source/drains can be used to apply uni-axial strain, further improving performance.

SEMATECH is pursuing a potentially defect-free alternative – mono layer doping (MLD). The III-V surface is terminated by a monolayer of dopant atoms, in this case sulfur, provided from a solution of ammonium sulfide. Other dopant solutions have been demonstrated and the concept could even be extended to ALD or plasma-based schemes. The wafer is then capped with dielectric and annealed. The dopant diffuses into the semiconductor and is activated. MLD has a number of advantages for the formation of ultrashallow junctions: low cost, high uniformity and low damage. Existing toolsets can be used to minimize costs. By terminating the surface with a monolayer of dopant, the exact dopant concentration is known and is self-limiting, allowing for accurate and repeatable junction resistivity. Perhaps most importantly, in contrast to ion implantation, MLD does not introduce lattice damage. Damage determines junction leakage current density, which in turn, controls the device off-state leakage. High mobility III-V materials generally have a smaller band-gap than silicon, which makes them particularly susceptible to high junction leakage, hence, damage-free junctions are a priority in this material system.

The extension sheet resistance as a function of depth for various doping densities. The mobility was calculated for each doping density using the empirical model from calibrated by data from internal measurements. For a 5nm junction depth, a N_d $>3\times10^{19}cm^{-3}$ is required to achieve sheet resistance

<~200Ω/sq. Although the maximum doping density achieved in III-Vs is approximately an order of magnitude lower than silicon, in the 1019 to 1020 cm^{-3} range, the superior electron mobility >1500cm^2/Vs even when highly doped allows low resistance shallow extensions at modest doping density. High N_d is not only attractive to reduce access resistance, but is also required to reduce contact resistance (R_c). When device dimensions and pitch are scaled aggressively Rc becomes dominant. When the access length (L_{ext}) is reduced, so is the access resistance, but when the contact length (L_c) is reduced, R_c is increased. At an L_c of 30nm, an R_c of ~1Ωμm^{-2} is required to meet the parasitic series source drain resistance targets put forth by the ITRS.

Hetero-integration on Silicon

A additional challenge facing III-V MOSFETs is hetero-integration on a silicon platform. For III-V materials to be considered seriously, the enormous momentum behind silicon manufacturing must be harnessed. III-V MOSFET process flows must use standard silicon technology and must not contaminate silicon lines. Additionally, III-Vs must be hetero-integrated on a large diameter silicon platform to be economically viable. In conjunction with growth partners, SEMATECH is pursuing different technologies to enable high quality III-V growth on Si substrates.

There is a long history of III-V buffer technology development to enable III-Vs to be grown on Si. Although thin buffer thickness <400nm would be required for a manufacturable co-integration scheme, thick buffer technology is a convenient interim technique to demonstrate the feasibility of hetero-integration and manufacturing compatibility.

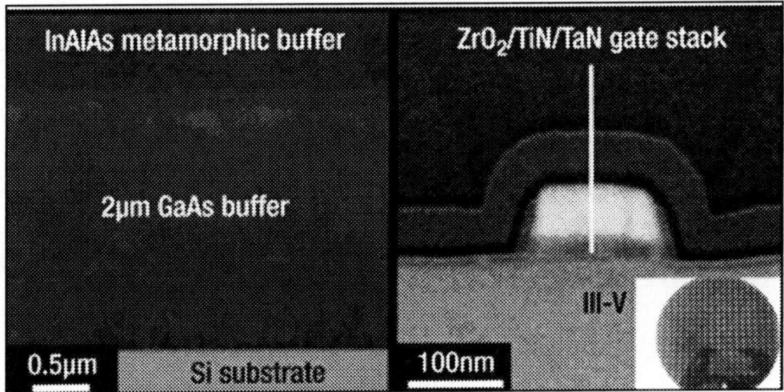

Figure : a) TEM cross-section of the III-V on Si buffer; b) SEM cross-section of a III-V L_g = 140nm device on 200mm Si wafer, with inset photograph of completed III-V on silicon device wafer.

A 53 % InGaAs buried HEMT structure was grown on a 4ˆ off-cut (100) Si wafer, with a 2μm GaAs buffer and a standard metamorphic InAlAs buffer. The mobility was measured to be ~ 5500cm2/Vs at a carrier concentration of

$3.1 \times 10^{12} cm^{-2}$. A TEM cross-section of the buffer from which a crude estimation of areal defect density was calculated to be $1.7 \times 10^{8} cm^{-2}$.

By means of similar buffer technology, SEMATECH has run a full device flow through a standard 200mm silicon process flow, including high-k, metal gate, source drain implantation and metal one processes. Systematic contamination monitoring was carried out after each step using TXRF analysis. A carefully designed processes flow, controlling tool contamination in a standard silicon fab line is under development. A SEM cross-section of a 140nm metal gate complete with high-k dielectric and silicon nitride spacers, fabricated on a 200mm production line using silicon compatible process flows. The inset is a photograph of the completed III-V on 200mm Si wafer.

MOSFET OPERATION

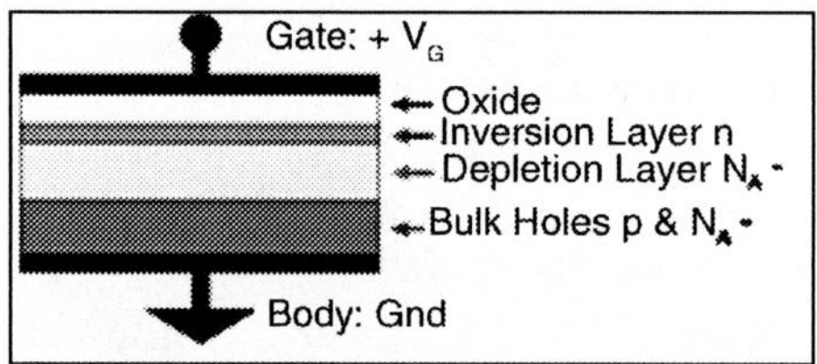

Figure: Metal–oxide–semiconductor structure on p-type silicon

METAL–OXIDE–SEMICONDUCTOR STRUCTURE

A customary metal–oxide–semiconductor (MOS) structure is obtained by growing a layer of silicon dioxide (SiO_2) on top of a silicon substrate and depositing a layer of metal or polycrystalline silicon (the latter is commonly used). As the silicon dioxide is a dielectric material, its structure is equivalent to a planar capacitor, with one of the electrodes replaced by a semiconductor. When a voltage is applied across a MOS structure, it modifies the distribution of charges in the semiconductor. If we consider a p-type semiconductor (with N_A the density of acceptors, p the density of holes; $p = N_A$ in neutral bulk), a positive voltage, V_{GB}, from gate to body creates a depletion layer by forcing the positively charged holes away from the gate-insulator/semiconductor interface, leaving exposed a carrier-free region of immobile, negatively charged acceptor ions. If V_{GB} is high enough, a high concentration of negative charge carriers forms in an inversion layer located in a thin layer next to the interface between the semiconductor and the insulator. Unlike the MOSFET, where the inversion layer electrons are supplied rapidly from the source/drain electrodes, in the MOS capacitor they are produced much more slowly by thermal generation through carrier generation and recombination centers in the depletion region. Conventionally, the gate voltage at which the volume density of electrons in the inversion layer is the same as the volume density of holes in the body is called the threshold voltage.

This structure with p-type body is the basis of the n-type MOSFET, which requires the addition of an n-type source and drain regions.

MOSFET structure and channel formation

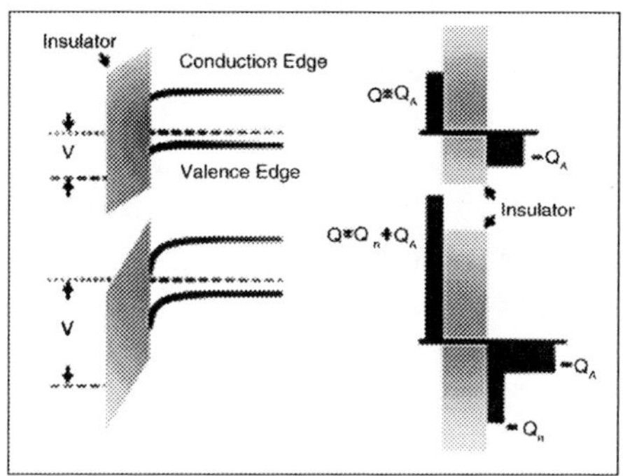

Figure: *Channel formation in nMOS MOSFET*: Top panels: An applied gate voltage bends bands, depleting holes from surface (left). The charge inducing the bending is balanced by a layer of negative acceptor-ion charge (right). Bottom panel: A larger applied voltage further depletes holes but conduction band lowers enough in energy to populate a conducting channel.

A metal–oxide–semiconductor field-effect transistor (MOSFET) is based on the modulation of charge concentration by a MOS capacitance between a body electrode and a gate electrode located above the body and insulated from all other device regions by a gate dielectric layer which in the case of a MOSFET is an oxide, such as silicon dioxide. If dielectrics other than an oxide such as silicon dioxide are employed the device may be referred to as a metal–insulator–semiconductor FET (MISFET). Compared to the MOS capacitor, the MOSFET includes two additional terminals (source and drain), each connected to individual highly doped regions that are separated by the body region. These regions can be either p or n type, but they must both be of the same type, and of opposite type to the body region. The source and drain (unlike the body) are highly doped as signified by a '+' sign after the type of doping.

If the MOSFET is an n-channel or nMOS FET, then the source and drain are 'n+' regions and the body is a 'p' region. If the MOSFET is a p-channel or pMOS FET, then the source and drain are 'p+' regions and the body is a 'n' region. The source is so named because it is the source of the charge carriers (electrons for n-channel, holes for p-channel) that flow through the channel; similarly, the drain is where the charge carriers leave the channel.

The tenancy of the energy bands in a semiconductor is set by the position of the Fermi level relative to the semiconductor energy-band edges. The sufficient gate voltage, the valence band edge is driven far from the Fermi level, and holes from the body are driven away from the gate. At larger gate bias still, near the semiconductor surface the conduction band edge is brought close to the Fermi level, populating the surface with electrons in an *inversion layer* or *n-channel* at the interface between the p region and the oxide. This conducting channel extends between the source and the drain, and current is conducted through it when a voltage is applied between source and drain. Increasing the voltage on the gate leads to a higher electron density in the inversion layer and therefore increases the current flow between the source and drain.

For gate voltages below the threshold value, the channel is lightly populated, and only a very small subthreshold leakage current can flow between the source and the drain.

When a negative gate-source voltage (positive source-gate) is applied, it creates a *p-channel* at the surface of the n region, analogous to the n-channel case, but with opposite polarities of charges and voltages. When a voltage less negative than the threshold value (a negative voltage for p-channel) is applied between gate and source, the channel disappears and only a very small subthreshold current can flow between the source and the drain.

The device may comprise a Silicon On Insulator (SOI) device in which a buried oxide (BOX) is formed below a thin semiconductor layer. If the channel region between the gate dielectric and a BOX region is very thin, the very thin channel region is referred to as an ultrathin channel (UTC) region with the source and drain regions formed on either side thereof in and/or above the thin semiconductor layer. Alternatively, the device may comprise a semiconductor on insulator (SEMOI) device in which semiconductors other than silicon are employed. Many alternative semiconductor materials may be employed. When the source and drain regions are formed above the channel in whole or in part, they are referred to as raised source/drain (RSD) regions.

Modes of Operation

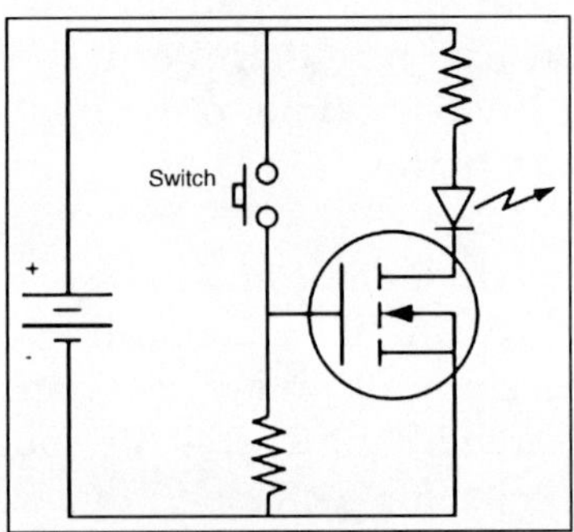

Figure: Example application of an N-Channel MOSFET. When the switch is pushed the LED lights up.

The operation of a MOSFET can be separated into three different modes, depending on the voltages at the terminals. In the following discussion, a simplified algebraic model is used that is accurate only for old technology. Modern MOSFET characteristics require computer models that have rather more complex behaviour. For an enhancement-mode, n-channel MOSFET, the three operational modes are:

Cutoff, subthreshold, or weak-inversion mode

When $V_{GS} < V_{\text{th}}$:

where Vth is the threshold voltage of the device.

According to the basic threshold model, the transistor is turned off, and there is no conduction between drain and source. A more accurate model considers the effect of thermal energy on the Boltzmann distribution of electron energies which allow some of the more energetic electrons at the source to enter the channel and flow to the drain. This results in a subthreshold current that is an exponential function of gate–source voltage. While the current between drain and source should ideally be zero when the transistor is being used as a turned-off switch, there is a weak-inversion current, sometimes called subthreshold leakage.

In weak inversion the current varies exponentially with gate-to-source bias V_{GS} as given approximately by:

$$ID \approx I_{D0} e \frac{V_{GS} - V_{th}}{nV_T},$$

where

I_{D0} = current at $V_{GS} - V_{th}$, the thermal voltage $V_T - kT/q$ and the slope factor n is given by

$$n = 1 + C_D / C_{OX},$$

with C_D = capacitance of the depletion layer and C_{OX} = capacitance of the oxide layer. In a long-channel device, there is no drain voltage dependence of the current once $V_{DS} >> V_T$, but as channel length is reduced drain-induced barrier lowering introduces drain voltage dependence that depends in a complex way upon the device geometry (for example, the channel doping, the junction doping and so on). Frequently, threshold voltage V_{th} for this mode is defined as the gate voltage at which a selected value of current I_{D0} occurs, for example, $I_{D0} = 1\ \mu A$, which may not be the same V_{th}-value used in the equations for the following modes.

Some micropower analog circuits are designed to take advantage of subthreshold conduction. By working in the weak-inversion region, the MOSFETs in these circuits deliver the highest possible transconductance-to-current ratio, namely: $gm/I_D = I/(nV_T)$, almost that of a bipolar transistor.

The subthreshold *I–V curve* depends exponentially upon threshold voltage, introducing a strong dependence on any manufacturing variation that affects threshold voltage; for example: variations in oxide thickness, junction depth, or body doping that change the degree of drain-induced barrier lowering. The resulting sensitivity to fabricational variations complicates optimization for leakage and performance.

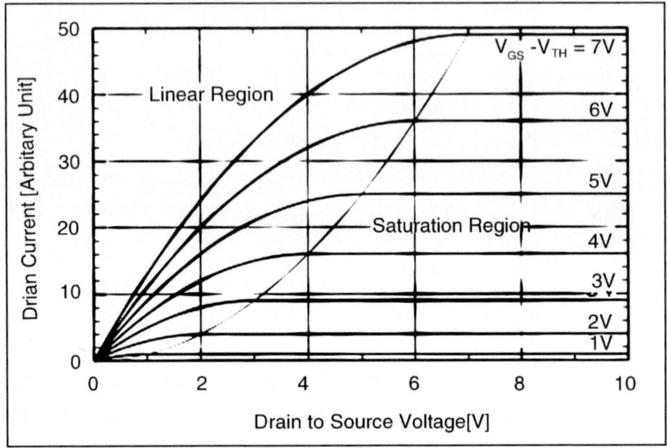

Figure: MOSFET drain current vs. drain-to-source voltage for several values of $V_{GS} - V_{th}$; the boundary between linear (Ohmic) and saturation (active) modes is indicated by the upward curving parabola.

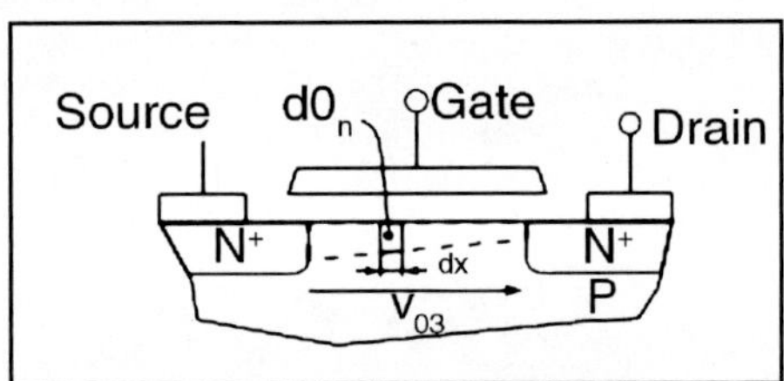

Figure: Cross section of a MOSFET operating in the linear (Ohmic) region; strong inversion region present even near drain

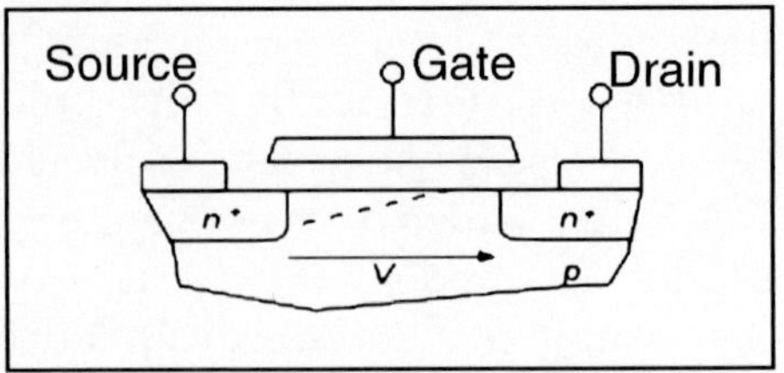

Figure: Cross section of a MOSFET operating in the saturation (active) region; channel exhibits pinch-off near drain

Triode mode or linear region (also known as the ohmic mode)

When $V_{GS} > V_{th}$ and $V_{DS} < (V_{GS} - V_{th})$

The transistor is turned on, and a channel has been created which allows current to flow between the drain and the source. The MOSFET operates like a resistor, controlled by the gate voltage relative to both the source and drain voltages. The current from drain to source is modelled as:

$$I_D = \mu_n C_{ox} \frac{W}{L}\left(\left(V_{GS} - V_{th}\right) V_{DS} - \frac{V_{DS}^2}{2} \right)$$

where μ_n is the charge-carrier effective mobility, W is the gate width, L is the gate length and C_{ox} is the gate oxide capacitance per unit area. The transition from the exponential subthreshold region to the triode region is not as sharp as the equations suggest.

Saturation or active mode;

When $V_{GS} > V_{th}$ and $V_{DS} > (V_{GS} - V_{th})$

The switch is turned on, and a channel has been created, which allows current to flow between the drain and source. Since the drain voltage is higher than the gate voltage, the electrons spread out, and conduction is not through a narrow channel but through a broader, two- or three-dimensional current distribution extending away from the interface and deeper in the substrate.

The onset of this region is also known as pinch-off to indicate the lack of channel region near the drain. The drain current is now weakly dependent upon drain voltage and controlled primarily by the gate–source voltage, and modelled approximately as:

$$I_D = \frac{\mu_n C_{ox}}{2} \frac{W}{L} \left(V_{GS} - V_{th}\right)^2 \left(1 + \lambda \left(V_{DS} - V_{DSsat}\right)\right)$$

The additional factor involving λ, the channel-length modulation parameter, models current dependence on drain voltage due to the Early effect, or channel length modulation. According to this equation, a key design parameter, the MOSFET transconductance is:

$$g_m = \frac{2I_D}{V_{GS} - V_{th}} = \frac{2I_D}{Vov},$$

where the combination $V_{ov} = V_{GS} - V_{th}$ is called the overdrive voltage, and where $V_{DSsat} = V_{GS} - V_{th}$ (which Sedra neglects) accounts for a small discontinuity in I_D which would otherwise appear at the transition between the triode and saturation regions.

Another key design parameter is the MOSFET output resistance r_{out} given by:

$$r_{out} = \frac{1}{\lambda I_D}.$$

r_{out} is the inverse of g_{DS} where $g_{DS} = \frac{\partial I_{DS}}{\partial V_{DS}}$. I_D is the expression in saturation region.

If λ is taken as zero, an infinite output resistance of the device results that leads to unrealistic circuit predictions, particularly in analog circuits.

As the channel length becomes very short, these equations become quite imprecise. New physical effects arise. For example, carrier transport in the active mode may become limited by velocity saturation. When velocity saturation dominates, the saturation drain current is more nearly linear than quadratic in V_{GS}. At even shorter lengths, carriers transport with near zero scattering, known as quasi-ballistic transport. In addition, the output current is affected by drain-induced barrier lowering of the threshold voltage.

Body Effect

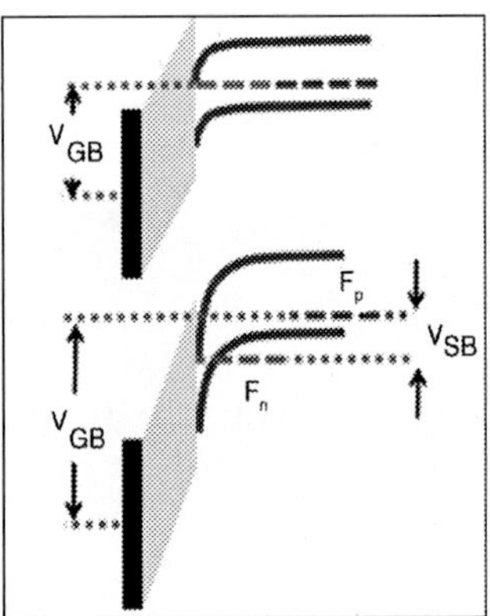

Figure: V_{SB} splits Fermi levels F_n for electrons and F_p for holes, requiring larger V_{GB} to populate the conduction band in an nMOS MOSFET

The tenancy of the energy bands in a semiconductor is set by the position of the Fermi level relative to the semiconductor energy-band edges. Application of a source-to-substrate reverse bias of the source-body pn-junction introduces a split between the Fermi levels for electrons and holes, moving the Fermi level for the channel further from the band edge, lowering the occupancy of the channel. The effect is to increase the gate voltage necessary to establish the channel. This change in channel strength by application of reverse bias is called the *body effect*.

Simply put, using an nMOS example, the gate-to-body bias V_{GB} positions the conduction-band energy levels, while the source-to-body bias V_{SB} positions the electron Fermi level near the interface, deciding occupancy of these levels near the interface, and hence the strength of the inversion layer or channel.

The body effect upon the channel can be described using a modification of the threshold voltage, approximated by the following equation:

$$V_{TB} = V_{T0} + \gamma\left(\sqrt{V_{SB} + 2\varphi_B}\right),$$

where V_{TB} is the threshold voltage with substrate bias present, and V_{T0} is the zero-V_{SB} value of threshold voltage, γ is the body effect parameter, and $2\varphi_B$ is the approximate potential drop between surface and bulk across the depletion layer when $V_{SB} = 0$ and gate bias is sufficient to insure that a channel is present. As this equation shows, a reverse bias $V_{SB} > 0$ causes an increase in threshold voltage V_{TB} and therefore demands a larger gate voltage before the channel populates.

The body can be operated as a second gate, and is sometimes referred to as the "back gate"; the body effect is sometimes called the "back-gate effect".

MOSFET SCALING

Over the precedent decades, the MOSFET has continually been scaled down in size; typical MOSFET channel lengths were once several micrometres, but modern integrated circuits are incorporating MOSFETs with channel lengths of tens of nanometers. Robert Dennard's work on scaling theory was pivotal in recognising that this ongoing reduction was possible. Intel began production of a process featuring a 32 nm feature size (with the channel being even shorter) in late 2009. The semiconductor industry maintains a "roadmap", the ITRS, which sets the pace for MOSFET development. Historically, the difficulties with decreasing the size of the MOSFET have been associated with the semiconductor device fabrication process, the need to use very low voltages, and with poorer electrical performance necessitating circuit redesign and innovation.

Reasons for MOSFET Scaling

Smaller MOSFETs are desirable for several reasons. The main reason to make transistors smaller is to pack more and more devices in a given chip area. This results in a chip with the same functionality in a smaller area, or chips with more functionality in the same area.

Since fabrication costs for a semiconductor wafer are relatively fixed, the cost per integrated circuits is mainly related to the number of chips that can be produced per wafer. Hence, smaller ICs allow more chips per wafer, reducing the price per chip.

In fact, over the past 30 years the number of transistors per chip has been doubled every 2–3 years once a new technology node is introduced. For example the number of MOSFETs in a microprocessor fabricated in a 45 nm technology can well be twice as many as in a 65 nm chip. This doubling of transistor density was first observed by Gordon Moore in 1965 and is commonly referred to as Moore's law.

It is as well expected that smaller transistors switch faster. For example, one approach to size reduction is a scaling of the MOSFET that requires all device dimensions to reduce proportionally. The main device dimensions are the transistor length, width, and the oxide thickness, each (used to) scale with a factor of 0.7 per node. This way, the transistor channel resistance does not change with scaling, while gate capacitance is cut by a factor of 0.7. Hence, the RC delay of the transistor scales with a factor of 0.7.

While this has been traditionally the case for the older technologies, for the state-of-the-art MOSFETs reduction of the transistor dimensions does not necessarily translate to higher chip speed because the delay due to interconnections is more significant.

Difficulties Arising due to MOSFET Size Reduction

Producing MOSFETs with channel lengths much smaller than a micrometre is a challenge, and the difficulties of semiconductor device fabrication are always a limiting factor in advancing integrated circuit technology.

Higher Subthreshold Conduction

As MOSFET geometries shrink, the voltage that can be applied to the gate must be reduced to maintain reliability. To maintain performance, the threshold voltage of the MOSFET has to be reduced as well. As threshold voltage is reduced, the transistor cannot be switched from complete turn-off to complete turn-on with the limited voltage swing available; the circuit design is a compromise between strong current in the "on" case and low current in the "off" case, and the application determines whether to favour one over the other. Subthreshold leakage (including subthreshold conduction, gate-oxide leakage and reverse-biased junction leakage), which was ignored in the past, now can consume upwards of half of the total power consumption of modern high-performance VLSI chips.

Increased Gate-oxide Leakage

The gate oxide, which serves as insulator between the gate and channel, should be made as thin as possible to increase the channel conductivity and performance when the transistor is on and to reduce subthreshold leakage when the transistor is off. However, with current gate oxides with a thickness of around 1.2 nm (which in silicon is ~5 atoms thick) the quantum mechanical phenomenon of electron tunnelling occurs between the gate and channel, leading to increased power consumption.

Insulators that have a larger dielectric constant than silicon dioxide, such as group IVb metal silicates *e.g.* hafnium and zirconium silicates and oxides are being used to reduce the gate leakage from the 45 nanometer technology node onwards. Increasing the dielectric constant of the gate dielectric allows a thicker layer while maintaining a high capacitance (capacitance is proportional to dielectric constant and inversely proportional to dielectric thickness). All else equal, a higher dielectric thickness reduces the quantum tunnelling current through the dielectric between the gate and the channel. On the other hand, the barrier height of the new gate insulator is an important consideration; the difference in conduction band energy between the semiconductor and the dielectric (and the corresponding difference in valence band energy) also affects leakage current level. For the traditional gate oxide, silicon dioxide, the former barrier is approximately 8 eV. For many alternative dielectrics the value is significantly lower, tending to increase the tunnelling current, somewhat negating the advantage of higher dielectric constant.

Increased Junction Leakage

To construct devices smaller, junction design has become more complex, leading to higher doping levels, shallower junctions, "halo" doping and so forth, all to decrease drain-induced barrier lowering. To keep these complex junctions in place, the annealing steps formerly used to remove damage and electrically active defects must be curtailed increasing junction leakage. Heavier doping is also associated with thinner depletion layers and more recombination centres that result in increased leakage current, even without lattice damage.

Lower Output Resistance

For analog operation, good gain requires a high MOSFET output impedance, which is to say, the MOSFET current should vary only slightly with the applied drain-to-source voltage. As devices are made smaller, the influence of the drain competes more successfully with that of the gate due to the growing proximity of these two electrodes, increasing the sensitivity of the MOSFET current to the drain voltage. To counteract the resulting decrease in output resistance, circuits are made more complex, either by requiring more devices, for example the cascode and cascade amplifiers, or by feedback circuitry using operational amplifiers.

Inferior Transconductance

The transconductance of the MOSFET decides its gain and is proportional to hole or electron mobility (depending on device type), at least for low drain voltages. As MOSFET size is reduced, the fields in the channel increase and the dopant impurity levels increase. Both changes reduce the carrier mobility, and hence the transconductance. As channel lengths are reduced without proportional reduction in drain voltage, raising the electric field in the channel, the result is velocity saturation of the carriers, limiting the current and the transconductance.

Interconnect Capacitance

Usually, switching time was roughly proportional to the gate capacitance of gates. However, with transistors becoming smaller and more transistors being placed on the chip, interconnect capacitance (the capacitance of the metal-layer connections between different parts of the chip) is becoming a large percentage of capacitance. Signals have to travel through the interconnect, which leads to increased delay and lower performance.

Heat Production

The ever-increasing density of MOSFETs on an integrated circuit creates problems of substantial localized heat generation that can impair circuit operation. Circuits operate more slowly at high temperatures, and have reduced reliability and shorter lifetimes. Heat sinks and other cooling techniques are now required for many integrated circuits including microprocessors. Power MOSFETs are at risk of thermal runaway. As their on-state resistance rises with temperature, if the load is approximately a constant-current load then the power loss rises correspondingly, generating further heat. When the heatsink is not able to keep the temperature low enough, the junction temperature may rise quickly and uncontrollably, resulting in destruction of the device.

Process Variations

By means of MOSFETS becoming smaller, the number of atoms in the silicon that produce many of the transistor's properties is becoming fewer, with the result that control of dopant numbers and placement is more erratic. During chip manufacturing, random process variations affect all transistor dimensions: length, width, junction depths, oxide thickness *etc.*, and become a greater percentage of overall transistor size as the transistor shrinks. The transistor characteristics become less certain, more statistical. The random nature of manufacture means we do not know which particular example MOSFETs actually will end up in a particular instance of the circuit. This uncertainty forces a less optimal design because the design must work for a great variety of possible component MOSFETs.

Modelling Challenges

Modern ICs are computer-simulated with the goal of obtaining working circuits from the very first manufactured lot. As devices are miniaturized, the complexity of the processing makes it difficult to predict exactly what the final devices look like, and modelling of physical processes becomes more challenging as well. In addition, microscopic variations in structure due simply to the probabilistic nature of atomic processes require statistical (not just deterministic) predictions. These factors combine to make adequate simulation and "right the first time" manufacture difficult.

FUNDAMENTALS OF MOSFETS FOR SWITCHING

The MOSFET (metal -oxide semiconductor field-effect transistor) is a near-ideal three-terminal component used for linear and switching applications. After a brief overview of the MOSFET's internal structure and operation, this course looks at critical and relevant device parameters to consider when using MOSFETs in their most common switching applications, such basic AC-line control, power-supply design, and motor control (via H-bridge and other topologies).

Among the parameters explored are on-resistance, switching time, and thermal performance. The course will also explore important associated issues such as the MOSFET driver and driver requirements (including isolated versus non-isolated drive), circuit protection, packaging, and paralleling MOSFETs, plus design issues and potential problems to take into aware of.

FUNDAMENTALS OF DESIGNING WITH MOSFET POWER SWITCHES

By means of power switches can be complex or even confusing for most electronic designers, especially for those who are not power management experts. In a broad range of applications such as portable electronics, consumer electronics, industrial or telecommunication systems, designers are increasingly working with power switches that can be used in a variety of ways including control, sequencing, protection, power distribution or even system supply turn-on management. Of course, each of these need power switching solutions with different characteristics.

This commentary summarizes important specifications and concepts that designers need to consider when using power switches in different applications, and reviews possible solutions to help designers select an optimized solution.

There are several ways to use power switches. The most popular are to:

- Control, distribute and sequence (*i.e.,* turn on/off a power rail to enable a subsystem or distribute power to multiple loads)
- Protect against short circuits or any kind or any kind of over-current or over-voltage (USB current limiting, sensors protection, power rail short circuit protection)
- Manage the turn-on inrush current (*i.e.,* when charging a capacitor)
- Select power supplies (*i.e.,* muxing or ORing) or load sharing.

ON Resistance, Maximum Currents and Input Voltage Range

ON Resistance (rON), maximum continuous current, and input voltage range are always key characteristics to consider. These are the basic characteristics you need to study before looking at any device. Depending of the application, the designer can easily determine the current that needs to be switched, and at what voltage. Based on this information, a first selection can be made. Indeed, if you need a switch to pass 1.2 V or 36 V, two distinct product ranges can be identified.

ON resistance impacts the dropout you determination see across the switch. Designers must be cautious to understand what the maximum acceptable dropout is with regard to their particular application set-up (voltage, current). This can be calculated easily using Equation 1:

$$V_{DROP} = r_{ON} * I$$

where the dropout is V_{DROP}, the pass FET ON resistance is r_{ON}, and the current through the switch is I.

If the submission needs to switch a lot of current, or switch a low-voltage rail (like 1.0-V), then dropout needs to be minimized. Therefore, the ON resistance needs to be as low as possible, for example, as with the TPS2292x series featuring a 14-mΩ r_{ON} at 3.6-V.

However, if the current to be switched is small, the ON resistance won't be a key concern, and you can select a higher ON resistance device like one of the TPS2294x series, which is about 1 Ω. The ON resistance is a key contributor to the die size of a power switch device and, hence, to the cost of the device. You will want to look closely at this to select the most cost-effective solution possible.

In totalling to the maximum continuous current the designer targets to switch, another important characteristic is the maximum pulsed-current the switch can accept. In certain applications, the load requested most of the time consists of moderate, continuous currents. However, spikes are evident when a subsystem requests additional power. A good example is the GSM/GPRS transmit burst, which sinks up to 1.7 A during 576 µsec with a duty cycle of 12.5 per cent. Make sure the selected part can support such pulsed current.

Power Dissipation and Protection Features

Power dissipation is also an important characteristic to consider. During normal operation as a pass switch, power dissipation can be calculated considering the switch's ON resistance as well as the current being switched. You can easily calculate the maximum power dissipated through the device by using Equation 2:

$$P = I^2 \times r_{ON}$$

If the part's ON resistance is selected low enough, the power dissipation is small and has little effect on the part's operating temperature. However, be careful if you plan to use the switch to protect the rail against an over-current, or a short circuit as with USB ports, or fingerprint sensor protection. In this case, you must select a current-limited switch like the TPS22944.

If you are not using a current-limited switch, power dissipation can be a major issue for system reliability. For instance, a 0.9-Ω short applied to a non-current-limited-load-switch with a 3.3-V input voltage (switch ON resistance being ~100-mΩ like for the TPS22902) translates into a dissipated power as shown in Equation 3:

$$P = I_{SHORT}{}^{2} \times r_{ON} = (3.3/(0.9+0.1))^2 \times 0.1 = 3.3^2 \times 0.1 = 1.089W$$

Usually, this power dissipation is too high for most packages in the market, resulting in failure and reliability issues.

In the similar manner, the designer using a current-limited switch needs to make sure the package can support a short-circuit condition. If the part goes into current limit, maximum power dissipation occurs when the output is shorted-to-ground.

For devices like the TPS22945 featuring auto-restart time, $t_{RESTART}$, and the overcurrent blanking time, t_{BLANK}, the maximum average power dissipated is shown in Equation 4:

$$P(average) = \frac{t_{BLANK}}{t_{RESTART} + t_{BLANK}} \times V_{IN}(\max) \times L_{LM}(\max) = \frac{10}{80+10} \times 5.5 \times 0.2 = 122mW$$

For devices that do not feature auto-restart loops like the TPS22944, a short on the output causes the part to operate in a constant-current state, dissipating a worst-case power level until the thermal shutdown activates. It then cycles in and out of thermal shutdown so long as the ON pin is active and the short is present.

Several current-limited switches exist in the market and the two main characteristics to look at are the current-limit minimum value (fixed current limit or programmable using an external resistor), and the current-limit accuracy and response time. In most applications the current-limit accuracy isn't a key concern because the device is used as a circuit breaker (*i.e.*, the switch is turned off in case of a short circuit). However, accuracy in some application like USB current limiting can be important, as the switch is used as a constant current source.

For applications where it is expected to switch large current or to face over currents, it is recommended that you select a device that features some kind of thermal protection. When the device temperature is identified as too high, most devices will activate the thermal shutdown, which will turn off the FET in order to protect the device itself against any potential thermal damage.

Aside from the current limit (or over-current protection, OCP), which is mandatory to protect against short circuits, other protection features like reverse current blocking can be interesting to consider.

Reverse current blocking (also known as reverse voltage protection) is mandatory when designers are trying to design a power selector (ORing), or to make some load sharing.

An example of power switches configured to supply a load from two potential power sources (*i.e.*, a DC input and a battery):

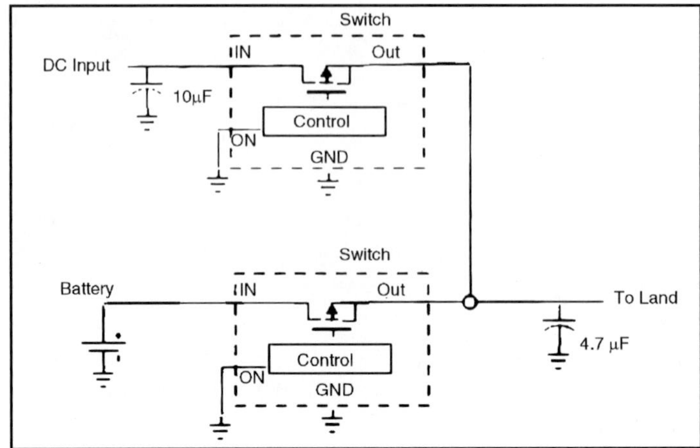

Figure : Dual-Source Power Selector

For a device that doesn't have reverse voltage protection, it is important that the input voltage of the pass FET stays higher than its output voltage. Otherwise the input will be clamped via the body diode of the FET, which will cause significant current to flow from the output to the input. The battery is a lithium-ion (Li-ion) battery at 4.2 V (max), and if the DC input is enabled and at 5.0 V, then a potential large current will flow from the load to the battery which, of course, is undesirable! A solution to make this work is to use a device featuring reverse-voltage protection. Reverse-current protection usually can be implemented using back-to-back FETs or by switching the back gate of a PMOS FET when a reverse voltage condition is detected. You will want to look at the reverse-voltage comparator trip point ($V_{OUT} - V_{IN}$ value threshold above which the reverse-current feature is activated), as well as the time from reverse-voltage condition to MOSFET turn off. One more protection that can be useful for some applications is the overvoltage protection (OVP). This feature protects the switch and the system, if an overvoltage is applied to the switch. It can be useful for example in some USB applications or in some battery applications.

Inrush Current Management

Another ordinary use of a power switch is to manage the inrush current when the system is being turned on. If the switch turns on without being controlled, a large inrush current is created that could result in a supply rail drop at the input of the switch. This could ultimately impact the system's entire functionality. When charging large output capacitances, inrushes will be large and need to be controlled and/or limited. The inrush current can be calculated using Equation 5:

$$I = C_{LOAD} * \frac{\partial V}{\partial t}$$

For example, with a C_{LOAD} = 1 μF, V = 3 V, and a rise time of 1 μsec, the inrush current could be as high as 3 A.

An simple way to avoid this inrush current is to slow-down the switch's rise time. This will slowly charge the output capacitor and reduce the current peak. In the example in Equation 5, a rise time of 200 μsec would result in a 15 mA inrush, which is acceptable.

In some cases, you may want to charge extremely large capacitors (several hundred of μF). A very slow rise time is usually recommended, but you can also select a switch with a high current limit. The device will go in current limiting at power on and the capacitor will be charged at the current limit value, which will be the maximum power dissipation capability of the power switch.

System Interoperability

Whenever selecting a power switch, system interoperability needs to be cautiously considered. For instance, when using a power switch to enable and disable loads to optimize power consumption in a portable application, the switch's control inputs must be compatible with general-purpose, low-voltage (1.8 V), GPIO is critical.

In addition, when turning off the switch, make sure that the floating output of the switch does not impact system performance. Therefore, some users may tie the power-switch output to ground when disabled with an additional transistor, or use an integrated device, which integrates this pull down-to-ground like in the TPS22902.

Another important point to check is the input and output capacitance used to design a stable system. Although an input capacitor is usually not required to stabilize power switches available in the market, it is considered good analog design practice to connect a 0.1 μF to 1 μF, low equivalent series resistance (ESR) capacitor across the input supply. This capacitor counteracts reactive input sources and improves transient response, noise and ripple rejection.

Depending on the load of the switch, you may want to consider some additional storage capacitors at the switch's output. If the switch doesn't have reverse-current blocking, then an input capacitor bigger than the output capacitor is strongly advised, otherwise the input will be clamped via the FET's body diode, which will cause significant current to flow from the output to the input.

ADVANCED MOSFETS CONCEPTS

Research MOSFETs with gate lengths of 15 nm or even 10 nm contain been fabricated. However, besides scaling the transistor to such extreme dimensions substantial improvements by the use of new materials and new transistor concepts are required. At small dimensions undesired short channel effects tend to become dominant, due to the distortion of the channel possible by the drain voltage. There are a number of alternative transistor concepts to reduce short channel effects and improve performance. Very promising is the use of SOI substrates instead of bulk Si wafers. SOI wafers are commercially produced either by wafer bonding and etch back (BESOI) or by ion implantation. The latter process, SIMOX (separation by implantation of oxygen) process involves implantation of a high dose of oxygen into a heated silicon substrate and subsequent annealing at very high temperature, typically * 1250°C for 8 h. Generally, SOI transistors have smaller parasitic capacitances, smaller source/drain leakage, are more immune to soft errors caused by alpha particles and allow higher speed and lower power consumption than bulk transistors. The buried, perfectly insulating SiO2 layer with a typical thickness of more than a few 100 nm eliminates several leakage paths. However, the SOI substrate is more expensive and the poor thermal conductivity of the buried silicon dioxide may create a heat problem. Nevertheless, many different transistor concepts are presently under investigation comparison with a standard bulk transistor. A transistor fabricated on a SOI substrate where the silicon region under the gate is partially depleted. Under gate bias, the potential of this silicon region may float and cause undesirable threshold shifts.

This problem is eliminated when the silicon layer on the silicon dioxide is made so thin that the region becomes fully depleted. A fully depleted transistor with an additional improvement, elevated source and drain contacts, which is an efficient measure to reduce the source and drain resistance. A cross-section TEM micrograph of a 50 nm gate length transistor with a thin silicon layer of 30 nm. The source/drain contacts are raised in order to lower the series resistance. This is made by selective deposition of silicon on the contact areas before the formation of the self-aligned silicide contacts. A comparison of transistors with and without raised contacts reveals substantial improvements. The *I*on drive current could be increased by 20 to 30 % for n- and p-channel devices, respectively. This results in a larger *I*on/*I*off ratio (order of 106) and steeper subthreshold slopes.

One more concept is *silicon on nothing* (SON) where a region under the gate is removed and possibly refilled with a dielectric material. A buried Si-Ge alloy layer is selectively underetched in a mesa structure to realise the SON structure. The advantage of this is that no expensive SOI wafers are necessary and the buried insulating layer is manufactured only at selected areas. This approach appears promising for SoC. Excellent performance can be achieved when the Si-channel layer is kept very thin (10 - 20 nm).

Alternatively, ultrathin body (UTB) transistors can be made on SOI. The structure is identical to the fully depleted SOI transistor, except for the very small thickness of the Si body which will be less than 10 nm. This concept allows the fabrication of very densely packed, high performance transistors with record frequencies. Scaling the linear dimension of a transistor by 30% reduces the area by 50%, thus doubling the number of transistors per unit area becomes feasible. When the gate length is reduced to about 20 nm, frequencies in the THz range seem achievable. In addition to the described concepts transistors with double gates (DG) are investigated, where two symmetric gates are arranged in a planar or in a vertical configuration as illustrate. A thin silicon ridge serves as the heart of the transistor with symmetric gates on both sides. The direction of the current flow from the source to drain can be horizontal or vertical. In the FinFET type source and drain are beside the gate region and the current flows along the silicon ridge. The name stems from the silicon fin which forms the basic building block of the transistor. The advantage of the vertical DG device is that the physical gate length can be defined by layer growth or by ion implantation and diffusion and not by lithography. However, all DG transistors are difficult to fabricate, in particular, the planar DG transistor, where a single crystalline silicon channel layer has to be grown on top of a buried gate stack. The two gates have to be perfectly aligned, which is very difficult task in view of the extremely small dimensions (1-3 (10 nm). The vertical DG transistors are somewhat easier to fabricate, nevertheless the technological problems, such as the etching of the nanostructures and the growth or deposition of a high quality gate dielectric on the vertical side walls, is a major challenge Both the DG and the UTB devices rely on the thickness of the silicon channel to obtain optimum gate control on the channel, suppression of short channel effects and minimization of leakage. The basic structure of these transistors, the corresponding band diagrams under inversion conditions and the charge distribution near the oxide layer. In a double gate configuration two symmetric gates produce two inversion layers on both sides of the silicon layer.

ELECTRON TUNNELLING SEEN IN REAL TIME

The strong electric field from an intense laser pulse can cause electrons to "tunnel" away from an atom in just one billion-billionth of a second. Now physicists in Germany are the first to observe this well-known quantum-mechanical process as it proceeds in real time. The breakthrough paves the way for a new technique that can prove short-lived states of atoms or molecules, giving the first direct insight into the dynamics of electron tunnelling (*Nature* 446 627).

A laser pulse consists of a small number of electric-field oscillations that can pull the outer electrons of atoms away from their binding nuclei. At the peaks of these oscillations, the pull can be so great that the outer electrons are given a chance to escape (or tunnel) from the atom, even though they don't have enough energy to overcome the attractive pull of the nucleus.

But this process occurs so quickly that current instruments can only see the final ionized atom, and not any intermediate states.

Ferenc Krausz and colleagues from the Max-Planck Institut für Quantenoptik have now found a way around instrument limitations by probing atoms with two pulses of different-wavelength light that are carefully-timed to take snapshots of the tunnelling process.

Tunnelling

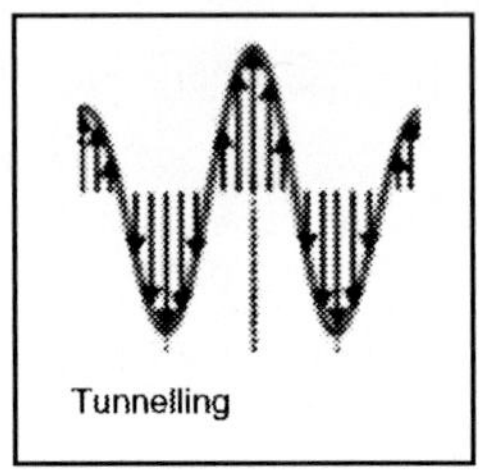
Tunnelling

The technique involves inquisitive atoms that have very tightly bound electrons – in this case those of neon – which do not readily ionize with a laser pulse alone. The atoms must first be prepared by an exciting pulse, which sends some of the electrons to outer regions of the atom where tunnelling can be induced by an ionizing pulse.

This two-stage process is where the advantage lies: if the exciting pulse has a much shorter wavelength than the ionizing laser pulse, it can be switched on at many points in the ionizing pulse's cycle, and only at these points will electrons be in a position to tunnel. Then, by noting when in the ionizing laser's cycle the electrons tunnel, a picture of how the electrons leave the atom can gradually be reconstructed. In perform the two pulses must be synchronized to within several billion-billionths of a second to achieve any sort of accuracy. To get over this hurdle, Krausz and his team used the same infrared laser to make both pulses. First, they shone the laser though a gas jet to produce a pulse of extreme (very short wavelength) ultraviolet light. This pulse then travelled to the sample of neon atoms with the original laser pulse delayed behind. Finally, using mirrors to carefully alter the delay, the physicists could chart the tunnelling of the electrons with a resolution of less than one femtosecond (10^{-15} s). According to 40-year-old quantum theory, the probability of an electron tunnelling through the potential barrier should increase stepwise with every successive peak of the ionizing pulse. Now Krausz and his team are the first to be able to confirm this is exactly the case.

The light-induced tunnelling technique may now be used to provide further observations of electron movement, giving scientists unprecedented insights in areas such as microelectronics and biological imaging. "There may be issues with transient [short-lived] states that we've never been able to capture," Jonathan Marangos, an expert in sub-femtosecond technology, told *Physics Web*. "This technique will enable us to do that."

ELECTRON TUNNELING FROM ATOMS

Now as a valley stabilizes the position of a body by gravity, the electric force attracting electrons to the atomic nucleus stabilizes them within a tiny volume of space, allowing the stable existence of atoms, the building blocks of our world. This stabilizing effect can also be represented by a valley, which physicists call a potential. According to the laws of quantum physics, microscopic particles can penetrate the potential wall confining their location like a wave. If the wall is sufficiently thin, the wave may reach its other side, *i.e.* the particle may overcome the potential barrier without climbing it. This phenomenon has been dubbed tunnelling and constitutes one of the most striking implications of quantum physics. Tunnelling of particles through some binding potential is commonplace in the microscopic world: it is believed to be responsible for nuclear as well as electronic phenomena, but – because of its awesome rapidity – it has never been observed in real time. This unsatisfactory state of matters has changed profoundly, when attosecond metrology permitted capturing electrons in the act of their tunnelling through the potential binding them to the atomic core under the influence of laser light. The experiment provides the first direct insight into the attosecond-scale history of electron tunnelling and reveal how light-field-induced tunnelling can be exploited for real-time observation of intra-atomic or intra-molecular motion of electrons.

RESONANT TUNNELLING DIODES: THEORY OF OPERATION AND APPLICATIONS

Tunnelling diodes (TDs) have been widely studied for their importance in achieving very high speed in wide-band devices and circuits that are beyond conventional transistor technology. A particularly useful form of a tunnelling diode is the Resonant Tunnelling Diode (RTD). RTDs have been shown to achieve a maximum frequency of up to 2.2 THz as opposed to 215 GHz in conventional Complementary Metal Oxide Semiconductor (CMOS) transistors. The very high switching speeds provided by RTDs have allowed for a variety of applications in wide-band secure communications systems and high-resolution radar and imaging systems for low visibility environments. In this paper, the theory of operation of RTDs will be explained. Next, the tradeoffs in the optimization of this technology will be discussed followed by some current circuit applications of RTDs.

Theory of Operation

Tunnelling diodes provide the same functionality as a CMOS transistor where under a specific external bias voltage range, the device will conduct a current thereby switching the device "on". However, instead of the current going through a channel between the drain and source as in CMOS transistors, the current goes through the depletion region by tunnelling in normal tunnelling diodes and through quasi-bound states within a double barrier structure in RTDs. A TD consists of a p-n junction in which both the n- and pregions are degenerately doped (>10^{19} cm^{-3}). There is a high concentration of electrons in the conduction band (EC) of the n-type material and empty states in the valence band (EV) of the p-type material. Initially, the Fermi level (EF) is constant because the diode is in thermal equilibrium with no external bias voltage.

When the forward bias voltage starts to increase, the EF will start to decrease in the p-type material and increase in the n-type material. Since the depletion region is very narrow (<10nm), electrons can easily tunnel through, creating a forward current. Depending on how many electrons in the n-region are energetically aligned to the empty states in the valence band of the p-region, the current will either increase or decrease. As the bias voltage continues to increase, the ideal diffusion current will cause the current to increase. When a reverse-bias voltage is applied, the electrons in the p-region are energetically aligned with empty states in the n-region causing a large reverse-bias tunnelling current.

RESONANT TUNNELING

A quantum healthy, in the general use of this term, is a potential structure which spatially confines the electron. According to quantum mechanics, an electron subjected to potential confinement has its energy quantized and a discrete energy spectrum would be expected for the electron system. However, the electron remains free to move in the perpendicular direction. This results in the creation of a two-dimensional electron gas of quasi-bound states.

Resonant tunnelling refers to tunnelling in which the electron transmission coefficient through a structure is sharply peaked about certain energies.

The emergence of these peaks can be qualitatively explained by introducing infinite walls as boundary conditions. It is usually possible to do this far from the quantum well itself. Then the calculation of the quantized energy levels in a quantum well of arbitrary shape is the solution of an eigenvalue problem. For electrons with an energy corresponding approximately to the virtual resonant energy level of the quantum well, the transmission coefficient is close to unity. That is, an electron with this resonant energy can cross the double barrier without being reflected. This resonance phenomenon is similar to that taking place in the optical Fabry-Perot resonator or in a microwave capacitively-coupled transmission-line resonator.

The outlined physics gives a simple theory of quantum transport through quantum wells from elementary quantum mechanics. However, for simulation, we model resonant tunnelling using open systems theory.

RESONANT TUNNELING DIODE STRUCTURE

A resonant-tunnelling diode requires a band-edge discontinuity at the conduction band or valence band to form a quantum well and, thus, necessitates heteroepitaxy. The most common combination used is GaAs-AlGaAs. The middle quantum-well thickness is typically around 5nm, and the barrier layers range from 1.5 to 5nm. Symmetry of the barrier layers is not required so their thickness can be different. A typical resonant tunnelling diode structure with analytical band edge model as used in simulations.

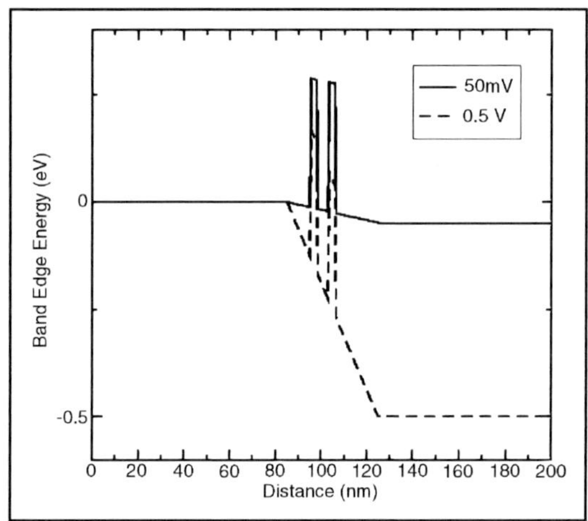

Figure : Conduction band edge of the RTD for different voltages. A linear voltage drop is assumed over a distance of 40 nm.

The well layer (GaAs) and the barrier layers (AlGaAs) are all undoped, and they are sandwiched between heavily doped, narrow energy-gap materials, which usually are the same as the well layer. Adjacent to the barrier layers are thin layers of undoped spacers to ensure that dopants do not diffuse to the barrier layers.

The region flanked by the two barriers defines a virtual quantum well since the electrons can escape the well confinement by tunnelling. The resonant tunnelling diode (RTD) is thus an open quantum system in which the electronic states are scattering states with a continuous distribution in energy space, rather than bound states with a discrete energy spectrum. Under these circumstances quasi-bound states (resonant states) are formed in the quantum well which accommodate electrons for a time that is characteristic for the double-barrier structure. So-called resonant tunnelling through the double-barrier structure occurs when the energy of the electrons flowing from the emitter coincides with the energy of the quasi-bound state,, in the quantum well. The effect of the external bias is to sweep the alignment of the emitter and quasi-bound states.

For many applications, negative differential resistance (NDR) devices should have a large peak current and a small valley current, where the latter is the minimum current following the peak current as the magnitude of the voltage increases. Therefore an important figure of merit for an NDR device such as the RTD is the peak to valley ratio (PVR). For good devices with thin AlAs barriers, PVRs close to 4:1 and peak current densities in excess of $10^5 A/cm^2$ may be obtained at 300K, although not in the same structures, since there is usually a trade-off between these two parameters in terms of device design.

Because tunnelling is inherently a very fast phenomenon that is not transit-time limited, the resonant-tunnelling diode is considered among the fastest devices ever made. On the other hand, using resonant-tunnelling diodes it is more difficult to supply high current and the output power of an oscillator is limited. Integration of RTDs with MOSFETs provides high speed operation due to the inherently fast tunnelling process, and negative differential resistance regime (NDR) that provides at least two stable operating points (*i.e.*, multiple valued logic) when combined with MOSFETs.

The resonant tunnel devices for logic applications include resonant tunnel transistors (RTT) and hybrid devices incorporating resonant tunnelling diodes and one or more FETs (RTD-FET). RTD designs can offer a reduction in circuit component count by up to 40% when compared with the equivalent CMOS logic family.

The major problem is the extreme sensitivity of device characteristics to the thickness of the tunnelling well as the tunnelling current depends exponentially on the thickness of the tunnel barrier. The challenge to the process engineer is to match device properties across the wafer.

Overall, the resonant tunnelling devices may be useful for certain niche applications requiring high speed and low dynamic range, provided the manufacturing issues associated with uniformity of the tunnelling barrier can be resolved. Thus, RTDs are on the verge of commercialisation. Potential practical applications are high-speed microwave systems and novel digital logic circuits.

COHERENT TUNNELING

In this description the electron does not experience any phase-coherence breaking events throughout the structure, it is based on the Schrödinger equation. We refer to this model as "global coherent tunnelling picture". It neglects any scattering processes. Note that transport in the absence of scattering is referred to as ballistic transport. Ballistic transport through a device can be assumed when the device length is short compared with the mean free path of the electron.

When no voltage is applied electrons are injected from the left and the right, and due to the symmetry of the device, no current results, as should be the case in equilibrium. With a positive bias applied to the right contact relative to the left, the Fermi energy on the left is pulled through the resonant level E_0. As the Fermi energy passes through the resonant energy, a large current flows due to the increased transmission from left to right. It reaches a local maximum, called peak current. With higher bias, the current ceases to flow, when E_0 falls below the conduction-band edge. The result is a marked decrease of the current with increasing voltage, giving rise to a region of negative differential resistance.

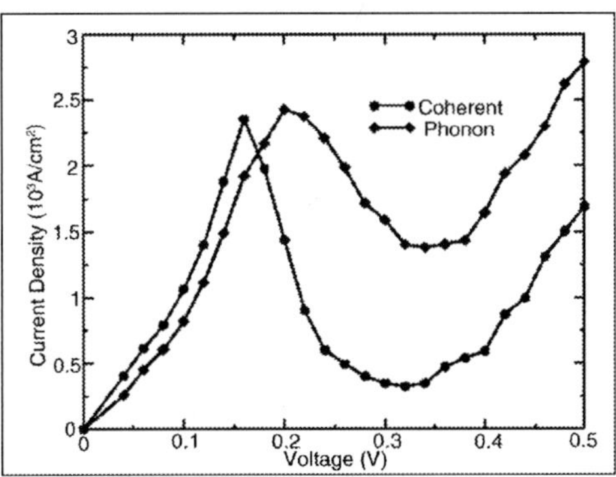

Figure : Influence of phonon scattering on the I/V characteristics of an RTD

The minimum present following the peak current as the voltage increases is denoted valley current. The nonzero valley current is mainly due to thermionic emission over the barriers, and it has a large temperature dependence. Another small but conceivable contribution is due to tunnelling of electrons through higher quantized levels. At larger bias the current increases again as particles acquire enough kinetic energy.

INTERBAND RESONANT TUNNELING DIODES

Resonant interband tunnelling diodes (RITDs) unite the structures and behaviours of both *intraband* resonant tunnelling diodes (RTDs) and conventional *interband* tunnelling diodes, in which electronic transitions occur between the energy levels in the quantum wells in the conduction band and that in the valence band. Like resonant tunnelling diodes, resonant interband tunnelling diodes can be realised in both the III-V and Si/SiGe materials systems.

III-V RITDs

In the III-V materials system, InAlAs/InGaAs RITDs with peak-to-valley current ratios (PVCRs) higher than 70 and as high as 144 at room temperature and Sb-based RITDs with room temperature PVCR as high as 20 have been obtained. The main drawback of III-V RITDs is the use of III-V materials whose processing is incompatible with Si processing and is expensive.

Si/SiGe RITDs

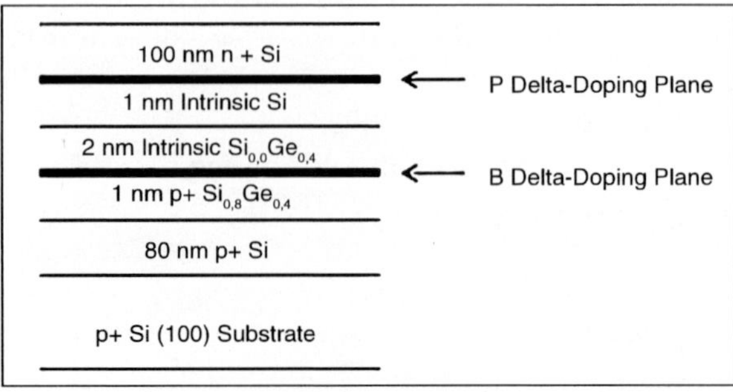

Figure: Typical structure of a Si/SiGe resonant interband tunnelling diode

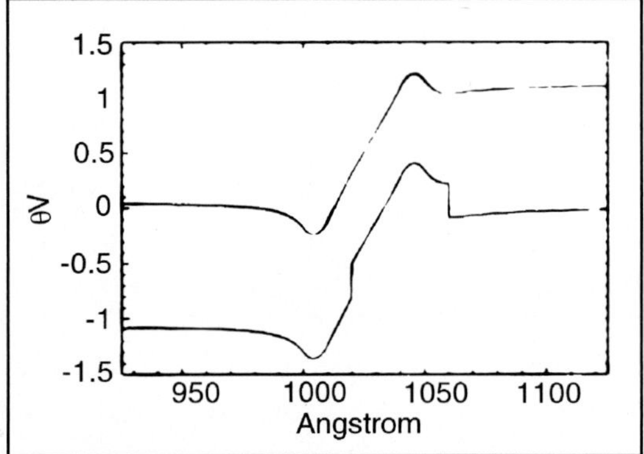

Figure: Band diagram of a typical Si/SiGe resonant interband tunnelling diode calculated by Gregory Snider's 1D Poisson/Schrodinger Solver.

In Si/SiGe materials system, Si/SiGe resonant interband tunnelling diodes have also been developed which have the potential of being integrated into the mainstream Si integrated circuits technology.

Structure

The five key points to the design are:

- An intrinsic tunnelling barrier,
- Delta-doped injectors,
- Offset of the delta-doping planes from the heterojunction interfaces,
- Low temperature molecular beam epitaxial growth (LTMBE), and
- Postgrowth rapid thermal annealing (RTA) for activation of dopants and reduction of density of point defects.

Presentation and Performance

A minimum PVCR of about 3 is needed for typical circuit applications. Low current density Si/SiGe RITDs are suitable for low-power memory applications, and high current density tunndel diodes are needed for high-speed digital/mixed-signal applications. Si/SiGe RITDs have been engineered to have room temperature PVCRs up to 4.0. The same structure was duplicated by another research group using a different MBE system, and PVCRs of up to 6.0 have been obtained. In terms of peak current density, peak current densities ranging from as low as 20 mA/cm^2 and as high as 218 kA/cm^2, spanning seven orders of magnitude, have been achieved. A resistive cut-off frequency of 20.2 GHz has been realised on photolithography defined SiGe RITD followed by wet etching for further reducing the diode size, which should be able to improve when even smaller RITDs are fabricated using techniques such as electron beam lithography.

Applications

In addition to the realization of integration with Si CMOS and SiGe heterojunction bipolar transistors that is discussed in the next section, other applications of SiGe RITD have been demonstrated using breadboard circuits, including multi-state logic.

Integration with Si/SiGe CMOS and Heterojunction Bipolar Transistors

Integration of Si/SiGe RITDs with Si CMOS has been demonstrated. Vertical integration of Si/SiGe RITD and SiGe heterojunction bipolar transistors was also demonstrated, realizing a 3-terminal negative differential resistance circuit element with adjustable peak-to-valley current ratio. These results indicate that Si/SiGe RITDs is a promising candidate of being integrated with the Si integrated circuit technology.

SINGLE ELECTRON DEVICES FOR LOGIC APPLICATIONS

Scaling down of electronic device sizes has been the fundamental strategy for improving the performance of ultra-large-scale integrated circuits (ULSIs). Metal-oxide-semiconductor field-effect transistors (MOSFETs) have been the most prevalent electron devices for ULSI applications, and thus the scaling down of the sizes of MOSFETs has been the basis of the development of the semiconductor industry for the last 30 years. However, in the early years of the 21st century, the scaling of CMOSFETs is entering the deep sub-50 nm regime. In this deep-nanoscaled regime, fundamental limits of CMOSFETs and technological challenges with regard to the scaling of CMOSFETs are encountered. On the other hand, quantum-mechanical effects are expected to be effective in these small structured devices. Therefore, in order to extend the prodigious progress of LSI performance, it is essential to introduce a new device having an operation principle that is effective in smaller dimensions and which may utilize the quantum- mechanical effects, and thus provide a new functionality beyond that attainable with CMOSFETs.

Single-electron devices are promising as new nanoscaled devices because single-electron devices retain their scalability even on an atomic scale and, moreover, they can control the motion of even a single electron. Therefore, if the single-electron devices are used as ULSI elements, the ULSI will have the attributes of extremely high integration and extremely low power consumption. In this respect, scalability means that the performance of electronic devices increases with a decrease of the device dimensions. Power consumption is roughly proportional to the electron number transferred from voltage source to the ground in logic operations. Therefore, the utilization of single-electron devices in ULSIs is expected to reduce the power consumption of ULSIs. In this chapter, firstly, the operation and operation principle of single-electron devices is briefly explained. Then, the advantages and disadvantages of single-electron devices over conventional MOSFETs are discussed. Next, the analytical device model of a single-electron transistor, which is a typical functional single-electron device, for circuit simulation is derived and the techniqueology of designing logic circuits with single-electron transistors.

SINGLE ELECTRON TRANSISTOR

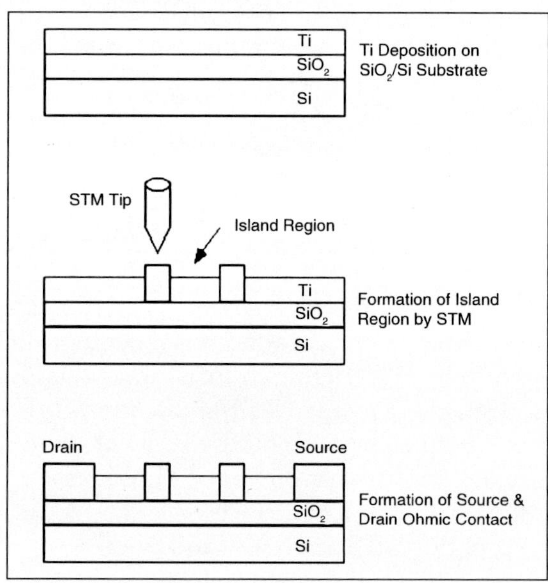

Figure : Fabrication of the Ti/TiOx SET by the STM nano-oxidation process.

The single electron transistor or SET is type of switching device that uses controlled electron tunnelling to amplify current. A SET is made from two tunnel junctions that share a common electrode. A tunnel junction consists of two pieces of metal separated by a very thin (~1 nm) insulator. The only way for electrons in one of the metal electrodes to travel to the other electrode is to tunnel through the insulator. Since tunnelling is a discrete process, the electric charge that flows through the tunnel junction flows in multiples of e, the charge of a single electron.

Room temperature operation of single electron memory has been realised by the use of self-organized, small-size structures on thin poly-silicon films. However, it is difficult to control the size and structure of the SET island with the spontaneous size formation fabrication technique.

Earlier, we demonstrated an artificial pattern formation technique based on the scanning tunnelling microscope (STM) which avoids the control problems in self- organized structures. Using this technique, we have succeeded in fabricating an SET. The SET operates at room temperature, showing a clear Coulomb staircase with a ~150 mV period at 300 K.

A account of the STM nano-oxidation process. A 3 nm thin titanium (Ti) metal film is deposited on a 100 nm thermally oxidized SiO_2/n-Si substrate. The Ti surface was oxidized by anodization through the water adhered to the surface of the Ti from the atmosphere, using the STM tip as a cathode, forming nanometer size Ti oxide (TiOx) lines. The barrier height of the TiOx/Ti junction has been found to be 285 meV for the electron from the temperature dependence of the current. The relative permittivity of the TiOx has been determined as e_r = 24 from the electric field dependence of the TiOx barrier height.

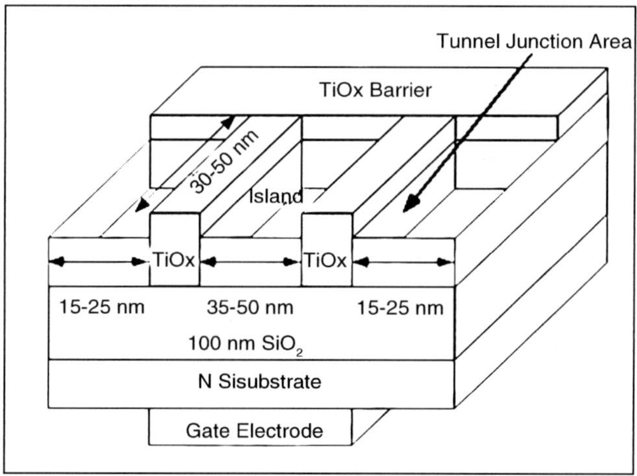

Figure: Schematic of a single electron transistor

A schematic illustration of the SET made by the STM nano- oxidation process. At both ends of the 3 nm thick Ti layer we formed the source and drain ohmic contacts, and on the back side of the n-Si substrate, we formed the gate ohmic contact. At the centre region of the Ti layer, we formed the island region, surrounded by two parallel, narrow TiOx lines, that serve as tunnelling junctions for the SET, and two large TiOx barrier regions. An atomic force microscopy (AFM) image of the island region of a fabricated SET.

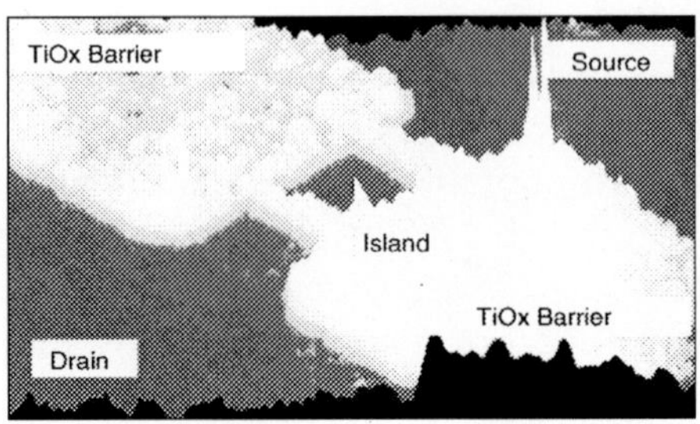

Figure: AFM image of a single electron transistor made by the STM nano-oxidation process

Typical sizes of the TiOx lines are 15-25 nm widths and 30-50 nm lengths. Typical island sizes are 30-50 nm by 35-50 nm. The most important feature of this structure is the small tunnel junction. The junction area corresponds to the cross section of the TiOx line, and is as small as 2-3 nm (the thickness of the Ti layer) by 30-50 nm (the length of the TiOx line). The deposited Ti layer is as thin as 3 nm, and the surface of the Ti layer is naturally oxidized to a depth of ~1 nm. Thus, the intrinsic Ti layer thickness is considered to be less than 3 nm. Owing to this small tunnelling junction area, the tunnel capacitance becomes as small as 10^{-19} F, which allows the SET to be operated at room temperature.

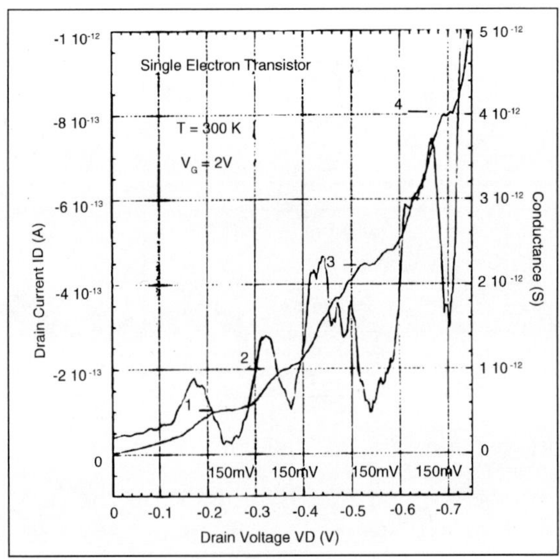

Figure: Drain current v. drain voltage characteristics of the SET at 300 K

The use up current-voltage characteristics of the SET were measured at room temperature. The gate bias was set to 2 V. The solid lines shows the current of the SET, and the dashed line shows the conductance of the SET. Between the drain bias of 0 V and -0.75 V, four clear Coulomb staircases with a ~150 mV period are observed. The conductance oscillates with the increase of the drain bias with almost the same 150 mV period. The lower peaks of the conductance oscillation correspond to the flat regions of the current of the Coulomb staircase.

The Coulomb staircase, may be attributed to the asymmetrical structure of the two tunnelling junctions. One TiOx tunnelling junction has a width of 18 nm, while the other junction is 27 nm wide. Due to this difference in junction widths, each tunnelling junction has different values of conductance and capacitance, which produces the Coulomb staircase.

The height of the Coulomb steps becomes larger with larger applied drain bias. This may be attributed to the increase of the tunnelling probability of the electron through the TiOx tunnelling barrier. The Fowler-Nordheim tunnelling current increases as the applied drain bias lowers the height of the TiOx tunnelling barrier.

The drain current v. gate bias characteristics with 150 mV drain bias at room temperature exhibit clear current oscillations with a period of ~460 mV, implying a periodic Coulomb oscillation of the current. The tunnelling capacitance (C_t) and gate capacitance (C_g) could be roughly estimated from the period of the Coulomb staircase and oscillation. Their values were found to be $C_t = \sim 3.6 \times 10^{-19}$ F and $C_g = \sim 3.5 \times 10^{-19}$ F. These estimated values of the capacitances coincide well with the calculated capacitances from the SET's structural parameters. These results confirm the existence of Coulomb blockade phenomena at room temperature, and are due to the small dimensions of the SET island formed by the STM nano-oxidation process.

In conclusion, we have succeeded in fabricating a room temperature-operable single electron transistor using the STM nano-oxidation process. The SET shows a Coulomb staircase with periods of 150 mV at a temperature of 300 K. The Coulomb gap and staircase observed at high temperatures are attributed to the small tunnelling junction area made by the STM nano-oxidation process. The fabrication process is quite easy and could be applicable to many kinds of devices.

SIMULATION OF SINGLE ELECTRON DEVICES

Logical solutions of the carrier transport in single-electron circuits are only possible for circuits with few junctions such as the double junction or SET transistor, or symmetric devices such as a one-dimensional array of tunnel junctions. For more complex circuits and for the consequences of co-tunnelling the obvious thing to do is numerical simulation. A SET circuit may be described with a Master Equation (ME) which is a conservation law for all probabilities of states a circuit can occupy. A ME is a standard means of describing stochastic processes. Basically there are two distinct ways for a numerical solution of a ME. Either one tries to solve the ME directly, that is to solve the equation for the probability density function, or one simulates the stochastic process by letting one or more particles jump from state to state according to the transition probabilities. From a sufficient set of samples, the desired statistical properties and quantities can be calculated. We refer to the former as the ME approach and to the latter as the Monte Carlo (MC) technique.

Only few publications deal with the simulation of single-electron devices. N. Bakhvalov *et al.* was the first to follow a MC approach, and E. Ben-Jacob *et al.* suggested a ME technique, as an appropriate technique applicable to single-electron devices. From time to time simulation results were published, but implementation details, limitations and assumptions of employed simulators were not always obvious. S. Roy explained in more detail his MC analysis tools which were developed from the study of linear arrays of tunnel junctions. He briefly explains an interesting semi-automatic approach to determine stable and instable regions of zero temperature operation. It is based on the critical voltage technique. L. Fonseca *et al.* developed a ME simulator called SENECA which is explained in. They focused especially on co-tunnelling and how to simulate it correctly and efficiently. M. Kirihara *et al.* describe briefly the MC simulation technique and give simulation results of basic logic circuits. H. Fukui *et al.* and S. Amakawa *et al.* also focused on logic circuits and studied in particular single-electron inverters. They developed similar tools to the one S. Roy describes, namely the calculation of stable and instable operation regions for zero temperature and a MC simulator. R. Chen developed the single-electron MC simulator MOSES which is described briefly in.

REFERENCES

- Awasthi, K P : *Advanced Magnetic Nanostructures*, Cyber Tech Publications, Delhi, 2008.
- Datta, S. : *Electronic Transport in Mesoscopic Systems*,Cambridge University Press, Cambridge, 1995.
- Drexler E.: *"Engines of Creation : The Coming Era of Nanotechnology."* Anchor Books New York, 1986.
- Kapoor, Manish : *BioMEMS and Biomedical Nanotechnology*, Oxford Book Company, Delhi, 2012.
- Julian W. Gardner, and Vijay K. Varadan: *Microsensors, MEMS and Smart Devices*, Wiley, NY, 2001.
- Mathur, K P : *Nanoscience and Technology*, Rajat, Delhi, 2007.

6

Carbon Nanotubes Devices

INTRODUCTION

Carbon nanotubes (CNTs) are an allotrope of carbon. They take the form of cylindrical carbon molecules and have novel properties that make them potentially useful in a wide variety of applications in nanotechnology, electronics, optics and other fields of materials science. They exhibit extraordinary strength and unique electrical properties, and are efficient conductors of heat. Inorganic nanotubes have also been synthesized. Nanotubes are members of the fullerene structural family, which also includes buckyballs. Whereas buckyballs are spherical in shape, a nanotube is cylindrical, with at least one end typically capped with a hemisphere of the buckyball structure.

Their name is derived from their size, since the diameter of a nanotube is on the order of a few nanometers (approximately 50,000 times smaller than the width of a human hair), while they can be up to several millimetres in length. There are two main types of nanotubes: single-walled nanotubes (SWNTs) and multi-walled nanotubes (MWNTs)

Nanotubes are members of the fullerene structural family, which also includes the spherical buckyballs, and the ends of a nanotube may be capped with a hemisphere of the buckyball structure. Their name is derived from their long, hollow structure with the walls formed by one-atom-thick sheets of carbon, called graphene. These sheets are rolled at specific and discrete ("chiral") angles, and the combination of the rolling angle and radius decides the nanotube properties; for example, whether the individual nanotube shell is a metal or semiconductor. Nanotubes are categorized as single-walled nanotubes (SWNTs) and multi-walled nanotubes (MWNTs). Individual nanotubes naturally align themselves into "ropes" held together by van der Waals forces, more specifically, pi-stacking.

Applied quantum chemistry, specifically, orbital hybridization best describes chemical bonding in nanotubes. The chemical bonding of nanotubes is composed entirely of sp^2 bonds, similar to those of graphite. These bonds, which are stronger than the sp^3 bonds found in alkanes and diamond, provide nanotubes with their unique strength.

WHAT IS A CARBON NANOTUBE?

A Carbon Nanotube is a tube-shaped fabric, made of carbon, having a diameter measuring on the nanometer scale. A nanometer is one-billionth of a meter, or about one ten-thousandth of the thickness of a human hair. The graphite layer appears somewhat like a rolled-up chicken wire with a continuous unbroken hexagonal mesh and carbon molecules at the apexes of the hexagons.

Carbon Nanotubes have many structures, differing in length, thickness, and in the type of helicity and number of layers. Although they are formed from essentially the same graphite sheet, their electrical characteristics differ depending on these variations, acting either as metals or as semiconductors.

As a group, Carbon Nanotubes typically have diameters ranging from <1 nm up to 50 nm. Their lengths are typically several microns, but recent advancements have made the nanotubes much longer, and measured in centimetres.

Carbon Nanotubes can be categorized by their structures:

- Single-wall Nanotubes (SWNT)
- Multi-wall Nanotubes (MWNT)
- Double-wall Nanotubes (DWNT).

What are the Properties of a Carbon Nanotube?

The intrinsic mechanical and transport properties of Carbon Nanotubes make them the ultimate carbon fibres. The following tables compare these properties to other engineering materials.

Overall, Carbon Nanotubes show a unique combination of stiffness, strength, and tenacity compared to other fibre materials which usually lack one or more of these properties.

Thermal and electrical conductivity are also very high, and comparable to other conductive materials.

Table : Mechanical Properties of Engineering Fibres

Fibre Material	*Specific Density*	*E (TPa)*	*Strength (GPa)*	*Strain at Break (%)*
Carbon Nanotube	1.3 - 2	1	10 - 60	10
HS Steel	7.8	0.2	4.1	< 10
Carbon Fibre - PAN	1.7 - 2	0.2 - 0.6	1.7 - 5	0.3 - 2.4
Carbon Fibre - Pitch	2 - 2.2	0.4 - 0.96	2.2 - 3.3	0.27 - 0.6
E/S - glass	2.5	0.07/0.08	2.4/4.5	4.8
Kevlar* 49	1.4	0.13	3.6 - 4.1	2.8

Kevlar is a registered trademark of DuPont.

Table : Transport Properties of Conductive Materials

Material	*Thermal Conductivity (W/m.k)*	*Electrical Conductivity*
Carbon Nanotubes	> 3000	106 - 107
Copper	400	6 × 107
Carbon Fibre - Pitch	1000	2 - 8.5 × 106
Carbon Fibre - PAN	8 - 105	6.5 - 14 × 106

What are the Potential Applications for Carbon Nanotubes?

Carbon Nanotube Technology can be used for a wide range of new and existing applications:

- Conductive plastics
- Structural composite materials
- Flat-panel displays
- Gas storage
- Antifouling paint
- Micro- and nano-electronics
- Radar-absorbing coating
- Technical textiles
- Ultra-capacitors
- Atomic Force Microscope (AFM) tips
- Batteries with improved lifetime
- Biosensors for harmful gases
- Extra strong fibres.

How Does Nanocyl Produce Carbon Nanotubes?

Nanocyl uses the "Catalytic Carbon Vapour Deposition" technique for producing Carbon Nanotube Technologies. This proven industrial process is well known for its reliability and scalability. It involves growing nanotubes on substrates, thus enabling uniform, large-scale production of the highest-quality carbon nanotubes worldwide.

FULLERENES

Fullerenes consist of 20 hexagonal and 12 pentagonal rings as the basis of an icosohedral symmetry closed cage structure. Each carbon atom is bonded to three others and is sp2 hybridised. The C60 molecule has two bond lengths - the 6:6 ring bonds can be considered "double bonds" and are shorter than the 6:5 bonds.C60 is not "superaromatic" as it tends to avoid double bonds in the pentagonal rings, resulting in poor electron delocalisation. As a result, C60 behaves like an electron deficient alkene, and reacts readily with electron rich species. The geodesic and electronic bonding factors in the structure account for the stability of the molecule. In theory, an infinite number of fullerenes can exist, their structure based on pentagonal and hexagonal rings, constructed according to rules for making icosahedra.

A Mechanism for Fullerene Formation

Although many mechanisms have been described, only the "pentagon road" appears to explain high yields of C60. An extract taken from an address by Professor Richard Smalley to the Robert A. Welch Foundation 39th Conference on Chemical Research: Nanophase Chemistry: Self-Assembly of Fullerene Tubes and Balls.

"At the moment there is only one mechanism that appears to be consistent with yields of C60 as high as 30-40%. It is known as the Pentagon Road. In this view, the yield is high because clustering continues in a hot enough region to permit the growing clusters to anneal to the minimum energy path: one where the graphene sheet (a) is made up solely of pentagons and hexagons, (b) has as many pentagons as possible, while (c) avoiding structures where two pentagons are adjacent.

If the pentagon rule structures really are the lowest energy forms for any open carbon network, then one can readily imagine that high-yield synthesis of C60 may be possible. In principal, all one needs to do is adjust the conditions of the carbon cluster growth such that each open cluster has ample time to anneal into its favoured pentagon rule structure before it grows further. This path through the kinetics is the Pentagon Road, and constitutes a mechanism of graphene-sheet self-assembly that leads to C60 in very high yield.

The Pentagon Road must be the path that is followed in the high yield conditions. This requires that the cluster growth temperature be high enough to anneal the open structures as they grow so that they follow the pentagon road, but that the temperature not be so high as to permit the extensive rearrangement and mounting of high activation barriers necessary to convert the open pentagon road structure to the closed fullerene.

In the Krätschmer-Huffman (KH) experiment, carbon radicals are produced simply by slow evaporation of the surface of a resistively heated graphite rod. Here the carbon vapour density is far lower than that in pulsed laser vaporization, thus so too is the rate of clustering. Most critically, the rate of cooling of the condensing carbon vapour is much slower in the KH technique. By adjusting the pressure of helium buffer gas around the evaporating graphite rods, one has at least coarse control over this rate of cooling and the clustering rate. At too low a pressure the carbon radicals migrate far away from the hot region around the resistively-heated rod and are too cold when they grow to the region of C60. But with just the right pressure of helium, the clustering in the critical size range occurs at just the right distance from the hot source that the temperature is optimal for the growing clusters to follow the Pentagon Road.

After the KH technique was introduced, it was found at Rice that a simple ac or dc arc would produce C60 and the other fullerenes in good yield as well, and this is now the technique used commercially. Even though the vapourization mechanism of a carbon arc differs markedly from that of a resistively heated carbon rod (because it involves a plasma) the optimum helium pressure for C60 formation is found to be very similar in each case. It is not so much the vapourization technique that matters, but rather the conditions prevailing while the carbon vapour condenses. By adjusting the helium buffer gas pressure, one controls the rate of migration of the carbon vapour away from the hot graphite rod and thereby controls (at least crudely) the effective temperature and carbon radical density in the region where clusters in the size range near C60 are formed.

The Pentagon Road instrument is just one of many by which fullerenes can form, but it may be the only one capable of accounting for the single most telling fact of fullerene formation: that the overall C60 yield can be as high as 40% of all the carbon vapourized. The Pentagon Road explains this as a consequence of annealing the open graphene sheets to their optimal open structure (given by the Pentagon Rule) at a rate faster than their growth, while avoiding rearrangement to a closed fullerene prior to the size of C60. It is a case of picking the reaction conditions to favour a particular product."

PURIFICATION OF CARBON NANOTUBES

Purification of carbon nanotubes (CNTs) generally refers to the separation of CNTs from other entities, such as carbon nanoparticles, amorphous carbon, residual catalyst, and other unwanted species. The classic chemical techniques for purification have been tried, but they have not been found to be effective in removing the undesirable impurities. Three basic techniques have been used with varying degrees of success, namely:

- Gas-phase
- Liquid-phase and
- Intercalation techniques.

Centrifugal Concentration

Generally, a centrifugal separation is necessary to concentrate the single walled nanotubes in low-yield soot before the micro filtration operation, since the nanoparticles easily contaminate membrane filters. The advantage of this technique is that unwanted nanoparticles and amorphous carbon are removed simultaneously and the CNTs are not chemically modified. However 2-3 mol nitric acid is useful for chemically removing impurities.

Sonication to Cut Carbon Nanotubes into Smaller Segments

At the present possible to cut CNTs into smaller segments, by extended sonication in concentrated acid mixtures. The resulting CNTs form a colloidal suspension in solvents. They can be deposited on substrates, or further manipulated in solution, and can have many different functional groups attached to the ends and sides of the CNTs.

Gas Phase

The first successful technique for purification of nanotubes was developed by Thomas Ebbesen and coworkers. Following the demonstration that nanotubes could be selectively attached by oxidizing gases these workers realised that nanoparticles, with their defect rich structures might be oxidised more readily than the relatively perfect nanotubes. They found that a significant relative enrichment of nanotubes could be achieved this way, but only at the expense of losing the majority of the original sample.

A new gas-phase technique has been developed at the NASA Glenn Research Centre to purify gram-scale quantities of single-wall CNTs. This technique, a modification of a gas-phase purification technique previously reported by Smalley and others, uses a combination of high-temperature oxidations and repeated extractions with nitric and hydrochloric acid. This improved procedure significantly reduces the amount of impurities such as residual catalyst, and non-nanotube forms of carbon) within the CNTs, increasing their stability significantly.

Liquid Phase

The current liquid-phase purification procedure follows certain essential steps:

- Preliminary filtration- to get rid of large graphite particles;
- Dissolution in organic solvents and concentrated acids to remove fullerenes and catalyst particles.
- Centrifugal separation
- Microfiltration- and
- Chromatography to either separate multi walled nanotubes and unwanted nanoparticles or single walled nanotubes and the amorphous carbon impurities.

It is significant to keep the CNTs well-separated in solution, so the CNTs are typically dispersed using a surfactant prior to the last stage of separation.

ELECTRONIC PROPERTICS

Properties

Carbon nanotubes are the strongest and stiffest materials yet discovered in terms of tensile strength and elastic modulus respectively. This strength results from the covalent sp^2 bonds formed between the individual carbon atoms. In 2000, a multi-walled carbon nanotube was tested to have a tensile strength of 63 gigapascals (GPa). The individual CNT shells have strengths of up to ~100 GPa, which is in good agreement with quantum/atomistic models. Since carbon nanotubes have a low density for a solid of 1.3 to 1.4 g/cm^3, its specific strength of up to 48,000 $kN \cdot m \cdot kg^{-1}$ is the best of known materials, compared to high-carbon steel's 154 $kN \cdot m \cdot kg^{-1}$.

Under excessive tensile strain, the tubes will undergo plastic deformation, which means the deformation is permanent. This deformation begins at strains of approximately 5% and can increase the maximum strain the tubes undergo before fracture by releasing strain energy. Although the strength of individual CNT shells is extremely high, weak shear interactions between adjacent shells and tubes leads to significant reductions in the effective strength of multi-walled carbon nanotubes and carbon nanotube bundles down to only a few GPa's. This limitation has been recently addressed by applying high-energy electron irradiation, which crosslinks inner shells and tubes, and effectively increases the strength of these materials to ~60 GPa for multi-walled carbon nanotubes and ~17 GPa for double-walled carbon nanotube bundles.

CNTs are not almost as strong under compression. Because of their hollow structure and high aspect ratio, they tend to undergo buckling when placed under compressive, torsional, or bending stress.

Table: Comparison of mechanical properties

Material	*Young's modulus (TPa)*	*Tensile strength (GPa)*	*Elongation at break (%)*
SWNT^E	~1 (from 1 to 5)	13–53	16
Armchair SWNT^T	0.94	126.2	23.1
Zigzag SWNT^T	0.94	94.5	15.6–17.5
Chiral SWNT	0.92		
MWNT^E	0.2–0.8–0.95	11–63–150	
Stainless steel^E	0.186–0.214	0.38–1.55	15–50
Kevlar–29&149^E	0.06–0.18	3.6–3.8	~2

^{E}Experimental observation; ^{T}Theoretical prediction

The axial properties of the nanotube, whereas simple geometrical considerations suggest that carbon nanotubes should be much softer in the radial direction than along the tube axis. Indeed, TEM observation of radial elasticity suggested that even the van der Waals forces can deform two adjacent nanotubes. Nanoindentation experiments, performed by several groups on multiwalled carbon nanotubes and tapping/contact mode atomic force microscope measurement performed on single-walled carbon nanotube, indicated Young's modulus of the order of several GPa confirming that CNTs are indeed rather soft in the radial direction.

Hardness

Standard single-walled carbon nanotubes can withstand a pressure up to 24GPa without deformation. They then undergo a transformation to superhard phase nanotubes. Maximum pressures measured using current experimental techniques are around 55GPa. However, these new superhard phase nanotubes collapse at an even higher, albeit unknown, pressure. The bulk modulus of superhard phase nanotubes is 462 to 546 GPa, even higher than that of diamond(420 GPa for single diamond crystal).

Kinetic Properties

Multi-walled nanotubes are manifold concentric nanotubes precisely nested within one another. These exhibit a striking telescoping property whereby an inner nanotube core may slide, almost without friction, within its outer nanotube shell, thus creating an atomically perfect linear or rotational bearing. This is one of the first true examples of molecular nanotechnology, the precise positioning of atoms to create useful machines. Already, this property has been utilized to create the world's smallest rotational motor. Future applications such as a gigahertz mechanical oscillator are also envisaged.

Electrical Properties

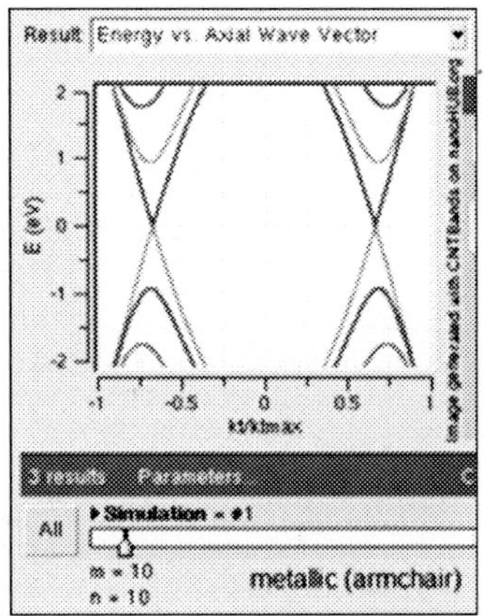

Figure: Band structures computed using tight binding approximation for (6,0) CNT (zigzag, metallic) (10,2) CNT (semiconducting) and (10,10) CNT (armchair, metallic).

Because of the symmetry and unique electronic structure of graphene, the structure of a nanotube strongly affects its electrical properties. For a given (n,m) nanotube, if $n = m$, the nanotube is metallic; if n " m is a multiple of 3, then the nanotube is semiconducting with a very small band gap, otherwise the nanotube is a moderate semiconductor. Thus all armchair ($n = m$) nanotubes are metallic, and nanotubes (6,4), (9,1), etc. are semiconducting.

However, this rule has exceptions, because curvature effects in small diameter carbon nanotubes can strongly influence electrical properties. Thus, a (5,0) SWCNT that should be semiconducting in fact is metallic according to the calculations. Likewise, *vice-versa*— zigzag and chiral SWCNTs with small diameters that should be metallic have finite gap (armchair nanotubes remain metallic). In theory, metallic nanotubes can carry an electric current density of 4×10^9 A/cm^2, which is more than 1,000 times greater than those of metals such as copper, where for copper interconnects current densities are limited by electromigration.

There have been reports of intrinsic superconductivity in carbon nanotubes. Many other experiments, however, found no evidence of superconductivity, and the validity of these claims of intrinsic superconductivity remains a subject of debate.

OPTICAL PROPERTIES

Wave Absorption

One of the more recently researched properties of multi-walled carbon nanotubes (MWNTs) is their wave absorption characteristics, specifically microwave absorption. Interest in this research is due to the current military push for radar absorbing materials (RAM) to better the stealth characteristics of aircraft and other military vehicles. There has been some research on filling MWNTs with metals, such as Fe, Ni, Co, etc., to increase the absorption effectiveness of MWNTs in the microwave regime. Thus far, this research has shown improvements in both maximum absorption and bandwidth of adequate absorption. The reason the absorptive properties changed when filled is that the complex permeability (μ_r) and complex permitivity (ε_r), shown in the equations below, have been shown to vary depending on how the MWNTs are called and what medium they are suspended in. The direct relationship between μ_r, ε_r, and the other system parameters that affect the absorption sample thickness, d, and frequency, f, is shown in the equations below, where Z_{in} is the normalized input impedance. As shown in the equation below, these characteristics vary by frequency. Because of this, it is convenient to set a baseline reflection loss (R.L.) that is deemed effective and determine the bandwidth within a given frequency that produces the desired reflection loss. A common R.L. to use for this bandwidth determination is –10 dB, which corresponds to a loss of over 90% of the incoming wave. This bandwidth is usually maximized at the same time as the absorption is. This is done by satisfying the impedance matching condition, getting $Z_{in} = 1$. In the work done at Beijing Jiaotong University it was found that Fe filled MWNTs exhibited a maximum reflection loss of –22.73 dB and had a bandwidth of 4.22 GHz for a reflection loss of –10 dB.

$$R.L.(dB) = 20\log_{10}\left[\frac{Z_{in}-1}{Z_{in}+1}\right]$$

$$Z_{in} = \sqrt{\frac{\mu_r}{\varepsilon_r}}\tanh\left[j\left(\frac{2\pi fd}{c}\right)\sqrt{\mu_r\varepsilon_r}\right]$$

Thermal Properties

All nanotubes are expected to be very good thermal conductors along the tube, exhibiting a property known as "ballistic conduction", but good insulators laterally to the tube axis. Measurements show that a SWNT has a room-temperature thermal conductivity along its axis of about 3500 $W \cdot m^{-1} \cdot K^{-1}$; compare this to copper, a metal well known for its good thermal conductivity, which transmits 385 $W \cdot m^{-1} \cdot K^{-1}$. A SWNT has a room-temperature thermal conductivity across its axis (in the radial direction) of about 1.52 $W \cdot m^{-1} \cdot K^{-1}$, which is about as thermally conductive as soil. The temperature stability of carbon nanotubes is estimated to be up to 2800 °C in vacuum and about 750 °C in air.

Defects

As by means of any material, the existence of a crystallographic defect affects the material properties. Defects can occur in the form of atomic vacancies. High levels of such defects can lower the tensile strength by up to 85%. An important example is the Stone Wales defect, which creates a pentagon and heptagon pair by rearrangement of the bonds. Because of the very small structure of CNTs, the tensile strength of the tube is dependent on its weakest segment in a similar manner to a chain, where the strength of the weakest link becomes the maximum strength of the chain.

Crystallographic defects also affect the tube's electrical properties. A common result is lowered conductivity through the defective region of the tube. A defect in armchair-type tubes (which can conduct electricity) can cause the surrounding region to become semiconducting, and single monoatomic vacancies induce magnetic properties.

Crystallographic defects strongly affect the tube's thermal properties. Such defects lead to phonon scattering, which in turn increases the relaxation rate of the phonons. This reduces the mean free path and reduces the thermal conductivity of nanotube structures. Phonon transport simulations indicate that substitutional defects such as nitrogen or boron will primarily lead to scattering of high-frequency optical phonons. However, larger-scale defects such as Stone Wales defects cause phonon scattering over a wide range of frequencies, leading to a greater reduction in thermal conductivity.

One-dimensional Transport

Because of the nanoscale dimensions, electrons propagate only along the tube's axis and electron transport involves many quantum effects. Because of this, carbon nanotubes are frequently referred to as "one-dimensional".

Toxicity

The toxicity of carbon nanotubes has been an important question in nanotechnology. Such research has just begun. The data are still fragmentary and subject to criticism. Preliminary results highlight the difficulties in evaluating the toxicity of this heterogeneous material. Parameters such as structure, size distribution, surface area, surface chemistry, surface charge, and agglomeration state as well as purity of the samples, have considerable impact on the reactivity of carbon nanotubes. However, available data clearly show that, under some conditions, nanotubes can cross membrane barriers, which suggests that, if raw materials reach the organs, they can induce harmful effects such as inflammatory and fibrotic reactions. A study led by Alexandra Porter from the University of Cambridge shows that CNTs can enter human cells and accumulate in the cytoplasm, causing cell death.

Results of rodent studies collectively show that regardless of the process by which CNTs were synthesized and the types and amounts of metals they contained, CNTs were capable of producing inflammation, epithelioid granulomas (microscopic nodules), fibrosis, and biochemical/toxicological changes in the lungs. Comparative toxicity studies in which mice were given equal weights of test materials showed that SWCNTs were more toxic than quartz, which is considered a serious occupational health hazard when chronically inhaled. As a control, ultrafine carbon black was shown to produce minimal lung responses.

The needle-like fibre shape of CNTs is similar to asbestos fibres. This raises the idea that widespread use of carbon nanotubes may lead to pleural mesothelioma, a cancer of the lining of the lungs or peritoneal mesothelioma, a cancer of the lining of the abdomen (both caused by exposure to asbestos). A recently published pilot study supports this prediction. Scientists exposed the mesothelial lining of the body cavity of mice to long multiwalled carbon nanotubes and observed asbestos-like, length-dependent, pathogenic behaviour that included inflammation and formation of lesions known as granulomas. Authors of the study conclude:

This is of considerable importance, because research and business communities continue to invest heavily in carbon nanotubes for a wide range of products under the assumption that they are no more hazardous than graphite. Our results suggest the need for further research and great caution before introducing such products into the market if long-term harm is to be avoided.

According to co-author Dr. Andrew Maynard:

This revise is exactly the kind of strategic, highly focused research needed to ensure the safe and responsible development of nanotechnology. It looks at a specific nanoscale material expected to have widespread commercial applications and asks specific questions about a specific health hazard. Even though scientists have been raising concerns about the safety of long, thin carbon nanotubes for over a decade, none of the research needs in the current U.S. federal nanotechnology environment, health and safety risk research strategy address this question.

Although further research is required, the available data suggests that under certain conditions, especially those involving chronic exposure, carbon nanotubes can pose a serious risk to human health.

SYNTHESIS OF CARBON NANOTUBES

Carbon nanotubes are one of the most intriguing new materials with extraordinary properties being discovered in the last decade. The unique structure of carbon nanotubes provides nanotubes with extraordinary mechanical and electrical properties. The outstanding properties that these materials possess have opened new interesting researches areas in nanoscience and nanotechnology. Although nanotubes are very promising in a wide variety of fields, application of individual nanotubes for large scale production has been limited. The main roadblocks, which hinder its use, are limited understanding of its synthesis and electrical properties which lead to difficulty in structure control, existence of impurities, and poor processability. This book makes an attempt to provide indepth study and analysis of various synthesis techniques, processing techniques and characterization of carbon nanotubes that will lead to the increased applications of carbon nanotubes.

What is Carbon Nanotube Synthesis?

What is Carbon Nanotube Synthesis, and What are its Implications for Modern Science?

Whereas the mystery and implications for science surrounding carbon nanotubes in modern science have gained them immense popularity with scientists and researchers in myriad fields. The age old question -what is carbon nanotube synthesis? is quickly gaining ground in modern science as well. And its implications may far outweigh and outreach those of just carbon nanotubes, since this process of synthesis may very well open doors for science and for the use of carbon nanotubes that no one ever thought possible.

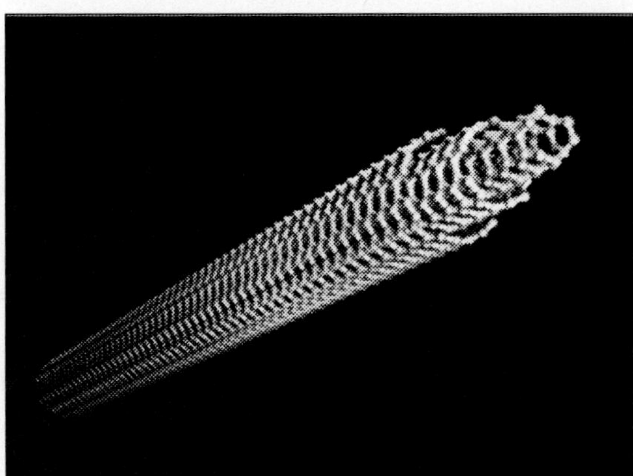

Carbon nanotubes are essentially constructed of a lattice work sheet of graphite that is rolled into a cylindrical shape. These carbon nanotubes are feather light in their construction and much stronger than many other properties used in not only nanotechnology but also in electronics, optics, additional areas of materials science, architecture and even the construction of motor vehicles, space vehicles and other materials useful in a wide array of traditional fields of research and modern living.

Carbon nanotubes are also known to have any unique electrical properties and have even been tested and proven as super efficient conductors of heat, otherwise referred to as thermal conductors. It is only their potential to have some toxic properties that has limited their widespread use and distribution for construction and other purposes.

Carbon nanotubes come in a variety of forms, including single walled carbon nanotubes, double walled carbon nanotubes and multi walled carbon nanotubes. Each type of carbon nanotubes has its own set of unique properties that make it useful for different areas of science. For instance, the single walled carbon nanotube has especially strong electric properties that the other varieties lack.

Carbon nanotube synthesis is incredibly important to the research of the potential uses of carbon nanotubes because nanotube synthesis is the technique of production used to create carbon nanotubes. The faster the rate of synthesis, or production, the more carbon nanotubes available for researches and scientists to use in testing for the various possible uses of carbon nanotubes in standard modern day living, current science and potential future applications.

Even though carbon nanotube synthesis can occur naturally through flame synthesis, the most common techniques of used in modern science include laser ablation, arc discharge and chemical vapour deposition.

The arc discharge technique of nanotube synthesis was initially developed by accident. In 1991, an arc discharge that was meant for producing fullerenes, produced, along with the fullerenes, carbon nanotubes. In 1992, the experiment was reproduced by two researchers at the Fundamental Research Laboratory of the NEC.

This process involves high temperatures for discharge which cause the negative electrode in the arc to force sublimate the carbon nanotubes contained within. Additionally, this process has a yield by weight of more than 30 per cent and it is able to produce both multi walled (including double walled) and single walled carbon nanotubes. This technique for carbon nanotube synthesis is most popular because it was the first technique discovered and used. The carbon nanotubes produced through this technique of synthesis often have few defects, if any, and lengths of more than 50 micrometers are able to be produced.

But arc discharge is not the only type of nanotube synthesis available to researchers and scientists. Other techniques have been developed in order to provide a wide variety of acceptable outcomes—acceptable outcomes being defined as producing usable carbon nanotubes in large quantities with minimal defects. One of the additional techniques of nanotube synthesis is laser ablation technique. When scientists use laser ablation synthesis technique for carbon nanotubes, both a pulsing laser and an inert gas are paired together in a tightly sealed reactor with high temperatures. The pulsed laser is used to vapourize the graphite as the gas is allowed to slowly seep into the reactor chamber. The graphite vapour condenses as the reactor surfaces of the chamber cool down and nanotubes form out of the vapours.

In the laser ablation technique, scientists most often produce single walled carbon nanotubes and the yield is approximately 70 per cent by weight. Although this technique yields much more by weight than the other two available techniques, it is much more expensive to use and many scientists and researchers opt in favour of the less expensive techniques. The concluding popular technique of carbon nanotube synthesis is the chemical vapour desposition technique. Although this technique has been used to synthesize carbon since 1959, the chemical catalytic vapour deposition technique of producing carbon nanotubes was not first used until 1993, nearly forty years after its first use for synthesis and two years after the first nanotubes were synthesized using the arc discharge technique.

This technique was further developed in 2007 by University of Cincinnati researchers and at that time began producing carbon nanotubes at lengths of 18 millimetres, much longer than those produced by the arc discharge technique at only fifty micrometers.

In chemical vapour desposition, a substrate made of iron, cobalt, nickel or a combination of any of these metals is heated to precisely 700 degrees Celsius inside a chemical reactor. Process gases such as hydrogen, nitrogen or ammonia along with a carbon gas like methane, ethanol or acetylene, is leaked into the reactor chamber. The meeting of the gases at the substrate cause a reaction that breaks the carbon gases apart and forces the particles to the sides of the chamber, where the carbon nanotubes are formed. This technique of nanotube synthesis is the most popular for commercial production because it allows for nanotubes to be grown from and on a specified substrate.

The most important factors relating to nanotube synthesis are how fast the nanotubes can be grown, how many can be made at any one time, how much the process to be used will cost and most importantly, how few structural defects will be present in the new carbon nanotubes that have been synthesized. All of these factors determine the future of carbon nanotubes and the development of better technique for synthesis will determine the future application of carbon nanotubes in everything that surrounds us.

COMMON SYNTHESIS TECHNIQUES

Arc Discharge and Laser Ablation

The primary technique that was successfully used to synthesize CNTs in small quantities was the arc discharge technique. Opposing anode and cathode terminals made of 6-mm and 9-mm graphite rods respectively are placed in an inert environment (He or Ar at ~500 Torr). A strong current, typically around 100 A (DC or AC), is passed between the terminals generating arc-induced plasma that evaporates the carbon atoms in the graphite. The nanotubes grow from the surface of these terminals. A catalyst can be introduced into the graphite terminal. Although MWNTs can be formed without a catalyst, it has been found that SWNTs can only be formed with the use of a metal catalyst such as iron or cobalt. A process called Laser Ablation, first developed in 1995, uses a similar principle to produce nanotubes. Carbon is evaporated at high temperatures from a graphite target using a powerful and focused laser beam. In the most basic laser ablation technique, a 1.25-cm diameter graphite target is placed in a 2.5-cm diameter, 50-cm long quartz tube in a furnace controlled at 1200°C and filled with 99.99% pure argon to a pressure of 500 Torr. A pulsed Nd:Yag laser[W] beam at 250mJ (10 Hz) is focused using a circular lens and the beam is swept uniformly across the graphite target surface. The nanotubes, mixed with undesired amorphous carbon, are collected on a cooled substrate at the end of the chamber.

Both of these techniques have limited potential for scale-up. Solid graphite must be evaporated at >3000°C to source the carbon needed, the nanotubes produced are in an entangled form, and extensive purification is required to remove the amorphous carbon and fullerenes that are naturally produced in the process.

Chemical Vapour Deposition

Chemical Vapour Deposition (CVD) has the highest potential for mass production of carbon nanotubes. It is a versatile technique that can produce bulk amounts of defect-free CNTs at relatively low temperatures.

The generalized process for producing CNTs using CVD is outlined here. A substrate material (*e.g.* alumina, quartz), is cleaned in preparation for the catalyst deposition. A porous substrate may be desired, so electrochemical etching with a hydrofluoric acid/methanol solution may be performed. Nanotubes can grow at a higher rate on a porous substrate, suggesting that carbon can diffuse through the porous substrate layer and feed growing nanotubes.

A catalyst (*e.g.* iron, nickel) is deposited onto the substrate by thermal evaporation. The furnace is raised to a temperature between 500-1200°C and a hydrocarbon gas such as acetylene, ethylene, or carbon monoxide is slowly pumped into the reactor. At these high temperatures carbon dissociates from the feedstock molecules and diffuses onto the catalyst. The atoms arrange themselves into a sheet of nanotubes on the substrate, combined with impurities such as amorphous carbon, fullerenes, as well as the catalyst material. In most cases these impurities must be removed using a purification step. An acid treatment followed by sonification[W] is popular.

Growth Mechanisms

The enlargement mechanisms of carbon nanotubes is not fully understood. Many different models of carbon nanotube growth have been proposed, and the most accepted model which is consistent with solid-phase growth is described here. Small carbon atoms diffuse around a large metal catalyst atom which is assumed to have a spherical shape. The carbon nucleates on the sides of the metal, forming a tube structure. The location of the catalyst material during formation depends on the strength of the interatomic forces between the metal atoms and the substrate: if the metal is weakly bonded to the substrate it may remain in place as a nanotube forms vertically from the substrate (this mechanism is entitled *base growth*); otherwise the metal may be carried at the tip as the tube grows beneath (entitled *tip growth*).

Catalyst

The catalyst provides a base around which the nanotubes can grow and the material chosen for CVD synthesis can wildly vary the yield and quality of CNTs produced. Conventionally, the catalysts are transition-metal nanoparticles (usually iron, nickel, molybdenum) deposited on a support material such as alumina or silica. However vapourized catalysts have also been used such that the nanotubes form in the gaseous phase and are condensed on cold surfaces. The catalyst becomes deactivated when it is coated by undesirable amorphous carbon, stopping the CNT growth process. An ideal catalyst material should have large surface area and pore volume (properties that must not sinter out at high temperatures), as well as strong metal-support interactions.

The average diameter of the carbon nanotubes produced by CVD is directly related to the size of the catalyst particles. This relationship can be illustrated by the growth mechanisms. As the nanotubes tend to grow perpendicular to the surface of the catalyst, the diameter of the particle dictates the diameter of the nanotube. Thus CVD provides direct control over the size of carbon nanotubes, making it a viable option for materials processing.

Carbon Feedstock

In CVD the carbon is supplied by a hydrocarbon gas. The reactivity of the molecule is an important factor determining which hydrocarbon to use. The reactivity is dictated by the strength of the covalent bonds between the carbon and the hydrogen atoms - that is, a more reactive molecule requires less energy to decompose into its components. This decomposition is achieved by elevating the temperature, providing the molecule with kinetic energy to help break the bonds. Thus, ethylene (C_2H_4) and acetylene (C_2H_2) are usually chosen over methane and carbon monoxide as the carbon feedstock because of their higher reactivity. Benzene, methanol, and ethanol have also been successfully used for various experiments.

Purification

The carbon nanotubes can be separated from impurities and catalyst particles using a variety of techniques. The most common post-production steps are:

- Dispersion, followed by
- Acid reflux and finally
- Micro-filtration.

Dispersion is performed by sonication of the nanotubes with a surfactant, such as Triton X100 (a type of detergent). A solution of 0.01ml of Triton X100 and 0.08 litres of water per gram of carbon is used to seperate the nanotubes from the impurities. The acid reflux treatment is often done using 0.4 litres of nitric acid per gram of dispersion product over 10 hours. This step is slow, and therefore requires several processing lines to maintain productivity. Filtration is performed through a polytetrafluoroethylene (PTFE) membrane over approximately 20 minutes, but may require many runs to achieve the desired purity. On average, about 90% of the nanotube yield can be recovered after the purification process.

Ordered Nanotubes by CVD

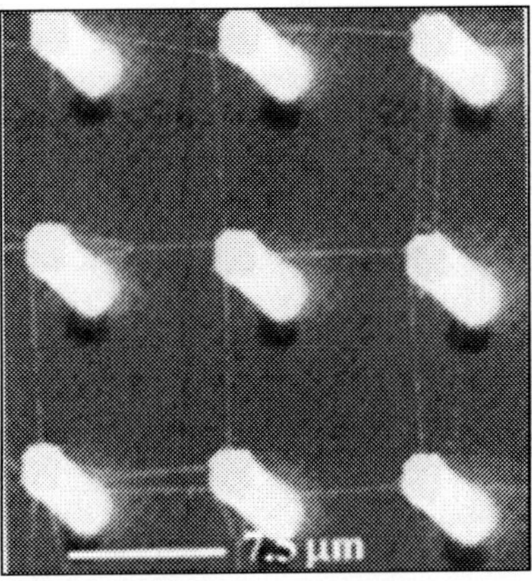

Figure: A CNT array with suspended SWNTs between pillars

Current advances have shown that nanotubes can be synthesized in an ordered manner. Catalysts can be deposited onto a porous silicon substrate in a pattern through a shadow mask[W] so that nanotubes can only nucleate at specific sites. The result is known as a 'nanotube forest' because the nanotubes grow vertically from the catalyst sites, with each tube held up by van der Waals forces[W] from surrounding tubes. Directed growth of SWNTs has also been demonstrated. Catalyst material was printed onto the tops of silicon pillars selectively, and after CVD, SWNTs were shown to grow suspended from pillar to pillar. The growth of carbon nanotubes in a controlled way is an essential skill to master in order to effectively produce CNT-based nanotechnologies.

OPTIMIZATION OF CVD SYNTHESIS

As CVD shows the most promise for mass production of carbon nanotubes, extensive research has been done to improve the yield and energy efficiency of the synthesis process. Manufacturing efficiency can be quantified in a very simple way:

Efficiency (manufacturing) = value added time/total time

The value added time encompasses all the time in a process that is put into creating a marketable product. To increase the efficiency either time must be used more productively, or the process must be completed in less time overall. In addition, increasing the yield of useful nanotubes increases the efficiency as the same amount of time adds more value to the manufacturing process. A more efficient process also produces less waste, as time must be spent consolidating the waste material post-production.

By extension the overall cost of the synthesis process is decreased with increasing efficiency. The ultimate goal is to minimize the cost so that nanotubes can be used in a variety of applications.

Water Assisted CVD ("Supergrowth")

A means is limited in its ability to nucleate CNTs. The reason that the catalysts become deactivated is that amorphous carbon naturally forms around the catalyst over time, smothering the available nucleation sites. Models of catalyst decay have shown strong correlation to models of radioactive decay[W]. Therefore the height of the forest H(t) under set growth conditions can be described by exponential decay,

$$H(t) = \beta\tau(1 - exp(- t/\tau))$$

where β is the initial growth rate, τ is the catalyst lifetime, and $\beta\tau$ intuitively gives the maximum height of the forest. Adding a small amount of water vapour (in the ppm range) to the ambient atmosphere can increase the lifetime of the catalyst material. Remarkably this exponential decay model has predicted catalyst behaviour very precisely. The catalyst lifetime, and therefore the height of the forest, can be maximized at a specific water-ethylene ratio. Recent experiments even show that this optimum ratio is independent of growth time and temperature, which points to the fact that water is changing a fundamental aspect of the carbon nanotube growth process.

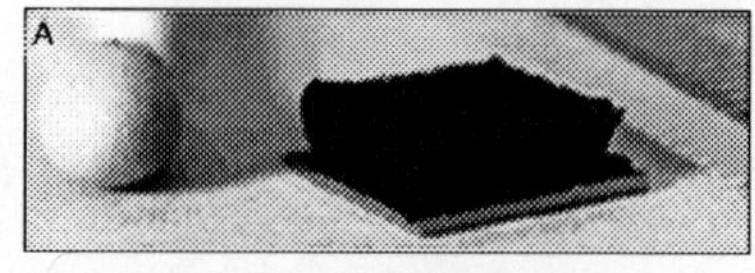

Figure: CNT forest produced through water-assisted CVD

Even though the effect of water is poorly understood, it has been hypothesized that water etches amorphous carbon without etching the nanotubes. The effect is that nucleation sites are kept free of amorphous carbon so that the catalysts are not inactivated while nanotubes are unaffected, fostering increased growth. Supergrowth has generated super-pure (99.98% purity) vertically aligned SWNT forests in the millimetre range – this is an unprecedented achievement in ordered CNT synthesis.

CVD with No Catalyst

Catalyst materials can encourage nanotube growth; however they must be purified out at the end of the synthesis process. It has been shown that, remarkably, an external catalyst is not necessarily needed to produce carbon nanotubes. This eliminates the need for purification and thus increases the efficiency of the process. A simple technique has been developed by Fadour *et al.* to grow CNTs on etched stainless steel. At an operating temperature of about 700-850°C this technique shows promise for increasing synthesis efficiency by eliminating the need for a pretreated catalyst and by depositing nanotubes directly onto a steel substrate. The process is performed as follows: substrate cleaning for 30 minutes with acetone in an ultrasonic bath, substrate etching in 35-38% hydrochloric acid, substrate heat treatment at 850°C, acetylene injection at 700°C and hold in nitrogen gas, followed by cool down.

Hot-Filament CVD

One of the main sources of inefficiency in any processing system is the need to sustain elevated temperatures for synthesis, so reducing the operating temperature of the CVD process would greatly enhance energy efficiency. Chemical Vapour Deposition has been demonstrated at low temperatures (~550°C) with the assistance of a heated filament (a process termed HF-CVD). A filament is pre-heated in a near-vacuum atmosphere to about 2000°C, and is used to heat the inflowing gases in the CVD chamber. The addition of this filament enables synthesis of highly pure nanotubes, but increasing the temperature above 750°C was shown to result in the deposition of undesired amorphous carbon. This alteration to the conventional CVD process has the potential to greatly enhance synthesis efficiency.

Plasma Enhanced CVD

The nucleation speed of carbon nanotubes can be dramatically increased by ionizing the incoming gases into plasma[W] phase. Plasma-Enhanced Chemical Vapour Deposition (PECVD) allows carbon to diffuse more readily and increases the yield of nanotubes. The plasma can be generated by direct current across two terminals, or more commonly by exciting the gas through wire coils with a high frequency current (13.56 MHz is the standard in industry). Although strong currents are required to generate the plasma, it has been shown that the operating temperature in the furnace can be reduced by half or more. The same experiment showed that applying an electric field causes an alignment among the growing nanotubes and the strength of the field is proportional to the degree of alignment. In addition, when NH_3 is pumped in with the hydrocarbon gas, the rate of nanotube nucleation is increased. This is due to the fact that NH_3 etches amorphous carbon most strongly at bias voltages above -300V. These results are promising as they provide a well-established technique for increasing synthesis efficiency by producing higher yields of aligned CNTs at relatively low temperatures.

CARBON NANOTUBES OUTPERFORM COPPER NANOWIRES AS INTERCONNECTS

To better understand and more precisely measure the key characteristics of both copper nanowires and carbon nanotube bundles, the researchers used advanced quantum-mechanical computer modelling to run vast simulations on a high-powered supercomputer. It is the first such study to examine copper nanowire using quantum mechanics rather than empirical laws. After crunching numbers for months with the help of Rensselaer's Computational Centre for Nanotechnology Innovations, the most powerful university-based supercomputer in the world, the research team concluded that the carbon nanotube bundles boasted a much smaller electrical resistance than the copper nanowires. This lower resistance suggests carbon nanotube bundles would therefore be better suited for interconnect applications.

"With this study, we have provided a road map for accurately comparing the performance of copper wire and carbon nanotube wire," said Saroj Nayak, an associate professor in Rensselaer's Department of Department of Physics, Applied Physics, and Astronomy, who led the research team. "Given the data we collected, we believe that carbon nanotubes at 45 nanometers will outperform copper nanowire."

The research results will be featured in the March issue of Journal of Physics: Condensed Matter. Because of the nanoscale size of interconnects, they are subject to quantum phenomena that are not apparent and not visible at the macroscale, Nayak said. Empirical and semi-classical laws cannot account for such phenomena that take place on the atomic and subatomic level, and, as a result, models and simulations based on those models cannot be used to accurately predict the behaviour and performance of copper nanowire. Using quantum mechanics, which deals with physics at the atomic level, is more difficult but allows for a fuller, more accurate model.

"If you go to the nanoscale, objects do not behave as they do at the macroscale," Nayak said. "Looking forward to the future of computers, it is essential that we solve problems with quantum mechanics to obtain the most complete, reliable data possible."

The dimension of computer chips has shrunk dramatically over the past decade, but has recently hit a bottleneck, Nayak said. Interconnects, the tiny copper wires that transport electricity and information around the chip and to other chips, have also shrunk. As interconnects get smaller, the copper's resistance increases and its ability to conduct electricity degrades. This means fewer electrons are able to pass through the copper successfully, and any lingering electrons are expressed as heat. This heat can have negative effects on both a computer chip's speed and performance.

Researchers in both industry and academia are looking for alternative materials to replace copper as interconnects. Carbon nanotube bundles are a popular possible successor to copper, Nayak said, because of the material's excellent conductivity and mechanical integrity. It is generally accepted that a quality replacement for copper must be discovered and perfected in the next five to 10 years in order to further perpetuate Moore's Law – an industry mantra that states the number of transistors on a computer chip, and thus the chip's speed, should double every 18-24 months.

Nayak said there are still many challenges to overcome before mass-produced carbon nanotube interconnects can be realised. There are still issues concerning the cost of efficiency of creating bulk carbon nanotubes, and growing nanotubes that are solely metallic rather than their current state being of partially metallic and partially semiconductor. More study will also be required, he said, to model and simulate the effects of imperfections in carbon nanotubes on the electrical resistance, contact resistance, capacitance, and other vital characteristics of a nanotube interconnect.

CARBON NANOTUBE FETS

Carbon nanotubes (NTs) are a new form of carbon with unique electrical and mechanical properties. They can be considered as the result of folding graphite layers into carbon cylinders and may be composed of a single shell–single wall nanotubes (SWNTs), or of several shells—multi-wall nanotubes (MWNTs). Depending on the folding angle and the diameter, nanotubes can be metallic or semiconducting. Simple theory also shows that the band gap of semiconducting NTs decreases with increasing diameter. These predictions have been verified in recent scanning tunnelling spectroscopy experiments. Their interesting electronic structure makes carbon nanotubes ideal candidates for novel molecular devices. Metallic NTs, for example, were utilized as Coulomb islands in single-electron transistors and, very recently, Tans and coworkers built a molecular field-effect transistor (FET) with a semi-conducting nanotube.

Carbon nanotubes are cylindrical sheets of one or more concentric layers of carbon atoms. Experiments have shown that the tubes can either have metallic or semiconducting properties. Their band structure depends on the position of the carbon atoms forming the tube. Particularly single-wall carbon nanotubes show superior electrical properties and are considered promising candidates for future nanoelectronic applications, either as interconnects or active devices. Semiconducting nanotubes can be used as active elements in field-effect transistor (FET) designs. Two possible applications of carbon nanotubes as transistor devices. Single-wall carbon nanotubes are ballistic conductors, so the current is governed by LANDAUER's equation. This, however, implies that the minimum resistance of a metallic nanotube is $h / 4q^2 \approx 6.5k\Omega.$. It is now generally accepted that the transport in the tubes is dominated by SCHOTTKY barriers at the metal contacts.

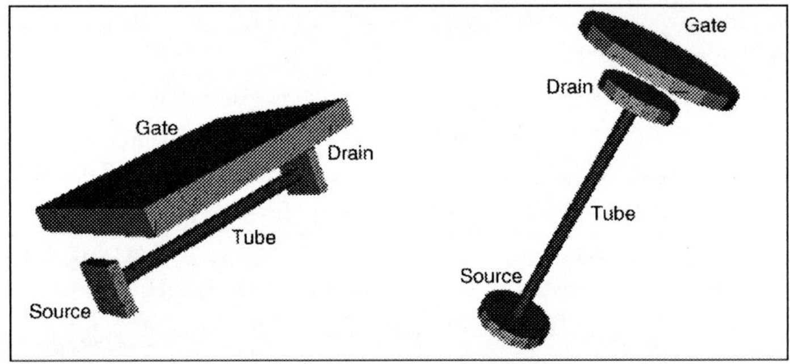

Figure : A lateral (left) and an axial (right) carbon nanotube FET.

Electronic structure of carbon nanotubes

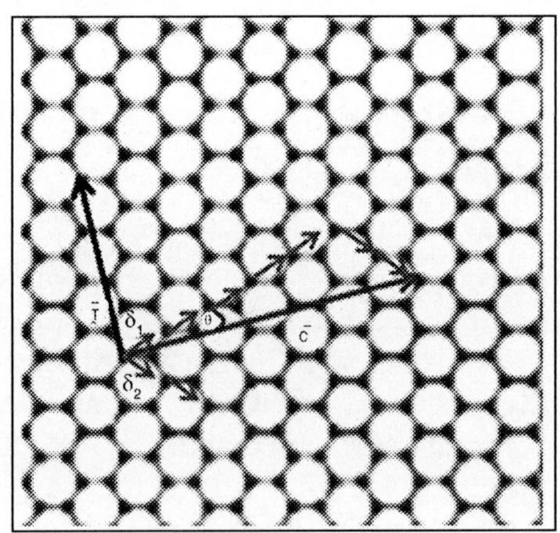

Figure: Graphene atomic structure with a translational vector T and a chiral vector ´h of a CNT

The exceptional electrical properties of carbon nanotubes arise from the unique electronic structure of graphene itself that can roll up and form a hollow cylinder. The circumference of such carbon nanotube can be expressed in terms of a chiral vector: $\hat{C}_h = n\beta_1 + m\beta_2$ which connects two crystallographically equivalent sites of the two-dimensional graphene sheet. Here n and m are integers and β_1 and β_2 are the unit vectors of the hexagonal honeycomb lattice. Therefore, the structure of any carbon nanotube can be described by an index with a pair of integers (n, m) that define its chiral vector.

In terms of the integers (n, m), the nanotube diameter d_t is given by: $d = \frac{\sqrt{3}a_{C-C}\sqrt{m^2 + mn + n^2}}{\pi}$ and the chiral angle θ is given by:

$$\theta = \tan^{-1}\left(\frac{\sqrt{3}n}{2m+n}\right).$$

The differences in the chiral point of view and the diameter cause the differences in the properties of the various carbon nanotubes. For example, it can be shown that an (n, m) carbon nanotube is metallic when $n = m$, has a small gap (*i.e.* semi-metallic) when $n - m = 3i$, where i is an integer, and is semiconducting when $n - m \neq 3i$. This is due to the fact that the periodic boundary conditions for the one-dimensional carbon nanotubes permit only a few wave vectors to exist around the circumference of carbon nanotubes. Metallic conduction occurs when one of these wave vectors passes through the K-point of graphene's 2D hexagonal Brillouin zone, where the valence and conduction bands are degenerate. For the semiconducting carbon nanotubes, there is a diameter dependency on bandgap. For example, according to a single-particle tight-binding description of the electronic structure, $E_g = \gamma\left(\frac{2a}{d_t}\right)$ where γ is the hopping matrix element, and a is the carbon-carbon bond distance.

Motivations for Transistor Applications

A carbon nanotube's bandgap is directly affected by its chirality and diameter. If those properties can be controlled, CNTs would be a promising candidate for future nano-scale transistor devices. Moreover, because of the lack of boundaries in the perfect and hollow cylinder structure of CNTs, there is no boundary scattering. CNTs are also quasi-1D materials in which only forward scattering and back scattering are allowed, and elastic scattering mean free paths in carbon nanotubes are long, typically on the order of micrometers. As a result, quasi-ballistic transport can be observed in nanotubes at relatively long lengths and low fields. Because of the strong covalent carbon-carbon bonding in the sp^2 configuration, carbon nanotubes are chemically inert and are able to transport large amounts of electric current.

In theory, carbon nanotubes are also able to conduct heat nearly as well as diamond or sapphire, and because of their miniaturized dimensions, the CNTFET should switch reliably using much less power than a silicon-based device.

NANOTUBE FOR MEMORY APPLICATIONS

Carbon nanotubes have successfully been made into a variety of nanoscale circuit components, including transistors, inverters, and switches. Now, a pair of scientists has made a rough, yet promising, flash memory device out of carbon nanotubes. The device is a long way from a finished, marketable product, but it nonetheless represents a significant step in the drive to incorporate carbon nanotubes into mainstream electronics. Flash memory devices are currently used to store data in many types of electronic items, including digital cameras, USB memory sticks, and cell phones. Flash memory is considered a "non-volatile" form of memory, meaning it can retain data without a constant supply of power.

Comparison with other Types of Non-Volatile Memory

Compared other NVRAM ("Non-Volatile RAM") technologies, NRAM has several advantages. The most common form of NVRAM today is FLASH RAM. In Flash Memory, each cell resembles a MOSFET transistor with a control gate (CG) modulated by a floating gate (FG) interposed between the CG and the FG. The FG is surrounded by an insulating dielectric, typically an oxide. Since the FG is electrically isolated by the surrounding dielectric, any electrons placed on the FG will be trapped on the FG which screens the CG from the channel of the transistor and modifies the threshold voltage (VT) of the transistor. By writing and controlling the amount of charge placed on the FG, the FG controls the conduction state of the MOSFET FLASH device depending on the VT of the cell selected. The current flowing through the MOSFET channel is sensed to determine the state of the cell forming a binary code where a 1 state (current flow) when an appropriate CG voltage is applied and a 0 state (no current flow) when the CG voltage is applied.

After being written to, the insulator traps electrons on the FG, locking it into the 0 state. However, in order to change that bit the insulator has to be "overcharged" to erase any charge already stored in it. This requires higher voltage, about 10 volts, much more than a battery can provide. FLASH systems include a "charge pump" that slowly builds up power and releases it at higher voltage. This process is not only very slow, but degrades the insulators as well. For this reason FLASH has a limited lifetime, between 10,000 and 1,000,000 "writes" before the device will no longer operate effectively.

NRAM potentially avoids all of these issues. The read and write process are both "low energy" in comparison to Flash (or DRAM for that matter due to "refresh"), meaning that NRAM can result in longer battery life in conventional devices. It may also be much faster to write than either, meaning it may be used to replace both. A modern cell phone will often include FLASH memory for storing phone numbers and such, DRAM for higher performance working memory because FLASH is too slow, and additionally some SRAM in the CPU because DRAM is too slow for its own use. With NRAM all of these may be replaced, with some NRAM placed on the CPU to act as the CPU cache, and more in other chips replacing both the DRAM and FLASH.

NRAM is one of a variety of new memory systems, many of which claim to be "universal" in the same fashion as NRAM – replacing everything from Flash to DRAM to SRAM. The merely alternative memory currently ready for commercial use is ferroelectric RAM (FRAM or FeRAM). FeRAM adds a small amount of a ferro-electric material in an otherwise "normal" DRAM cell, the state of the field in the material encoding the bit in a non-destructive format. FeRAM has all of the advantages of NRAM, although the smallest possible cell size is much larger than for NRAM. FeRAM is currently in use in a number of applications where the limited number of writes in Flash is an issue. The FeRAM read operation is inherently destructive, requiring a restoring write operation afterwards.

Other more speculative memory systems include MRAM and PRAM. MRAM is based on a grid of magnetic tunnel junctions. Key to MRAM's potential is the way it reads the memory using the tunnel magnetoresistive effect, allowing it to read the memory both non-destructively and with very little power. Unfortunately, the 1st generation MRAM, which utilized field induced writing, reached a limit in terms of size, which kept it much larger than existing Flash devices. However, two new MRAM techniques are currently in development and hold the promises of overcoming the size limitation and making MRAM more competitive even with Flash memory. The techniques are Thermal Assisted Switching (TAS), which is being developed by Crocus Technology, and Spin Torque Transfer (STT) on which Crocus, Hynix, IBM, and several other companies are working.

PRAM is based on a technology similar to that in a writable CD or DVD, using a phase-change material that changes its magnetic or electrical properties instead of its optical ones. The PRAM material itself is scalable but requires a larger current source.

REFERENCES

- Cerofolini, G. F.: *Nanoscale Devices*, Springer-Verlag, Berlin, 2009.
- Ljubisa Ristic: *Sensor Technology and Devices*, Artech House, Delhi, 1994.
- Reddy, K. Krishna : *Nanotechnology in Electronics*, Campus Books International, Delhi, 2007.

7

Molecular Electronics

Molecular electronics has witnessed increased interest in recent years, triggered by the forecast that silicon technology might reach its scalability limits in a few years. In order for molecular electronics to become a valuable alternative to silicon technology, it will not be sufficient to fabricate molecular electronic devices with outstanding characteristics, but appropriate circuit and architectural solutions will also be needed. While a lot of effort has been dedicated to the demonstration of electronic functionalities of single molecules and organic films, research at the circuit and system level is still in its infancy.

Investigation on particular molecules or nanotube-based devices promises to keep Moore's law alive once miniaturization of silicon-based structures becomes impractical. As first proposed by Aviram and Ratner, one can imagine to squeeze entire nonlinear circuit elements (such as diodes or transistors) into single molecules. In principle, such devices could be significantly faster and smaller than end of the roadmap solid-state electron devices. Despite an enormous progress on the experimental characterization of single-molecule conduction in the last years, only few device concepts have emerged and it is still unclear whether individual molecular devices could be integrated into a larger-scale computing circuit.

In any case, we can be sure that molecular circuits and architectures will be very different from what we are used to in today's systems. Several architectures, which would be suitable for the realization of electronic logic circuits or memory cells based on molecular devices, have been proposed. One possibility is to synthesize complex molecules whose arms can be separately contacted to provide the same electrical input/output of conventional logic gates. Such fascinating idea, suggested by Ellenbogen and Love, is unfortunately extremely challenging from the chemical-synthesis point of view, and it has not been realised up to now. Another possibility is to create a programmable interconnected network of nanoparticles and molecular entities ("nanocells") : conducting metallic nanoparticles are randomly deposited on a substrate (which can be a silicon one) and subsequently bridged via molecular connections, to create electrical pathways between previously patterned metallic leads. The molecular linkers should exhibit non linear IV characteristics, in the form of negative differential resistance or hysteretic behaviour and the network is connected to a limited number of input/output pins at the edges of the nanocell, which could then be accessed, configured and programmed from the edges. Simulations have demonstrated the capability of the nanocell to act as logic gate.

Integration with CMOS technology will certainly be the first step for devices based on single molecules, with solutions that follow a road that has been called "More than Moore", in an attempt to extend the standard chip functionalities already offered by silicon technology. Such hybrid systems would benefit from the speed and reliability of CMOS, while offering at the same time the versatility and intrinsically nanometer footprint of molecular devices. Two architectures which are in principle compatible with silicon technology are the "Quantum Cellular Automata" and the "cross bar",. While the latter, employs solutions based on standard interconnections, the former relies on a highly innovative approach. In fact, devices and their interconnections can no longer be separated in a straightforward way in single-molecule electronics and it is still not clear if the paradigm of classical circuit theory for building a complex system can still be followed. It is very timely to explore alternative ideas to interconnect molecular building blocks inside the molecular circuit and the molecules to the outer world without metal wires, using electric or magnetic fields, too. One very interesting hybrid architecture is the semiconductor/nanowire/molecular one ("CMOL"), proposed by Likharev and coworkers. The basic difference with respect to cross bars is that in CMOL the interface between CMOS circuitry and nanowires is provided by pins distributed all over the circuit area. Thus, a much larger integration should be achieved, with relaxed requirement concerning alignment of the wire interconnects.

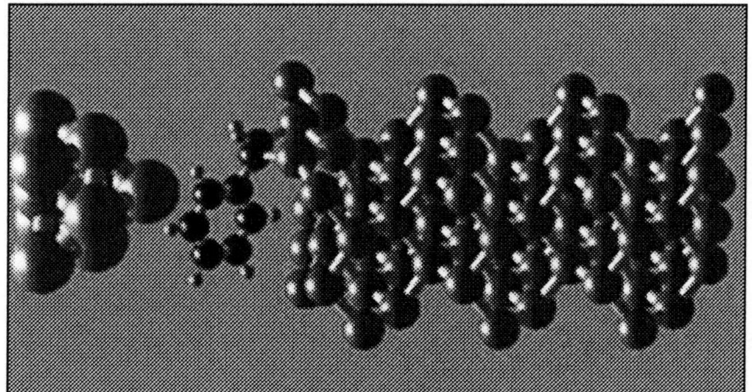

Besides single molecule structures, molecular electronics refers also to the employment of organic materials as substitutes for the solid-state layers of traditional semiconductor devices. Since their carrier mobility is relatively low, organic devices cannot compete with silicon-based circuits in high-speed, high-performance computing applications, but can become competitive in the market of large area devices, because their process technology is potentially inexpensive and circuits can be realised on virtually any substrate. We will nearby our theoretical analysis on organic transistors, inverters and ring oscillators, describing the simulation tools used and its applications to device and circuit modelling.

In the areas of chemical sensing, light emission and light sensing, organic materials could make far more versatile devices as their semiconductor counterparts, because of the wide selection of possible sensor and light-emitting materials. Due to their versatility in terms of fabrication and performance, organic films could also be integrated with silicon circuitry.

MOLECULAR SCALE ELECTRONICS

Recently, molecular electronics-based computation has involved attention because it addresses the decisive in a dimensionally scaled structure: ultradense and molecular scale. The important scaling factor gained from molecular-scale devices implies eye-opening comparisons: a contemporary computer utilizes ~10^{10} silicon-based devices, whereas one could prepare ~10^{23} devices in a single beaker using routine chemical syntheses. An additional driving factor is the potential to utilize thermodynamically driven directed self-assembly of components such as chemically synthesized interconnects, active devices, and circuits. This is a novel technological approach for post-VLSI electronic systems and can conceivably lead to a new era in ultradense electronic systems. This approach for spontaneously assembling atomic scale electronics attacks the interconnection and critical dimension control problems in one step and is implicitly atomic scale. Concurrently, the approach utilizes inherently self-aligned batch processing techniques that address the fabrication limitations of conventional VLSI.

Molecular materials for electronic and optoelectronic applications have been realize for fairly some time. In addition to uses such as liquid crystal displays, devices such light emitting diodes, lasers, transistors, and sensors have been demonstrated. The distinction between these (essentially "bulk") applications and molecular scale electronics is not just one of size, but of concept: the design of a molecule that itself is the active element.

Molecules be proposed as vigorous electronic devices as early as 1973, when Aviram and Ratner proposed that unimolecular rectification, or asymmetrical electrical conduction, should occur through the molecular orbitals of a single D-σ-A molecule by "through-bond tunneling". D in an electron donor with low ionization potential, and A is an electron acceptor with high electron affinity, and s is a covalent "bridge." The excited zwitterionic state D^{+}-σ-A^{-} would be relatively accessible from the ground neutral state D-σ-A while the opposite zwitterion D^{-}-σ-A^{+} would lie several eV higher and would be inaccessible. In solid-state language, the system is an asymmetric multilevel resonant tunneling structure. The molecular systems have superior analogies to solid state systems. Instead of the Fermi levels of the solid state, one deals with the highest occupied and lowest unoccupied molecular orbitals (HOMO and LUMO, respectively) of molecules. Instead of metal and interconnects and degenerate contacts, one uses conjugated linear polymeric systems. Instead of "doping" to modify Fermi levels, one modifies the electron affinity and ionization potential of molecules by the chemical substitution. By designing in the molecular orbitals, one has the equivalent of bandgap engineering.

Used for molecular-scale electronics to come of age, fabrication and measurement techniques had to reach the atomic scale. The advent of atomic imaging techniques, such as the scanning tunneling microscope (STM) and the atomic force microscope (AFM) have given us an atomic view of molecular placement, fabrication, and self meeting. Nanofabrication techniques have created interconnects small enough to reliably contact molecules. A better understanding of electronic transport at the atomic level has developed over the last decade, and theoretical models of conduction through such systems are beginning to develop. All these advances have led to the first electrical measurements of molecular systems. Among these are conductivity measurements of molecules by STM, the first measurement of electronic conduction through a single "molecular wire" and the primary molecular diodes. The measurement of accuse transport in single organic molecules, and the determination of their conductance, has been a long sought objective. Such measurements are experimentally challenging and intriguing since one can test the validity of transport approximations at the molecular level. A conceptually simple pattern would be to connect a single molecule between metallic contacts. Such a metal- molecule-metal configuration would current the molecular embodiment of a system analogous to a quantum dot with the potential barriers of the semiconductor structure replaced by the contact barrier of the molecule/metal interface. Although making nanofabricated planar contacts at the molecular scale is challenging, it has been done and candidate molecular conductors have been absorbed into the gaps. An absence of observable conductivity in these systems may be due to inefficient electron transfer, undesirably large contact potentials, conjugation-breaking due to substrate interactions, or unfavorable absorption configurations. Thus, a number of alternative approaches contain be attempt and realize.

NOTION OF MOLECULAR ELECTRONICS

The notion 'molecular electronics' has been used more frequently since the 1970s and summarizes a series of physical phenomena and ideas for their application in connection with organic molecules, oligomers, polymers, organic aggregates and solids. The properties studied in this field were connected to optical and electrical phenomena, such as optical absorption, fluorescence, nonlinear optics, energy transport, charge transfer, electrical conductance, and electron and nuclear spin-resonance. The final goal was and is to build devices which can compete or surpass some aspects of inorganic semiconductor devices. For example, on the basis of organic molecules there exist rectifiers, transistors, molecular wires, organic light emitting diodes, elements for photovoltaics, and displays. With respect to applications, one aspect of the organic materials is their broad variability and the lower effort and costs for their processability.

The step from microstructures to the investigation of nanostructures is a big challenge also in this field and has lead to what nowadays is called molecular electronics in its narrow sense. In this field the subjects of the studies are often single molecules, *e.g.* single molecule optical spectroscopy, electrical conductance, *i.e.* charge transport through a single molecule, the influence of vibrational degrees of freedom, etc. A challenge here is to provide the techniques for addressing in a reproducible way the molecular scale. In another approach small molecular ensembles are studied in order to avoid artefacts from particular contact situations. The recent development of the field is presented in.

In this Focus Issue we present new results in the field of 'molecular electronics', both in its broad and specialized sense.

One of the basic questions is the distribution of the energy levels responsible for optical absorption on the one hand and for the transport of charge on the other. A still unanswered question is whether the Wannier exciton model applies in which the excitation is distributed over several molecules or whether a good description is given by the Frenkel exciton model with the electron and the whole being localized at the same molecular unit.

In organic semiconductors the charge transport usually occurs on the basis of holes because of the presence of many defects giving rise to a localization of the electrons. It is therefore a challenge to produce materials with both positive and negative mobile charge carriers. In the 1990s V M Agranovich introduced the idea of hybrid excitons, *i.e.* of nanostructured materials consisting of both organic and inorganic semiconductors. At the interface between the organic and inorganic parts new excitons can appear, being a superposition of both Frenkel and Wannier excitons and having both the high oscillator strength of the Frenkel and the large optical nonlinearity of the Wannier exciton. The problem is to find optimum combinations of the organic and inorganic parts to enable the hybrid structure concept to work.

Micro-cavities also play an important role in the investigation of organic materials resulting in a new state (polariton) as the superposition of a photon and an exciton because of the large exciton–photon interaction. A similar excitation arises because of the interaction between plasmons and photons. A special geometrical shape of a nano-cavity increases the interaction between the electromagnetic radiation and a dipole sitting in the cavity. The communication between vibronic degrees of freedom and electronic excitations plays an important role for various phenomena such as nonlinear processes, the question of coherence, information on the shape of a potential hypersurface, etc.

With the help of femtosecond laser pulses, detailed information on such vibrations can be obtained. Also of great importance is the investigation of the energy transfer in artificial light-harvesting systems, *e.g.* in dendrimers. Finally the combination of experimental and theoretical investigations allows for a comparison of the spectra of two molecules with the same backbone (tetracene and rubrene).

The transport of charge through a molecule occurs possibly in a stationary, but at any rate in a non-equilibrium situation. The study of dissipation in such situations requires special approaches, both in theory and in experiments. One key issue is the understanding of the role of the microscopic phenomena such as the excitation of vibrational modes and their macroscopic outcome, *i.e.* the dissipation. This topic is addressed in several contributions both theoretically and experimentally.

From the theoretical side, for the investigation of the heat production during the electron transfer, non-equilibrium Green's functions have been utilized. In another contribution a combination of the non-equilibrium Green's function technique together with the density functional technique has been developed for the calculation of the elastic and inelastic electronic transport. To calculate the transport of indistinguishable particles a path integral Monte Carlo approach has been put forward. In single-molecule transistors the gate-voltage dependence on the Kondo temperature and an accompanying strong Coulomb blockade can be explained by taking into account a strong electron–vibron interaction including anharmonicities of the molecular potential surface.

The transport of charges is heavily influenced by disorder. The case of static disorder is investigated for linear chains, carbon nanotubes and graphene ribbons. Finally it is shown that the charge transport through a single energy level coupled to a localized vibrational mode and two leads shows hysteretic effects which could possibly be used in a memory device. For applications the control of the current through a molecular junction is considered theoretically. Two possible mechanisms are discussed: the control via coherent destruction using predefined ultrafast laser pulses and the formation of laser pulses using optimal control theory.

A collection of contributions is dedicated to the study of electronic transport through molecules using various techniques ranging from scanning-tunnelling techniques via controllable break-junctions to printing techniques and molecular networks. The molecules under study can be classified into two main groups. On the one hand the functionalization of aromatic or alkane molecules with thiols is used for establishing chemical binding to metal electrodes; in the other set of experiments fullerenes are used as model systems for studying the influence of orientation and heat dissipation.

Molecular conducting networks are important under various aspects. Such networks are formed by an array of gold nanoparticles connected by conjugated molecular chains with one or two thiol ends and conduction investigations are performed under various conditions. A careful investigation of the experimental conditions is necessary when comparing conductance measurements using the break junction technique. For example there are results demonstrating that several molecular junctions are formed in parallel between the electrodes.

Other experiments use different materials for the junctions and measurements are performed at various temperatures. The transport of charge through an alkane-monolayer is investigated using micro-transfer printing to establish contacts without shorts. It is shown that both tunnelling between the electrodes and transport through the states of the molecules contribute to the conductance. C_{60} plays a role in various fields of molecular electronics. One aspect is the conductance through the molecule as a function of the orientation of the molecule on the surface. It is found that there is a strong orientation dependence of the transport on Au(111) surfaces and that it is almost independent on a Cu(100) surface.

A further important phenomenon is the heating and cooling of C_{60} during charge transport depending on the surface of C_{60} adsorption. The differences are ascribed to the amount of charge transfer into C_{60} upon adsorption on different surfaces.

In summary, in the field of molecular electronics new materials and structures are developed and investigated, both with respect to a basic understanding of the materials and their compositions and with respect to possible applications in electronics. While in the field of molecular electronics on the microscale the techniques are well established, they still need to be refined in the field of nano-molecular electronics. Nevertheless, both subfields share some of the most challenging questions: *e.g.* the problems of charge and energy transport, of excitations and the formation of new quasi-particles. Another question is the role of vibrational degrees of freedom, where on the one hand one has to cope with the unavoidable effect of heat dissipation. On the other hand, vibrational excitations are intimately connected to the individual molecule under study and thus offer the possibility to be used in functional devices based on intrinsic molecular properties.

This Focus Issue represents a snapshot of the state of the art of this emerging field in the first half of 2008. We expect that the fast development which the research has undergone in recent years will even speed up in the near future.

MOLECULAR MATERIALS FOR ELECTRONICS

APPLICATIONS OF MOLECULAR ELECTORNICS

Molecular electronics is the branch of nanotechnology which deals with the applications and structure of nano building blocks that are used in electronic circuit manufacturing and design. It is sometimes called as molecular electronics. Everyone the main electronic fabrications are supported by molecular electronics. It has extensive range of applications in the occupation areas of chemistry, physics, electronics and nanoelectronics, technology, artificial aptitude and medical equipements. Approximately all the fabricated chips in the smart machinery that is used on large scale has molecular electronic concerned in its construction. Meant for instance resistors and transistors that are used in producing electricity, capacitors in space crafts, automation circuits of robots, strategic plant temperature handlers and CT scan for displaying the infected areas of body.STM and AFM both have molecular electronics involved in their construction. In the field of chemistry it is used to observe the chemical reactions in stimulated models of nuclear reactors, and as well for measuring the acidic and reactive properties of individual constituent. The thought of molecular electronics is not original in the field of technology and science, its core concepts were originated with the concept of nanoelectronics. In 1997 the first ever speculation for the molecular electronics was given by Mark Reed and co-workers. The significant discipline to provides us way for dealing with all the electronic characteristics of conductors, insulators and semi conductors molecular electronics has two sub disciplines

- Molecular materials for electronics
- Molecular scale electronics.

Together of them collectively give high tech and ultra intelligent devices to be used in various type of applications. This field is progressing rapidly and supporting nanoelectronics in arrange to develop excellent applications.

MOLECULAR ELECTRONIC WORKS

The molecular electronics works at the smallest level involving single molecules, its characteristics and sub properties. It deals with the substance which are considered to be un explorable in the field of electronics. It too implements all the main laws of electronics in the practical opinion basis so that original implementation of theoretical frame work be able to seen. A lot of major components in the field of chemistry, physics, biological science and electronics are the gift of molecular electronics. With the help of this wonderful technology the building blocks of structures can be used in strong and complex fabrications of integrated circuits. It be able to control the molecular scale properties of person atom of matter according to the use. The direct measurement of material properties is one of the very important feature of molecular electronics which cannot be found in any other technology.

Disadvantages of Molecular Electronics

- Molecular electronics is not at all cost effective complete fabricated components developed by it are quite expensive and their maintenance cost is also high.
- The components developed because of high manufacturing cost, its components are not readily available.
- Specialized engineers and scientists are required to handle and control the risks factors of molecular electronics.

CURRENT PROGRESS AND FUTURE PERCEPTION

Current progress can be seen in the shape of experimental and theoretical learn of molecular electronics with the advancement in the nanosciences ,this very concept is stepping up and up to facilitate the world. The most recent foundation of molecular electronics are scanning tunneling microscope and atomic force microscope which has high molecular fabrications and circuits and are also able to measure and control its function accordingly. It can be unspecified that in the near future molecular electronics would be flourishing on broader scale and commonly used instruments that are used in many work areas would be having molecular electronic circuits and fabrications in them.

MOLECULAR ELECTRONICS TECHNOLOGY

The field of molecular electronics seeks to use individual molecules to perform functions in electronic circuitry now performed by semiconductor devices. Individual molecules are hundreds of times smaller than the smallest features conceivably attainable by semiconductor technology. Because it is the area taken up by each electronic element that matters, electronic devices constructed from molecules will be hundreds of times smaller than their semiconductor-based counterparts. Moreover, individual molecules are easily made exactly the same by the billions and trillions. The dramatic reduction in size, and the sheer enormity of numbers in manufacture, are the principal benefits offered by the field of molecular electronics.

At the heart of the semiconductor industry is the semiconductor switch. Because semiconductor switches can be manufactured at very small scales, and in combination can be made to perform all desired computational functions, the semiconductor switch has become the fundamental device in all of modern electronics. California Molecular Electronics Corporation's Chiropticene® Switch is a switchable device that goes beyond the semiconductor switch in size reduction. This switch is a single molecule that exhibits classical switching properties.

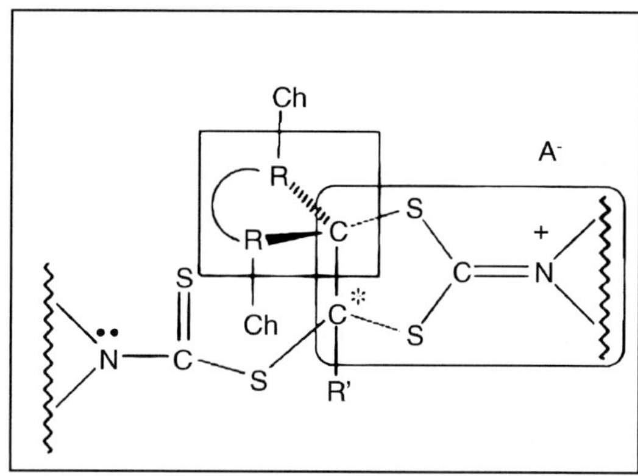

Chiropticene® Molecular Switch Design
C = Carbon Atom
S = Sulfur Atom
N = Nitrogen Atom
R = Proprietary Substituents
Ch = Chromophoric Groups
A = Anion (variable structure)

The investigate and development effort at California Molecular Electronics Corporation has resulted in the preparation of a novel Chiropticene molecular switch design.

Prochiral

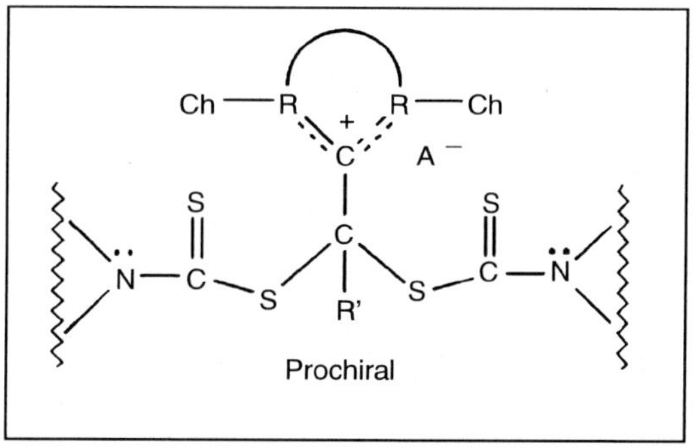

Although its detailed composition remains proprietary, the perfected molecule incorporates the following key features required for a viable, practical molecular switch:

- Spiro Carbon Atom: The two ring systems, shown in shadow boxes, are forced to lie in mutually perpendicular planes yielding enhanced optical rotation and high signal strength.
- Chromophore Activation: Prudent selection of the chromophoric groups (Ch) provides optical tuning to match laser input signals.
- Cation Control: By molecular engineering the proprietary substituents (R), we can determine the stability of the cation (+) and thus the speed of the switch.

- Optical Carbon Atom: The carbon atom that changes chirality (C*) is not directly involved in the switch mechanism thereby affording optical stability.

PHYSICAL PRINCIPLE OF FUNCTIONING MOLECULAR ELECTRONIC DEVICES

The phenomenon of the electronic-structure instability in a one-dimensional electronic system was theoretically predicted by Peierls in 1955 in his book *Quantum Theory of Solids*. According to his assumption, a one-dimensional (or low-dimensional) metal with a half-filled band is structurally unstable against the lattice modulation and distortion and opening the dielectric gap leading to the transition of the metal state to a semiconductor or dielectric state. In other words, a simple one-dimensional lattice, which consists of atoms with one electron per lattice unit and equal distances between the neighbouring atoms can be distorted and transformed into a lattice with short and long distances between neighbouring atoms. The important characteristics of this phenomenon is the gap value at a zero temperature. The gap value depends on many factors. In particular, one of them is temperature. It leads to the phase transition which is known as Peierls transition. We can obtain for the phase transition temperature the following relation (the gap value is equal $1{,}76T_p$, where T_p is temperature of the phase transition. By now, extensive experimental and theoretical studies of this phenomenon have been carried out. In the first place, these investigation have been connected with problems of superconductivity.

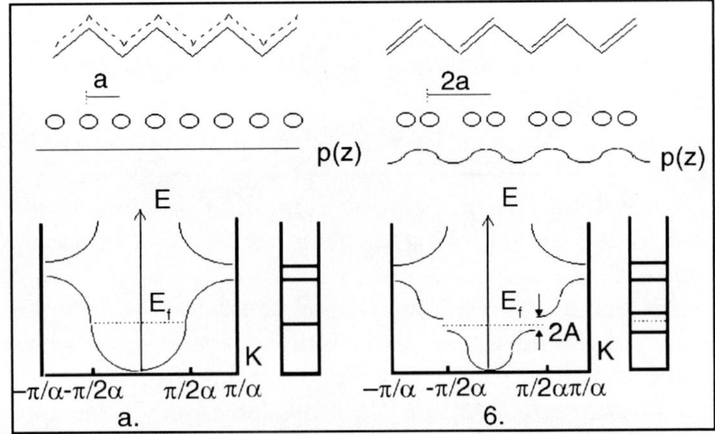

Classical examples of the Peierls systems are polyacetylene and related polyconjugated polymers. Polyacetylene has clearly alternating double and single bonds. The difference between their lengths is 0.06-0.08 Å. It is known, polyacetylene can be easily transformed from dielectric state to metallic state by doping it with positive or negative ions. Electrical conductivity of the polyacetylene samples can be obtained depending on doping concentration. The doping effect can be explained in different ways. From a chemistry perspective, doping is described as oxidation or reduction reactions, in band theory, as a change in band packing. From a low-dimensional physics perspective, a result of doping is caused by a change in the dimension of electronic systems. In other words, intercalation of the conjugate molecular systems by ions leads to the excitation of one-dimensional electron wave functions, thereby changing the dimension of this system. We believe that such perturbation of an electron system may be induced not only by ion doping, but also by applying an external electrical or magnetic field, electromagnetic radiation, pressure or actions of chemical compounds.

Therein lies our universal suggestion for using this phenomenon in Molecular Electronics. From the point of view of molecular electronics, we must be interested in other reasons for changing the gap value and conductivity, respectively (especially with external fields: electrical, magnetic and so on). For this purpose we will analyse factors which define the gap value and observe conditions of the phenomenon of electronic instability in the given system. In other words, we must find ways of controlling the physical properties, especially conductivity, such molecular systems without changing the temperature.

The amplitude of gap depends on a series of factors; from a molecular electronics stand point the most important for us are:

- The change of the charge per lattice period;
- The change of the relative difference of the transfer integrals for the short and long bonds;
- The variation of the transfer integrals between neighbouring one-dimensional molecular systems;
- The disorder of the molecular units of one-dimensional systems or guest molecules or ions which are located between neighbouring systems.

In order to observe the electron instability, we have to use a one-dimensional molecular system with a value of no less than 50-100 Å. In other words, the dimension of future molecular electronic devices will be about 50 Å and more. From the point of view of Electronics, it is desirable, that this variation should be made by an external electrical field and electrical conductivity could be controlled. When designing devices it is necessary for the state of one dimensional electronic system to be entered into the instability region of the Peierls's quasi-transitions.

BASIC IDEAS OF DESIGNING ELECTRONIC DEVICES

To design molecular electronics devices let us consider polyconjugated molecular systems, although it is applicable for other families of one-dimensional systems with different type of organization. These may be organic, organometalic and inorganic compounds. The electronic state of a one-dimensional system can be changed not only with intercalation mechanisms, but also via certain active molecular units, which lie in close vicinity to the van der Waals interaction or chemically bound with one.

To exemplify let us consider design devices, which perform the function of an amplifier (a molecular transistor). There are many ways of designing a molecular transistor. First of all we shall consider ways of changing the charge of a single element in a one-dimensional system. It can be reached at the expense of donor or acceptor molecules connected with an one-dimensional system or located in immediate proximity. In last case, we have the charge transfer system which consists of a polyconjugated molecular system and donor or acceptor molecules. The value of the transferred charge in such systems will depend on the value of an applied electrical field. In turn, it should result in change of conductivity of a one-dimensional system.

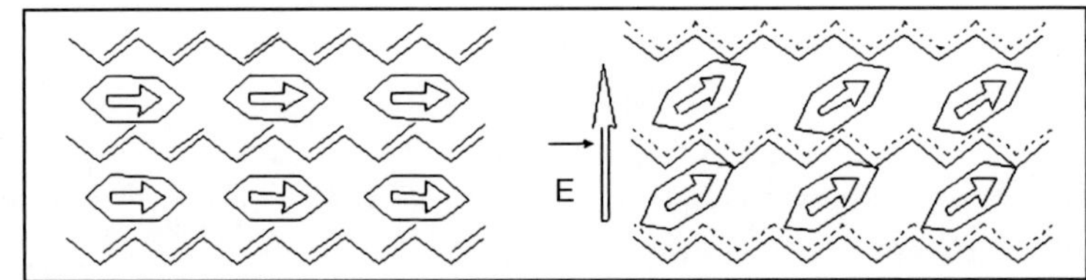

The quantitative characteristics of the initial and final value of transistor conductivity are determined by the number and characteristic of the special molecules and by their susceptibility to external fields or chemical reagents, as well as the character of electron transfer between the one-dimensional conjugate electron system and the electron system of the special molecules.

In another approach it is assumed that the device operations based on the interaction between separate one-dimensional molecular systems through the influence of the electrical field on a "guest" special molecule with a high polarizability and a large dipole moment located in the matrix consisting of one-dimensional molecular fragments. Under an internal electrical field (E), on the one hand, the high polarizability molecules affect the one-dimensional electronic system, on the other hand, the equilibrium orientation of the dipole moment of the "guest" molecules is changed. In both cases it is possible to expect the alteration of exchange interactions between neighbouring one-dimensional molecular systems, but with different relaxation times. As a result, the electrical conductivity of the compounds on a direction perpendicular to the internal field will be changed.

In revolve, disorder of the one-dimensional lattice can be obtained by a disorder of the molecular units of which this lattice consists. There can exist both a static and dynamic disorder. Moreover, it can be achieved by a disorder of guest molecules or ions which are located between neighbouring one dimensional systems. A fragment of a molecular structure in which the ordering of protons happens in an operation of an electrical field

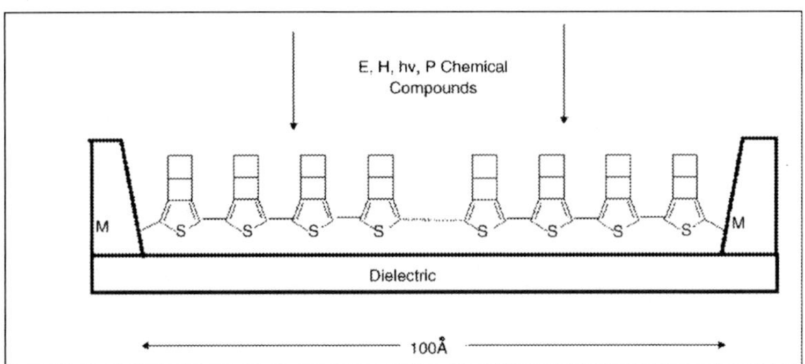

Figure: The Molecular transistors which using of polythiophenes with an activemolecular group which changes the transfer integrals for long donds.

On basis of similar molecular systems simple and effective field transistors can be created. This hypothetical molecular transistor, which using deviations of polythiophenes with an active molecular group which allow us to change the relative transfer integrals for the short and long bonds.

SIMULATION AND CIRCUIT DESIGN

Simulations

For purposes of psychiatry and as a means of setting clearer design goals, it is helpful to categorize simulations into four categories: parameter-driven, tactical, social process, and hybrid simulations. The categories of tactical and social process simulations come from Gredler 1994 though the interpretation below reflects my understanding of her concepts within the context of history education. I have added the categories of parameter driven and hybrid simulations to round out a basic taxonomy. Again, it is worth noting that these terms should not be applied too rigidly; they are intended simply to help in the process of analysing and designing simulations.

Parameter-Driven Simulations refer to those simulations that consist of a set of relationships between parameters. The user provides the initial parameters to the simulation and the simulation generates the results based on those parameters. This type of simulation is not interactive *per se*; rather, it runs a set of calculations based on user input and the system it is modelling. Parameter-driven simulations tend to be used for scientific or social scientific purposes. For example, one might use a parameter driven simulation to estimate the demographics of a given population with a given birth and death rate or to calculate the spread of a historical disease (or a prospective future outbreak).

Like all simulation types, parameter-driven simulations have strengths and weaknesses. They are beneficial because of their power to illustrate the potential effects of complex factors and even to predict. To the extent that a parameter-driven simulation accurately models the relationships between variables, it can provide reasonable predictions of outcomes. At the same time, they do not allow for participants to make decisions after the initial setup and, to this extent, may be less helpful students of history learn in dynamic fashion about the factors that shaped historical decisions and the potential effects of human decisions.

Key Steps in Designing a Simulation

Select a historical event or system for the simulation. There are a wide range of historical events and historical systems suitable for simulating. Since simulations are inherently models of processes it is important at the very beginning to recognize that simulations are not the best fit for all sorts of learning. In particular, simulations are not particularly efficient for helping students memorize a list of facts. This is not to say that students are unable to learn historical facts during the course of a simulation. A true simulation that allows participants room to make decisions within a set of historical factors, however, will by necessity focus to an extent on long term historical factors of cause and effect. It is well worth noting, however, that considering and determining those long term issues of causes and effect require higher order skills from students than memorization. The hallmarks of a viable topic for simulation design depend on the type of simulation to be designed. Hybrid social process/tactical simulations are best suited for historical events/systems where more than one individual (nation, social group, region, etc.) is in some form of conflict with other individuals and these groups have a set of goals and tools for reaching those goals.

Determine the key *processes* in the historical event or system. What will this simulation model for students? Since a simulation–particularly one for use by middle and high school students–cannot include all the processes involved in a historical event of system, the designer will have to choose which processes are crucial. Attempting to define the key processes at work before beginning to design the simulation gives the designer a standard for what to include and not include in the simulation. Determine the participants' roles. (Whom do the participants represent?) Will the participants of the simulation be representatives of historical groups (the governments of countries for example) or take on the roles of individuals (prosecutors and accused in witch trials).

Determine the participants' goals. (What are the participants trying to achieve?) It is critical that participants be aware of the goals for their roles. While simulations need not be competitive they function best when goal oriented. The decisions and actions of the players, ideally, will be guided by consideration of the goals in mind. Make sure the goals reflect faithfully the concerns and aspirations of the historical groups and individuals represented.

Determine the tools available to the participants. (What kinds of actions and decisions can participants make in order to achieve their goals?) Though all steps of simulation design are critical, a well considered set of roles and goals can founder if participants are not provided with a set of tools to reach those goals. Participants need to have different tactics available for achieving their goals; in general they also benefit from knowing what those tactics are, to a certain degree. The tools available to participants are a crucial distinction between role-playing activities and simulations. Activities that are accurately classified as role-plays are limited to discussion between characters. For an activity to qualify as a historical simulation–and this is a qualitative distinction, not a improvement or degradation of the role-play–participants must have actions/options available to them beyond sharing ideas and views. This is true even if those actions/options are verbal in nature.

ELECTRONIC CIRCUIT SIMULATION

Electronic circuit simulation uses mathematical models to replicate the behaviour of an actual electronic device or circuit. Simulation software allows for modelling of circuit operation and is an invaluable analysis tool. Due to its highly accurate modelling capability, many Colleges and Universities use this type of software for the teaching of electronics technician and electronics engineering programs. Electronics simulation software engages the user by integrating them into the learning experience. These kinds of interactions actively engage learners to analyse, synthesize, organize, and evaluate content and result in learners constructing their own knowledge.

Simulating a circuit's behaviour before actually building it can greatly improve design efficiency by making faulty designs known as such, and providing insight into the behaviour of electronics circuit designs. In particular, for integrated circuits, the tooling (photomasks) is expensive, breadboards are impractical, and probing the behaviour of internal signals is extremely difficult. Therefore almost all IC design relies heavily on simulation. The most well known analog simulator is SPICE. Probably the best known digital simulators are those based on Verilog and VHDL.

Some electronics simulators integrate a schematic editor, a simulation engine, and on-screen waveforms, and make "what-if" scenarios easy and instant. They also typically contain extensive model and device libraries. These models typically include IC specific transistor models such as BSIM, generic components such as resistors, capacitors, inductors and transformers, user defined models (such as controlled current and voltage sources, or models in Verilog-A or VHDL-AMS). Printed circuit board (PCB) design requires specific models as well, such as transmission lines for the traces and IBIS models for driving and receiving electronics.

Types

While there are strictly analog electronics circuit simulators, popular simulators often include both analog and event-driven digital simulation capabilities, and are known as mixed-mode simulators. This means that any simulation may contain components that are analog, event driven (digital or sampled-data), or a combination of both. An entire mixed signal analysis can be driven from one integrated schematic. All the digital models in mixed-mode simulators provide accurate specification of propagation time and rise/ fall time delays.

The event driven algorithm provided by mixed-mode simulators is general purpose and supports non-digital types of data. For example, elements can use real or integer values to simulate DSP functions or sampled data filters. Because the event driven algorithm is faster than the standard SPICE matrix solution, simulation time is greatly reduced for circuits that use event driven models in place of analog models.

Mixed-mode simulation is handled on three levels;

- With primitive digital elements that use timing models and the built-in 12 or 16 state digital logic simulator,
- With subcircuit models that use the actual transistor topology of the integrated circuit, and finally,
- With In-line Boolean logic expressions.

Exact representations are used mainly in the analysis of transmission line and signal integrity problems where a close inspection of an IC's I/O characteristics is needed. Boolean logic expressions are delay-less functions that are used to provide efficient logic signal processing in an analog environment. These two modelling techniques use SPICE to solve a problem while the third technique, digital primitives, use mixed mode capability. Each of these techniques has its merits and target applications. In fact, many simulations (particularly those which use A/D technology) call for the combination of all three approaches. No one approach alone is sufficient.

Another type of simulation used mainly for power electronics represent piecewise linear algorithms. These algorithms use an analog (linear) simulation until a power electronic switch changes its state. At this time a new analog model is calculated to be used for the next simulation period. This techniqueology both enhances simulation speed and stability significantly.

Complexities

Process variations occur in the manufacture of circuits in silicon and often circuit simulators do not take into account these variations that occur when the design is fabricated into silicon. These variations can be small, but taken together can change the output of a chip significantly.

Temperature variation can also be modelled to simulate the circuit's performance through temperature ranges.

CIRCUIT DESIGN

The process of circuit design can cover systems ranging from complex electronic systems all the way down to the individual transistors within an integrated circuit. For simple circuits the design process can often be done by one person without needing a planned or structured design process, but for more complex designs, teams of designers following a systematic approach with intelligently guided computer simulation are becoming increasingly common. In integrated circuit design automation, the term "circuit design" often refers to the step of the design cycle which outputs the schematics of the integrated circuit. Typically this is the step between logic design and physical design. Formal circuit design usually involves the following stages:

- Sometimes, writing the requirement specification after liaising with the customer
- Writing a technical proposal to meet the requirements of the customer specification
- Synthesising on paper a schematic circuit diagram, an abstract electrical or electronic circuit that will meet the specifications
- Calculating the component values to meet the operating specifications under specified conditions
- Performing simulations to verify the correctness of the design
- Building a breadboard or other prototype version of the design and testing against specification
- Making any alterations to the circuit to achieve compliance
- Choosing a technique of construction as well as all the parts and materials to be used
- Presenting component and layout information to draughtspersons, and layout and mechanical engineers, for prototype production
- Testing or type-testing a number of prototypes to ensure compliance with customer requirements
- Signing and approving the final manufacturing drawings
- Post-design services (obsolescence of components etc.)

Specification of Circuit Design

The process of circuit design begins with the specification, which states the functionality that the finished design must provide, but does not indicate how it is to be achieved. The initial specification is basically a technically detailed description of what the customer wants the finished circuit to achieve and can include a variety of electrical requirements, such as what signals the circuit will receive, what signals it must output, what power supplies are available and how much power it is permitted to consume. The specification can (and normally does) also set some of the physical parameters that the design must meet, such as size, weight, moisture resistance, temperature range, thermal output, vibration tolerance and acceleration tolerance.

As the design process progresses the designer(s) will frequently return to the specification and alter it to take account of the progress of the design. This can involve tightening specifications that the customer has supplied, and adding tests that the circuit must pass in order to be accepted. These additional specifications will often be used in the verification of a design. Changes that conflict with or modify the customer's original specifications will almost always have to be approved by the customer before they can be acted upon.

Correctly identifying the customer needs can avoid a condition known as 'design creep' which occurs in the absence of realistic initial expectations, and later by failing to communicate fully with the client during the design process. It can be defined in terms of its results; "at one extreme is a circuit with more functionality than necessary, and at the other is a circuit having an incorrect functionality". Nevertheless some changes can be expected and it is good practice to keep options open for as long as possible because it's easier to remove spare elements from the circuit later on than it is to put them in.

Design Process

The design process involves moving from the specification at the start, to a plan that contains all the information needed to be physically constructed at the end, this normally happens by passing through a number of stages, although in very simple circuit it may be done in a single step. The process normally begins with the conversion of the specification into a block diagram of the various functions that the circuit must perform, at this stage the contents of each block are not considered, only what each block must do, this is sometimes referred to as a "black box" design. This approach allows the possibly very complicated task to be broken into smaller tasks which may either by tackled in sequence or divided amongst members of a design team.

Each block is then consider, still at an abstract stage, but with a lot more focus on the details of the electrical functions to be provided. At this or later stages it is common to require a large amount of research or mathematical modelling into what is and is not feasible to achieve. The results of this research may be fed back into earlier stages of the design process, for example if it turns out one of the blocks cannot be designed within the parameters set for it, it may be necessary to alter other blocks instead. At this point it is also common to start considering both how to demonstrate that the design does meet the specifications, and how it is to be tested (which can include self diagnostic tools).

Finally the individual circuit components are chosen to carry out each function in the overall design, at this stage the physical layout and electrical connections of each component are also decided, this layout commonly taking the form of artwork for the production of a printed circuit board or Integrated circuit. This stage is typically extremely time consuming because of the vast array of choices available. A practical constraint on the design at this stage is that of standardization, while a certain value of component may be calculated for use in some location in a circuit, if that value cannot be purchased from a supplier, then the problem has still not been solved. To avoid this a certain amount of 'catalogue engineering' can be applied to solve the more mundane tasks within an overall design.

Costs Design

Proper design philosophy and structure incorporates economic and technical considerations and keeps them in balance at all times, and right from the start. Balance is the key concept here; just as many delays and pitfalls can come from ill considered cost cutting as with cost overruns. Good accounting tools (and a design culture that fosters their use) is imperative for a successful project. "Manufacturing costs shrink as design costs soar," is often quoted as a truism in circuit design, particularly for ICs.

Verification and Testing

Once a circuit has been designed, it must be both verified and tested. Verification is the process of going through each stage of a design and ensuring that it will do what the specification requires it to do. This is frequently a highly mathematical process and can involve large-scale computer simulations of the design. In any complicated design it is very likely that problems will be found at this stage and may involve a large amount of the design work be redone in order to fix them.

Testing is the real-world counterpart to verification, testing involves physically building at least a prototype of the design and then (in combination with the test procedures in the specification or added to it) checking the circuit really does do what it was designed to.

Ideas of Prototyping

Prototyping is a means of exploring ideas before an investment is made in them. Depending on the scope of the prototype and the level, prototypes can be built at any time during the project. Sometimes they are created early in the project, during the planning and specification phase, commonly using a process known as breadboarding; that's when the need for exploration is greatest, and when the time investment needed is most viable. Later in the cycle packaging mock-ups are used to explore appearance and usability, and occasionally a circuit will need to be modified to take these factors into account.

MEMS/MicroRobotics

MicroElectroMechanical Systems (MEMS) and Microrobotics research spans design techniqueologies, physical investigations and manufacturing techniques involving various microsensors, microactuators and other microsystems. Selected applications include inertial sensor suites for control and guidance, miniature wall-climbing robots using micro/nano-fibre adhesives, arrayed MEMS probe manipulators for tip-based nanomanufacturing, gas chemical sensor arrays for early warning systems, and ultra-compliant neural probes for brain-computer interfaces.

MEMS and Microrobotics RI faculty have developed strong multidisciplinary research programs on integrated MEMS design and fabrication (Fedder), micro and nanorobotic systems (Sitti), and implantable medical microsystems (Fedder, Weiss). This research links the RI to many departments throughout the University through active collaborations with faculty from across CIT and MCS as well as with Pitt and UPMC. These RI faculty members also have joint appointments in other entities. Expanded research in micro-inertial navigation technologies (Kelly and Fedder) is also planned to target new DOD initiatives.

The future for research in MEMS and microrobotics remains bright, with trends toward use of the technology to access and manipulate nano-scale phenomena as well as to enable emerging biomedical applications. RI faculty are taking leads in these areas in collaboration with faculty throughout the University. One of the challenges faced is to merge systems across macro-, micro-, and nano-scales. CMU's planned Nano/Bio/Energy Building (2014 expected date of completion) for nano- and biotechnologies will facilitate efforts in this area by providing state-of-the-art shared infrastructure for nano/microfabrication. In addition, a budding collaboration in multi-scale factory automation (Hollis, Fedder) is leveraging renewed efforts in reconfigurable, high-precision minifactories (Hollis).

RESEARCHERS STRIVE FOR COMPUTER-ON-A-CHIP WITH MEMS MEMORY

A microelectromechanical memory now in development could eventually put an entire computer system — CPU, RAM, I/O and hard drive — on a single chip, researchers at Carnegie Mellon University believe. If the researchers succeed — and even they and their supporters acknowledge potential pitfalls — the MEMS-based storage would offer faster access times at lower costs than existing disk-drive technology provides. For now, researchers are concentrating on building a memory big enough to be worthwhile — 2 Gbytes — yet small enough to integrate with the rest of the system. In the MEMS memory, hundreds of microelectromechanical probes each interrogate a small array of positions on a magnetic medium. Though the researchers have yet to construct a prototype, their work represents a new approach to the next stage of storage miniaturization, one that is also being pursued by companies like IBM and Hewlett-Packard.

"The mixture of ruggedness, low power consumption, low cost and high storage capacity will make these devices indispensable," said Richard Carley, a professor of electrical and computer engineering and director of the Centre for Highly Integrated Information Processing and Storage Systems at Carnegie Mellon (Pittsburgh).

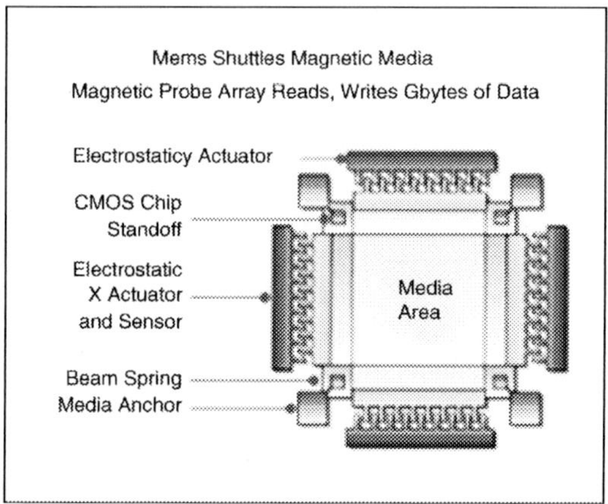

Carley said NASA has given Carnegie Mellon a research award to pursue the design of so-called integrated MEMS storage devices (IMSDs) for satellites. "One of the earliest drivers for IMSDs will be volume and mass critical applications," he said. "For example, planned low-earth-orbit microsatellites will need large amounts of storage capacity to buffer data as it is relayed around the world. However, even a micro-disk drive would be too large to fit into these satellites." Chuck Morehouse, director of the Information Storage Technology Laboratory (Palo Alto, Calif.), called Carnegie Mellon's work in the area some of the best in the world. He sees the market for the new systems, when they arrive, as being very different than for existing hard drives. "The MEMS-based devices will first go into use where portability is important — low weight, high data density, low power consumption, rugged, modest capacity — not the same capacity as desktop hard-drive storage."

Low Cost

The devices determination be useful, Carley said, in low-cost, 1- to 10-Gbyte applications. "For example, future intelligent cell phones will need large amounts of local storage for programs, personal data and downloaded applets, etc. E2PROM will be too expensive and disk drives will be too expensive, too bulky and too power hungry." Other examples of this kind of consumer product, he said, are digital cameras, an emerging class of smart toys, and smart VCRs.

Conventional chip-based memories rely on lithography improvements for miniaturization. Essentially, they consist of small circuits that have to be at least a few times the minimum lithographic line width in height and width. According to the International Technology Roadmap for Semiconductors, which forecasts progress in computing technology, lithography will not produce 2-Gbyte chips until at least 2011. Today's commercial process, at 180 nanometers, can only provide storage densities up to a few tens of megabytes per square centimeter.

But micromechanical-probe memories — which use a small cantilever tip to address a point on some sensitive material — can already reach useful densities for miniature hard disks, at least in theory. Probe tips can detect individual features as small as 30 nm. If this power is used effectively, the areal density could be 1 Gbyte/cm^2 or more. The question is how to exploit the technology in a fast, single-chip configuration.

The Carnegie Mellon team has come up with a scheme that it hopes will use the full area of the memory material, keep the speed high and accommodate various alignment problems. First, an array of read/write tips is fabricated onto the top of a chip that has the rest of the computer system laid down around and underneath it. Then a second wafer, with its bottom surface covered by some magnetic medium, is suspended above the probes.

To use every bit of the memory, the probe tip must be scanned over the same area within the memory as the probe circuit takes up within its array. Carnegie Mellon achieves this by scanning the entire medium (the top wafer, known as a "media sled") over the bottom. Moving the individual probes by the required scanning distance — 100 microns — would not be possible because of limitations on movement imposed by MEMS technology. As a result, the probes are essentially passive when addressing individual bits. All the necessary movement is performed by the medium.

The probe tips are also microactuated in two dimensions, however, to keep the system aligned. In particular, two distortions need to be compensated for: thermal variations in the chips and curvature of the two wafers, especially that carrying the medium.

The Carnegie Mellon researchers said a difference of 5° across an 8 × 8-mm media sled can produce an error of a few hundred nanometers, which must be removed. In one proposed design, comb drives provide the lateral displacement in the x-direction to compensate for this. In the y-direction, any errors are removed through electronic signal tracking. A second actuator, this time in the z-direction, compensates for curvature and mechanical instability. Carley said the MEMS devices offer a "much smaller physical size and mass/unit storage capacity than any other storage modality," and do so at the "lowest cost in the 1- to 10-Gbyte capacity range." Other advantages, he said, include high reliability (because of redundancy in the large probe-tip arrays), high data rate (through parallellism), low power consumption and environmental ruggedness.

As vocation progresses, technical challenges appear both numerous and inevitable. Though his company is working on a similar approach, HP's Morehouse sees the read-write mechanism, the MEMS microactuator and the packaging as potential problems. In addition, he said, "silicon MEMS uses silicon, which is expensive in large quantities. It will be difficult to amass large areas of storage."

Reliability Issues

Carley agreed that the read-write mechanism will be an issue. "The primary breakthrough needed to enable IMSDs is the development and demonstration of a reliable, robust probe-tip/media combination." He said it must be nonvolatile (a greater than 10-year lifetime), rewritable, have high areal storage density (more than 1,000 bits/micron^2) and be compatible with the MEMS manufacturing process.

In addition, he pointed to tight integration between MEMS and electronics as an important issue. "A system integration challenge for building efficient IMSDs is the integration of the control and channel electronics with the probe tip array." For example, Carley said, IBM's Millipede uses a 32 × 32 array of probe tips, but only 32 can be used at a time because of row column addressing, designed to reduce the number of I/O pads.

"Practical implementations of future systems would seem to demand integration of the probe tips into a standard electronics IC substrate," he said. "The work we are pursuing... proposes to build the magnetic probe tip array on top of a standard CMOS electronics wafer."

From a marketing standpoint, further disadvantages include cost, access time and data rates. "The cost per bit stored of IMSDs is likely to be significantly higher than that of hard-disk drives. Therefore, when the other advantages of IMSDs — size, ruggedness and low power, for example — are not important, and the total required storage capacity is high — more than 100 Gbytes — then IMSDs will not be a cost-effective solution," said Carley. And, he said, "For applications in which cost is no object, it will always remain the case that a large array of battery-backed-up RAM will offer better performance than an IMSD."

DESIGNING SYSTEMS WITH MEMS-BASED STORAGE

MEMS-based storage is an exciting new technology that could provide significant performance gains over current disk drive technology and at costs much lower than EEPROM technology. Based on MEMS (Micro-ElectroMechanical Systems), this non-volatile storage technology merges magnetic recording material and thousands of recording heads to provide storage capacity of 1-10 GB of data in under 1 cm^2 area with access times of under a millisecond and streaming bandwidths of over 50 Mbytes per second. Further, because MEMS-based storage is built using photolithographic IC processes similar to standard CMOS, MEMS-based storage has per-byte costs significantly lower than DRAM and access times an order of magnitude faster than conventional disks.

An additional interesting aspect of MEMS-based storage is its ability to incorporate both storage and processing into the same chip. Because MEMS-storage is CMOS-based, it is possible to integrate several microprocessors or hundreds of custom computational engines (*e.g.*, MPEG encode/decode, cryptography) directly with the storage device.

This integration will significantly improve performance, power consumption, and cost. More importantly, it will lay the foundation for a single computing brick that contains processing and both nonvolatile and volatile storage.

Developing these concepts is the central focus of a research centre at CMU called CHIPS. Although MEMS-based storage devices may still be several years away from commercialization, their potential impact in reducing the memory gap makes them an important technology for systems designers' consideration.

Therefore, we have begun the exploration process, seeking an understanding of how MEMS-based storage can improve application performance and how different MEMS device characteristics can fundamentally change the behaviour and design of storage systems. Our early results indicate that MEMS-based storage can significantly reduce application I/O stall times for a set of five file system and database workloads. The resulting speedups for these applications are around 5X, depending mainly on the ratio of computation to I/O.

Ongoing work refines the device models, explores how they should change system architectures and memory hierarchies, and investigates newly-enabled applications.

To ensure that our models of MEMS-based storage accurately reflect potential implementations, we work closely with a group of researchers who are actively building MEMS-based storage devices. This collaboration allows us to explore the system-level impact of various types of MEMS-based storage, evaluating which physical design trade-offs (*e.g.*, acceleration speed, velocity, size, capacity) are most important across a range of applications. In turn, our results feed back to the MEMS researchers, focusing their attention on design parameters that significantly impact system-level performance and avoiding optimizations that provide little real benefit.

MEMS-BASED STORAGE: EXTENDED OVERVIEW

Envision a world where gigabytes of storage, 1,000s of MIPS, and gigabit/second networking are merged into a single chip smaller than a quarter. This is the vision of CMU's Centre for Highly Integrated Information Processing and Storage Systems (CHIPS). CHIPS' goal is to revolutionize systems, creating low-cost embedded computers with multiple gigabytes of IC-based mass storage that will become a ubiquitous part of our everyday environment. IC-based mass storage devices will also enhance the security and archivability of data storage systems by enabling a tightly integrated coupling of storage and processing. Desktop and laptop computers architectures will evolve to incorporate IC-based mass storage into their memory hierarchies and exploit order-of-magnitude access times reduction, resulting in significant performance improvements. Even the capabilities of massively parallel computers will be enhanced under this vision as the small size of mass storage brings it closer to the processor, enabling dramatic performance improvements on applications ranging from data mining to fast FFTs.

The key to a true system-on-a-chip is the integration of gigabytes of non-volatile memory on a chip. To solve this problem, CMU researchers have turned to hybrid approaches that leverage the best of semiconductor memories and disk drives (IBM and HP have adopted similar approaches). From semiconductor memories, we adopt the wafer fabrication process to minimize unit costs. From disk drives, we adopt recording heads that use mechanical position to address data stored in a thin film material. For compatibility with silicon fabrication processes, we abandon the rotating disk paradigm in favour of using simple microelectromechanical systems (MEMS) to position probe tips over the storage media.

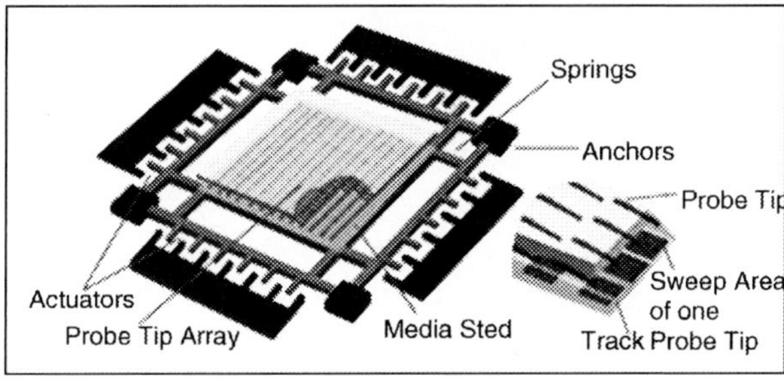

Figure : Prototype MEMS-based data storage system.

Figure depicts CMU's prototype MEMS-based data storage system. Similar to disk drives, the device has recording heads and a recording media surface that moves. However, the recording heads are actually MEMS probe tips that are fabricated in a parallel wafer-level manufacturing process. The

CMU prototype employs magnetic storage media much like that used by disk drives, but the media surface does not rotate; instead it translates in the X and Y directions to seek to the appropriate data. Data access is accomplished by moving the media at a constant velocity in the Y direction while data is read or written by the stationary probe tips.

This design avoids problems with stiction that occur in rotating bearings at very small geometries. This is critical as stiction problems can prevent precise nanometer position control because elements tend to move by alternatively sticking and slipping. This design also avoids the potential wear (to date, MEMS bearings have tended to have quite short lifetimes) that arises when micromechanical surfaces come into contact. The media for the CMU prototype is deposited on a large (8mm × 8mm × 500μm) square plate (the "media sled") and is held above the probe tip array by a network of springs. A force is applied to the sled using electrostatic actuators, though in principle electromagnetic or thermal actuators could be used. Unfortunately, such reciprocating motion is usually limited to a small fraction of the size of the structure. With typical motions being 10% or less of the suspension/actuator length, a single probe tip only "sweeps" 1% of the media sled. However, by using a large array of probe tips, all of the media area can be addressed as long as the media sled moves in X and Y by the pitch of the probe tip array. A large array of probe tips also provides a significant increase in data rate and reliability for the overall system.

Since the media surface is not perfectly flat and individual probe tip heights can vary across the probe-tip array due to both manufacturing variations and curvature of the media sled, nearly all MEMS-based storage approaches incorporate some form of tip height control. CMU's prototype provides for independent active control of the Z motion at every probe tip. Individual probe tips are placed on cantilevers that are electrostatically actuated to a fixed distance from the media surface using a local Z-positioning feedback loop.

Wiring the MEMS-based storage system's 6,400 probe tips' servo and channel electronics requires the electronics to be integrated directly into the same die as the probe tips. This integration greatly improves the bandwidth and sensitivity of the capacitive sensors that are integrated into the probe tips to determine their Z positions relative to the media. To achieve a highly integrated CMOS+MEMS process, we have developed a series of post-processing steps following a standard CMOS fabrication that turns conventional interconnect into released movable mechanical structure. Further, extensions to this integrated CMOS+MEMS process are being developed to fabricate the read/write probe heads. Further, best use of the media requires that the media sled move by at least the probe tip actuator pitch in X and Y. The target is a probe tip array with 100μm centres in X and Y; hence, the media actuator must move at least 50μm.

Of course, the ultimate success of MEMS-based data storage depends on its price and the performance gains in terms of speed, power, or robustness that it offers. Our simulation results show MEMS-based storage devices decrease average I/O service time an order-of-magnitude over disk drives (0.52 ms vs. 10.1 ms).

This translates into large reductions in application I/O stall time (*e.g.*, 0.3 sec. vs. 22.3 sec. on TPC-D #6). Moreover, MEMS-based storage's ability to rapidly power-down and its lower data-access power consumption creates an order-of-magnitude decrease in power consumption over a modern low-power disk drive (*e.g.*, 350 joules vs. 6000 joules for Netscape).

Given these performance improvements, there are many opportunities for MEMS devices in the storage hierarchy. Besides replacing disks, MEMS-based storage devices could serve as a non-volatile disk cache, absorbing write traffic at a much greater speed than conventional disk drives. Further, the cache could be explicitly exposed to and managed by software, allowing software to make customised allocation decisions based on the performance needs and access patterns of various data objects, such as metadata, small files, and files with real-time constraints (*e.g.*, video).

For a lot of "portable" applications such as notebook PCs, PDAs, and video camcorders, MEMS-based storage provides a more robust and lower-power solution. Unlike rotating storage, which cannot cope with device rotation (*e.g.*, rapidly turning a PDA) and is very sensitive to shock (*e.g.*, dropping a device), MEMS-based storage is much more immune to gyroscopic effects and can absorb much greater external forces. Further, MEMS-based storage creates a new low-cost entry point for modest-capacity applications in the 1-10 GB range. This is because disks' assemblies of mechanical components keep manufacturing costs from falling below a certain point, while MEMS-based storage rides the linear decline in IC manufacturing costs. With new applications aggressively creating massive amounts of data, we are also exploring how MEMS-based data storage devices can help solve data archival problems, including capacity, time to access data, and long-term data retrieval. For example, medical imaging generates gigabytes of data per patient, which, for cost reasons, is usually stored directly on tape. Write-once MEMS devices provides an attractive alternative to tape. With areal densities 10X greater than high-capacity tape, it should be cost-effective to build storage "bricks" that hold 1000s of MEMS devices. Each brick would hold petabytes of data that could be accessed in under 1 second. Further, by incorporating logic into the MEMS-based storage device, it would be possible to process data directly within the storage brick. With massive numbers of storage bricks there is massive computational parallelism available, creating the ultimate active disk.

Another application domain for MEMS-based storage is bulk non-volatile storage for embedded computers. Single-chip "throw-away" devices that store very large datasets can be built for such applications as civil infrastructure monitoring (*e.g.*, bridges, walls, roadways), weather or seismic tracking, and medical applications.

For example, one forthcoming application is temporary storage for microsatellites in very low earth orbit. Given that a satellite in a very low orbit moves very quickly, communications are only possible in very short bursts. Therefore, a low-volume, high-capacity, non-volatile storage device is required to buffer data. MEMS-based storage devices could also add huge databases to single-chip continuous speech recognition systems and be integrated into low-cost consumer or mobile devices. Such chips could be completely self-contained, with hundreds of megabytes of speech data, custom recognition hardware, and only minimal connections for power and I/O.

REFERENCES

- Bhatia, A.S. : *Nano Electronics*, Deep and Deep, Delhi, 2011.
- Datta, S. : *Electronic Transport in Mesoscopic Systems*,Cambridge University Press, Cambridge, 1995.
- Drexler, K. Eric: *Nanosystems: Molecular Machinery, Manufacturing, and Computation,* New York: Wiley, 1992.
- Héctor J. De Los Santos: *Introduction to Microelectromechanical (MEM) Microwave Systems*, Artech House, Delhi, 1999.
- Reddy, K. Krishna : *Nanotechnology in Electronics*, Campus Books International, Delhi, 2007.
- Remco J. Wiegerink, Miko Elwenspoek: *Mechanical Microsensors (Microtechnology and MEMS)*, Springer Verlag, Delhi, 2001.

8

Molecular and Biological Nanodevices

Organic materials have turn out to be quite widespread in electronic and optoelectronic applications for a variety of reasons that include lower costs, larger flexibility in designing materials with desired parameters, the possibility of fabricating circuit architectures and/or materials with parameters that are either better or different from those of semiconductor materials or, and last, but not least, the possibility to fabricate electronic components such as thin-film transistors and ring oscillators on almost any substrate. The fabrication of such devices on flexible polyetherether ketone film and even on paper has been recently reported. However, despite the number of published reports concerning organic nanoelectronic devices, they have not yet reached a breakthrough in applications, due, in part, to their general lower conductivity and mobility compared to semiconductor devices.

On the other hand, living cells face the same challenges as human technologies in that they must fabricate materials, convert, transmit, and make use of energy, generate motion, process and store information, and so on, based on principles that are sometimes totally different from the nanotechnologies and with goals that do not necessarily match those of humans. For example, the term "cost" certainly has a different meaning and the accuracy of cellular information processing is undoubtedly more important than speed. At the same time, the cell is not optimized for a specific function. Living cells have unique properties such as self-assembling, self-repairing, and self-replicating.

BIOLOGICAL MOLECULAR MOTORS

Biological molecular motors abound in natural systems and many of these motors are now readily available, having been isolated and purified, and have been studied extensively at both the biochemical level and the single-molecule level. These motors exist as both rotary motors (*e.g.* ATP synthase and the bacterial flagella motor) and linear tracking motors (*e.g.* kinesin), which provides an opportunity for use of the motors in a wide variety of different types of devices. In addition, nature's machines are often optimised for efficiency, which suggests that using what is already available may be much easier than de novo design and production of equivalent machines. Perhaps the most complex biological machine is the ribosome, which acts as the "protein factory" of the cell, continually synthesising proteins with extremely high fidelity in a complex that is no larger than 30–50 nm. This machine is seen by many as the model for future nanodevices in which a nanoscale machine can assemble atoms and molecules in a highly programmable manner (*i.e.* a model system on which to base the "molecular assembler"), to produce novel combinations and, consequently, new materials. Yet such a device remains, at present, only a long term goal and is unlikely to function without the use of biological molecular motors to provide much of the movement required to transport the building blocks supplied by some programmable template and to position those building blocks for chemical synthesis of new materials.

OVERVIEW OF TYPES OF BIOLOGICAL MOLECULAR MOTORS

Even the simplest organisms possess molecular motors:

- The bacteriophage (a virus that infects bacteria) possesses a rotary motor used to pack DNA into the bacteriophage head. This motor works rather like a cork and corkscrew, where the DNA is the corkscrew. Simple but elegant, the molecular motor, consisting of a 10 nm ring of proteins, compresses DNA into the phage head by reducing the volume occupied by the DNA approximately 6000 fold; the resulting internal pressure within the phage head is thought to be of the order of 60 atmospheres.
- Bacteria have rotary motors that drive the whip-like motion of their flagellae, which, in turn, provide the bacteria with a swimming movement.

  These motors provide one of the best examples of self-assembly in nature. They comprise ~40 proteins that are synthesised within the cell and then transported, through the self-assembled structure of the motor, to the appropriate site for assembly of the flagellum on the outside of the cell. This type of self-assembly is frequently observed in biological systems and provides a blueprint for the requirements of nanodevices— they will have to be able to self-assemble at precise locations to provide the required "bottomup" approach to nanotechnology.
- Muscle (myosin) is a typical linear motor in that it enables sliding of actin fibres along myosin fibres, although the motion at the heart of the myosin motor is in part a rotary motion that is

transmitted to the actin fibre as a linear motion through a long lever-arm. This lever-arm also amplifies the motion produced by the molecular motor from a few nanometres to 10 nm. Myosins are ubiquitous in the cell, with a wide range of functions ranging from control of balance in complex organisms such as man, through to cell division during mitosis.

- Kinesins may represent the most useful type of linear motors as nature already uses them for carrying objects around the cell. These motors travel along microtubules, which radiate around the cell in three dimensions, transporting their cargoes to various parts of it. The cargoes can be proteins, vesicles or organelles many times the size of the motor.
- Dyneins also act as ATP-driven molecular motors that are able to generate a force relative to microtubules. They are placed in three classes, which are primarily determined by subcellular localization—inner- and outer-flagella-arm dyneins and cytoplasmic dyneins. This localization matches the cellular activities, the first being the inner- and outer-flagella-movements and the second is the movement of cellular organelles.
- Another important group of motors are those that utilize DNA as their linear track. These include polymerases such as RNA polymerase, DNA polymerases, helicases and topoisomerases. One key area of interest generated by these molecular motors is single molecule DNA sequencing.
- Finally, an unusual, but maybe an all-important group of motors that move DNA are DNA translocases. Unlike the other motors, these enzymes do not simply run along the DNA track, but rather bind the DNA and pull the rest of it through the bound complex. This provides for a very flexible system because the motors usually have a specific recognition sequence on the DNA and they can produce useful work, because they create relative motion with respect to the surface on which the DNA is attached.

THEORY AND MODELLING OF BIOLOGICAL NANODEVICES

Bionanodevices are biomolecular complexes that self-assemble, recognize and control each other with the end of performing one overall function in living cells such as genetic expression, energy conversion, metabolism, signaling, and motion. Examples of bionanodevices are nucleosome and ribosome, ATPase, multienzyme complexes, actin and myosin, all systems the structures of which have recently been solved or are about to be solved. Combined computational-experimental studies have proven successful in explaining the structure—function relationship of proteins, but bionanodevices due to their size pose a great challenge to computational biologists that can be met only through advanced software and hardware.

Our group has build the tools needed for modelling bionanodevices, in particular, the molecular dynamics program NAMD that runs effectively on hundreds to thousands of processors and the molecular graphics program VMD.

The lecture will demonstrate how computational techniques support and complement observation as well as integrate physical descriptions from the electronic to the multi-protein level. It will focus on two bionanodevices, the purple membrane (PM) of Halobacteria and the photosynthetic unit of purple bacteria (PSU).

The PM of Halobacterium salinarium converts sun light into a chemiosmotic potential. The membrane is a crystalline hexagonal array that contains a single protein, bacteriorhodopsin, and ten lipids per protein. Recent observations resolved the structure of most of the components and through modelling we have combined the data to build the entire purple membrane. This hexagonally periodic, lamellar model has been hydrated and refined through NpT/PME molecular dynamics simulations. The resulting structure connects extracellular bulk water with water molecules and key side groups in the interior of the membrane proteins, permitting a seamless overall description of proton conduction and pumping in the PM, from intracellular to extracellular space. For the first time a complex cellular reaction in a bionanodevice can be accounted for in full atomic detail in its complete native environment.

Purple bacteria, like plants, fuel their metabolism with light energy and have developed for this purpose an efficient apparatus for harvesting sunlight, the PSU, key features of which had been conceptually established long ago. The atomic structure of all of its components have been solved, some only recently: the photosynthetic reaction centre, the light harvesting complexes LH-I and LH-II, and the bc1 complex that work together to absorb light and pump protons vectorially across the membrane, as well as the ATPase, that lets protons pass back converting their energy into synthesis of DNA. Our group has modelled in a series of studies during the past decade all individual protein components of the PSU and has pioneered simulations of large patches of membranes. We seek to eventually place all components into the integral bionanodevice and model it as it exists in the cell.

COMPREHENSIVE CHARACTERIZATION OF MOLECULAR INTERACTIONS BASED ON NANOMECHANICS

This vocation focuses on the development and testing of an instrument that measures the integral nanomechanics of molecular interactions. This device relies upon the unique ability of thin cantilevers to detect both the mass of the adsorbed molecules and nanomechanical changes on the cantilever interface, *e.g.* structural rearrangements. The mass is measured *via* the resonance frequency of the cantilever (dynamic mode). Structural changes are detected by static bending of cantilevers (static mode) as demonstrated recently. Here nanomechanical interaction changes generate a surface-stress difference between the asymmetrically functionalized cantilever interfaces forcing the beam to bend. The micro-fabricated cantilever arrays are actuated for a given frequency range and the response is recorded as amplitude and phase spectra for the individual cantilever sensors. These spectra are post-processed and various physical properties of the system can be extracted, such as the adsorbed mass (dynamic mode) and the static cantilever bending (static mode).

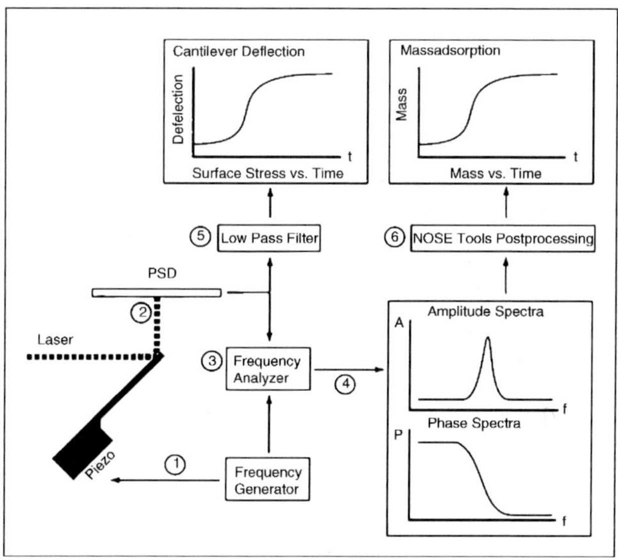

Figure : Schematic of the homemade measurement set-up for combined mode measurements.

An array of silicon cantilevers was mounted onto a piezo element. A sinusoidal excitation signal generated from a network analyser swept the requested frequency range vibrating the cantilevers. The laser beam deflection detection technique was used to monitor the response of individual cantilevers. A frequency analyser compared input- and output signals and continuously recorded amplitude and phase spectra as well as static deflection (bending) of the cantilevers. NOSETools software was used to analyse the spectra and extract the mass adsorption on the cantilever.

To validate the concept we measured molecular interactions between synthetic melittin and lipid vesicles. Melittin is the main component of the bee venom from the European honeybee and is responsible among other constituents for the hemolytic activity of this poison. The small peptide

(2.84 kDa) and its interactions with lipid membranes, using a combination of various biophysical techniques. The peptide binds spontaneously to lipid membranes, forms an α-helix, inserts into the membrane, aggregates and creates channels. While the exact pore formation mechanism is still uncertain it is known that the binding and channel-formation of melittin into vesicle bilayers involves nanomechanical changes. The vesicle mass increases and as visualized recently the insertion of the peptide leads to mechanical forces that subsequently push circumfluent membranes.

MAIN EXPERIMENTS AND THE RESULTS

The workflow of the main experiments and the results. The adsorption of lipid to the cantilever has to be controlled carefully since asymmetrical functionalization of the cantilevers was crucial for detection of the static cantilever bending (a single sided coating was not a prerequisite for the mass adsorption signal). This was achieved by a specific pre-functionalization of the sensor interfaces.

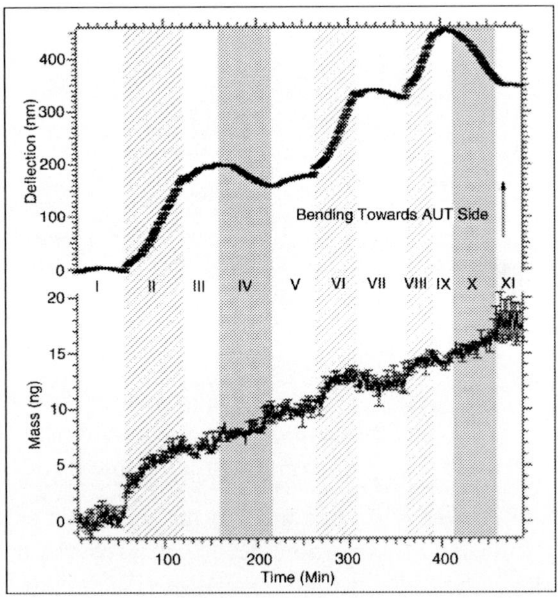

Figure : Combined mode measurements of vesicle and melittin adsorption on the cantilever sensor.

The optimistic controls were pre-functionalized in such a way that vesicles only bind to the upper cantilever surface and melittin does not bind at all. The lower graph displays the mass adsorption and the upper graph reveals the surface stress development measured simultaneously. Note that the surface stress represents the differential signal between the positively and negatively functionalized cantilevers (two cantilevers each).

Functionalization of Cantilevers

First, the upper side of the cantilever array was coated with a 20 nm gold layer onto a 3 nm Titanium adhesion layer. For the positive controls, the cantilevers were pre-functionalized by a self-assembled monolayer (SAM) of 11-Aminoundecan-1-thiol (AUT) on the gold-coated cantilevers using liquid-filled glass capillaries. This resulted in a positive charge selectively formed on the upper cantilever surface. The negative controls remained untreated. After the SAM formation, the complete array was immersed in casein to block "unspecific" binding of melittin and lipid to the silicon. A series of mass-adsorption control experiments were conducted to carefully direct the specific binding or blocking of lipid-vesicles. Double sided AUT-functionalized cantilevers bound approximately double the amount of lipids than single sided functionalized ones as shown in supplementary data S1. Therefore we conclude that the cantilevers pre-functionalized with an AUT SAM on the gold-coated cantilever promote specific binding of lipid vesicles. We also found that casein blocks efficiently the binding of lipid-vesicles and Melittin to silicon and gold. Note that this treatment of the cantilever not only promotes the specific binding of lipid vesicles to AUT pre-functionalized interfaces but also blocks by electrical repulsion the direct adsorption of the melittin peptide without preceding lipid-vesicles immobilization.

Binding Experiments

The simultaneously measured mass adsorption and surface stress for three vesicles (500 ng/ml lipid) and two melittin (1 μM) solution injections. The differential signal is shown obtained by the subtraction of the average of the negative controls from the average of the positive controls. The mass and deflection *changes* during injection of lipid or melittin solutions. We used lipid and buffer conditions known to procure the membrane insertion and channel formation as reported previously. After recording a baseline, vesicles were injected with a concentration of 500 ng/ml. A mass increase of 6.4±0.06 ng is observed. The surface stress difference between the AUT functionalized top-side and the casein passivated silicon bottom side of the cantilever leads to an upward bending (towards the AUT) of the cantilever by 185±1.2 nm during vesicle adsorption. After vesicle injection buffer was flushed through the measurement chamber again before melittin (1 μM) was injected once resulting in a mass increase of around 3.3±0.06 ng. Simultaneously the cantilever bent down by 14.6±1.1 nm. These surface stress changes are in close agreement with previously reported static mode measurements. The injection sequence was complemented with two additional vesicle exposures and a final melittin dose (X) exhibiting the same qualitative mass and deflection changes. Every vesicle and melittin injection was terminated by a buffer wash. This procedure removes not only weakly bound molecules but also ensures that the signal is not due to the liquid rheology.

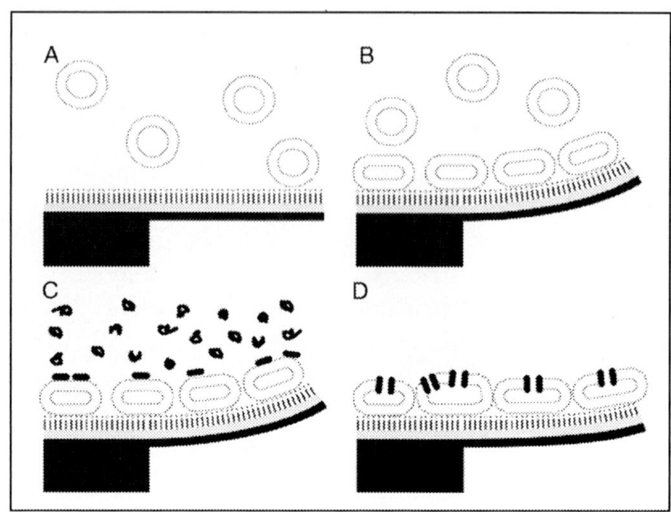

Figure : Molecular model of the nano-mechanical changes on the sensor interface explaining the data.

Without vesicles the pre-functionalized cantilever (AUT SAM on gold) is straight (a). Adsorption of vesicles on the cantilever surface bends the cantilever upwards driven by the interaction forces between the cantilever and the vesicles, which are flattened by this interaction (b). During the peptide injection, the melittin molecules first bind to the vesicle surface (c), and later insert into the membrane and form channels by oligomerization (d).

Conversation

The high lively range of this technique measuring large and small masses in a reproducible manner over several injections in the same experimental series. Furthermore, these experiments show that mass adsorption as well as static bending of the cantilever can be recorded simultaneously, as previously demonstrated for non-biological gas measurements and in liquids studying temperature changes. We interpret the mass changes during the injection of vesicles as lipids binding to the AUT functionalized side of the cantilevers. During the injection of melittin this mass change is attributed to the binding of the small peptide to the lipid vesicles. The change in deflection of the cantilever is construed as a result of the interaction through electrostatic forces between lipid-vesicles and the cantilever. During the melittin injection we interpret the change in deflection as melittin binding and insertion into the lipid bilayer thus forming channels. A schematic of the (simplified) molecular interpretation. The qualitative results are in excellent agreement with current models of the binding and melittin action on and in lipid bilayers. Interestingly, the mass changes follow exactly the injection of the adsorbents (vesicles, melittin) whereas deflection alterations are also observed during the injections of buffer. Such changes we interpret as global structural rearrangements taking place on the cantilever surface after vesicle or melittin binding respectively. All our data show that the lipid binds in the form of a vesicle layer on the AUT functionalized side of the cantilever. From the mass-adsorption we can determine a molecular protein to lipid ratio of 1:10 mol/mol (0.5w/w) for the first melittin injection. This is in the range of typical melittin to lipid ratios and the bilayer structure is reported to stay intact for this particlar mixture.

For the second melittin injection, a protein to lipid ratio of 1:3 mol/mol (0.8w/w) was measured. This ratio is reported to destabilize the lipid structure. Indeed, the static mode signal does not exhibit any change in deflection after the last melittin injection. The first melittin incubation. The cantilever shows an upward bending similar to the signal during vesicle adsorption but no mass changes occurs. This fact indicates that the observed deflection changes are due to structural rearrangements and not caused by electrostatic repulsion between the melittin peptides. The bending of the cantilever does not correlate linearly with the amount of the bound melittin. For the first melittin injection the relative deflection change is "4.7 nm/ng and for the second injection it is "20 nm/ng. This is expected for cooperative processes with many interaction sites involved. Following the first melittin injection a net increase in vesicle mass was observed with subsequent lipid injections. This is explained by the property of melittin to disturb lipid bilayers leading to association/fusion of lipid vesicles as reported previously.

A more quantitative discussion of the surface stress development observe includes the fact that static mode measurements do not only depend on the amount of absorbed (melittin) molecules, but also the nature of the adsorbents and their specific molecular interactions. Note that surface stress is an intensive dimension (in contrast to the measured mass) and is the result of the ensemble of interactions between the molecules adsorbed on the sensor surface. In our case, the vesicular structure of the lipid on the cantilever interface complicates the geometrical arrangement of the adsorbed melittin peptides. Only in-plane force components in the direction of the cantilever main-axis are contributing to the measured surface stress. Therefore the average mechanical lipid-expansion work of a melittin peptide contributing to the cantilever bending is significantly underestimated and requires extensive corrections (not presented here).

In a different way designed melittin binding experiments, confirming the results discussed above, are presented in Supplementary data S3. For the negative control reference, every second cantilever was pre-incubated with melittin. Interestingly, the mass measurement reveals that melittin bound to the positive as well to the negative control during the *in situ* binding experiments. However, the pre-incubated cantilevers did not exhibit any significant change in deflection (surface stress). This finding demonstrates that electrostatic interactions are not dominating the forces that lead to cantilever bending but rather an nanomechanical expansion of the lipid layer as reported before.

For the combined mode (static and dynamic recording) we experienced that the optimized thickness of the cantilever is about 1 μm. The sensitivity for static mode increases with lower spring constant but the sensitivity of the dynamic mode increases with higher frequencies (higher spring constants). In our experiments presented here we used cantilever-arrays with soft spring constants (0.02 N/m) but measured at higher modes (mode 13, 14 or 15) of vibration to increase the sensitivity. Using supported amphiphilic polymer or lipid (bi-) layers would even enhance functionalization efficiency and static mode information content. This would allow a more precise and quantitative interpretation of the static mode information. This technique has the potential to replace Langmuir monolayer assays with the advantages that in addition to the surface stress signal the number of adsorbed molecules could also be measured.

This work demonstrates for the first time that simultaneous and *direct* measurement of nanomechanical (structural) changes and mass adsorption can be performed on the same sensor platform in a liquid environment. Other techniques based on optical detection in combination with Quartz crystal mass balance (QCM) techniques were successfully applied for vesicle adsorption measurements but the different signals were recorded independently.

Conclusions and Outlook

Our results demonstrate firstly that this sensor can measure large ultrastructures and small peptides successively and secondly that the combined measurement of two intrinsic physical properties allows a comprehensive description of molecular interactions. In summary, the dynamic mode mass measurements provide binding information, which does not depend on the nature of the detected system, whereas the static mode provides information about the characteristics of the interactions system, *e.g.* global structural changes as demonstrated here. We strongly believe that the combined measurement will be established as a general tool to characterize molecular interactions.

Systems biology needs tools that not only detect binding partners, but also provide further information on structural changes to comprehend higher organizational levels. Cantilever sensors are perfectly suited for this purpose because the nano-mechanical measurement principle monitors both the binding of effector molecules to their partner and also the subsequent effect on a biological system *in vitro*. This sensor characteristic is unique and allows intriguing applications in nano-medicine as a new technique for drug screening and diagnostics.

ELECTRICAL CONDUCTION THROUGH MOLECULES

In recent years, several experimental groups have reported measurements of the current-voltage (I-V) characteristics of individual or small numbers of molecules. Even three-terminal measurements showing evidence of transistor action has been reported using carbon nanotubes as well as self-assembled monolayers of conjugated polymers. These developments have attracted much attention from the semiconductor industry who are actively looking for ways to progress from gigabit to terabit integration by complementing or even replacing presentday CMOS circuitry. There is great interest therefore from an applied point of view to model and understand the capabilities of molecular conductors. At the same time, this is also a topic of great interest from the point of view of basic physics. A molecule represents a quantum dot, at least an order of magnitude smaller than semiconductor quantum dots, which allows us to study many of the same mesoscopic and/or many-body effects at far higher temperatures.

SINGLE MOLECULES AS ELECTRIC CONDUCTORS

Minimum size, maximum efficiency: The use of molecules as elements in electronic circuits shows great potential. One of the central challenges up until now has been that most molecules only start to conduct once a large voltage has been applied. An international research team with participation of the Graz University of Technology has shown that molecules containing an odd number of electrons are much more conductive at low bias voltages. These fundamental findings in the highly dynamic research field of nanotechnology open up a diverse array of possible applications: More efficient microchips and components with considerably increased storage densities are conceivable.

One electron instead of two: Most stable molecules have a closed shell configuration with an even number of electrons. Molecules with an odd number of electrons tend to be harder for chemists to synthesize but they conduct much better at low bias voltages. Although using an odd rather than an even number of electrons may seem simple, it is a fundamental realization in the field of nanotechnology – because as a result of this, metal elements in molecular electronic circuits can now be replaced by single molecules. "This brings us a considerable step closer to the ultimate minitiurization of electronic components", explains Egbert Zojer from the Institute for Solid State Physics of the Graz University of Technology.

Molecules Instead of Metal

The inspiration for this basic research is the vision of circuits that only consist of a few molecules. "If it is possible to get molecular components to completely assume the functions of a circuit's various elements, this would open up a wide array of possible applications, the full potential of which will only become apparent over time. In our work we show a path to realizing the highly electrically conductive elements", Zojer excitedly reports the momentous consequences of the discovery.

Specific new perspectives are opened up in the field of molecular electronics, sensor technology or the development of bio-compatible interfaces between inorganic and organic materials: The latter refers to the contact with biological systems such as human cells, for instance, which can be connected to electronic circuits in a bio-compatible fashion via the conductive molecules.

ELECTRICAL CONDUCTANCE TO THE LIMIT

The unusual, often non-intuitive characteristics of single molecules may eventually be introduced into a broad range of microelectronics, suitable for applications including biological and chemical sensing electronic and mechanical devices.

Delicate molecular manipulations requiring patience and finesse are routine for Tao, whose research at Biodesign's Centre for Bioelectronics and Biosensors has included work on molecular diodes, graphene behaviour and molecular imaging techniques. Nevertheless, he was surprised at the outcome described in the current paper: "If you have a molecule attached to electrodes, it can stretch like a rubber band," he says. "If it gets longer, most people tend to think that the conductivity will decrease. A longer wire is less conductive than a shorter wire."

Indeed, diminishing conductivity through a molecule is commonly observed when the distance between the electrodes attached to its surface is increased and the molecule becomes elongated. But according to Tao, if you stretch the molecule enough, something unexpected happens: the conductance goes up — by a huge amount. "We see at least 10 times greater conductivity, simply by pulling the molecule."

As Tao explain, the intriguing result is a byproduct of the laws of quantum mechanics, which dictate the behaviour of matter at the tiniest scales: "The conductivity of a single molecule is not simply inversely proportional to length. It depends on the energy level alignment."

In the metal leads of the electrodes, electrons can move about freely but when they come to an interface — in this case, a molecule that sits in the junction between electrodes — they have to overcome an energy barrier. The height of this energy barrier is critical to how readily electrons can pass through the molecule. By applying a mechanical force to the molecule, the barrier is lowered, improving conductance.

"Theoretically, people have thought of this as a possibility, but this is a demonstration that it really happens," Tao says. "If you stretch the molecule and geometrically increase the length, it energetically lowers the barrier so electrons can easily go through. If you think in optical terms, it becomes more transparent to electrons." The reason for this has to do with a property known as force-induced resonant tunnelling. This occurs when the molecular energy moves closer to the Fermi level of the electrodes — that is, toward the region of optimal conductance. Thus, as the molecule is stretched, it causes a decrease in the tunnelling energy barrier.

For the experiments, Tao's group used 1,4'-Benzenedithiol, the most widely studied entity for molecular electronics. Further experiments demonstrated that the transport of electrons through the molecule underwent a corresponding decrease as the distance between the electrodes was reduced, causing the molecule's geometry to shift from a stretched condition to a relaxed or squeezed state. "We have to do this thousands of times to be sure the effect is robust and reproducible."

In addition to the discovery's practical importance, the new data show close agreement with theoretical models of molecular conductance, which had often been at variance with experimental values, by orders of magnitude. Tao stresses that single molecules are compelling candidates for a new types of electronic devices, precisely because they can exhibit very different properties from those observed in conventional semiconductors. Microelectromechanical systems or MEMS are just one domain where the versatile properties of single molecules are likely to make their mark. These diminutive creations represent a $40 billion a year industry and include such innovations as optical switches, gyroscopes for cars, lab-on-chip biomedical applications and microelectronics for mobile devices.

MOLECULAR CONDUCTANCE

Molecular Conductance (G = I/V), or the conductance of a single molecule, is a physical quantity in molecular electronics. Molecular conductance is dependent on the surrounding conditions (*e.g.* pH, temperature, pressure), as well as the properties of measuring device. Many experimental techniques have been developed in an attempt to measure this quantity directly, but theorists and experimentalists still face many challenges.

Recently, a great deal of progress has been made in the development of reliable conductance-measuring techniques. These techniques can be divided into two categories: molecular film experiments, which measure groups of tens of molecules, and single-molecule-measuring experiments.

Molecular Film Experiments

Molecular film experiments generally consist of the sandwiching of a thin layer of molecules between two electrodes which are used to measure the conductance through the layer. Two of the most successful implementations of this concept have been the bulk electrode approach and in the use of nanoelectrodes. In the bulk electrode approach, a molecular film is typically immobilized onto one electrode and an upper electrode is brought into contact with it allowing for a measure of current flow as a function of applied bias voltage. The nanoelectrode class of experiments, in creatively utilizing equipment such as atomic force microscope tips and small-radius wires, are able to perform the same sorts of current versus applied bias measurements but on a much smaller number of molecules as compared to bulk electrode. For instance, the tip of an atomic force microscope can be used as a top electrode and, given the nano-scale radius of curvature of the tip, the number of molecules measured is drastically cut. The difficulties encountered in these experiments have come mainly in dealing with such thin layers of molecules which often results in problems with short-circuiting the electrodes.

Single-molecule-measurement

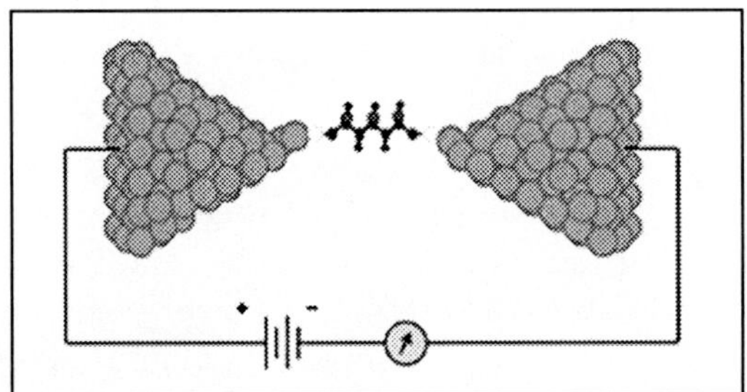

Figure: A molecule covalently connected to two electrodes.

More recently, single-molecule-measurement experiments have been developed that are bringing experimenters a better look at molecular conductance. These fall under the categories of scanning probe, which involves fixed electrode, and mechanically formed junction techniques. One example of a mechanically formed junction experiment involves using a movable electrode to make contact with and then pull away from an electrode surface coated with a single layer of molecules. As the electrode is removed from the surface the molecules that had bonded between the two electrodes begin to detach until eventually one molecule is connected. The atomic-level geometry of the tip-electrode contact has an effect on the conductance and can change from one run of the experiment to the next so a histogram approach is required. Forming a junction in which the precise contact geometry is known has been one of the main difficulties with this approach.

Applications

An important first step toward the goal of building electronic devices on the molecular level is the ability to measure and control the electrical current through an individual molecule. Based on the anticipated continuation of Moore's Law, which is expected to carry the miniaturization of transistors on integrated circuits into the atomic scale within the next 10 to 20 years, this goal of single-molecule-level circuit design is likely to become widespread throughout the semiconductor industry.

Other applications focus on the insight provided by these experiments in the area of charge transport, which is a recurrent phenomenon in many chemical and biological processes. This sort of insight gives researchers the ability to read the chemical information stored in a single molecule electronically, which can then be used in a wide variety of chemical and biosensor applications.

THE ELECTRIC PROPERTIES OF MOLECULES

Dipoles

A dipole is a pair of equivalent and opposite charges. Although overall the pair of charges is electrically neutral, a dipole does give rise to an electrostatic field because the charges are spatially separated. The magnitude of the dipole moment formed from two charges $+q$ and $-q$ separated by a distance r is defined as

$$\mu = qr$$

Polar neutral molecules have a permanent dipole moment because of the uneven distribution of charge. The magnitude of the dipole moment for most molecules is of the order of 10^{-30} C m and it is therefore often more convenient to express molecular dipole moments in units of Debyes, symbol D, where 1 D = 3.33564×10^{-30} C m. Conventionally, the direction of a dipole moment is taken as running from negative to positive. This can be rather confusing since some text books use the alternative Debye-Lewis notation in which the dipole appears to run from positive to negative.

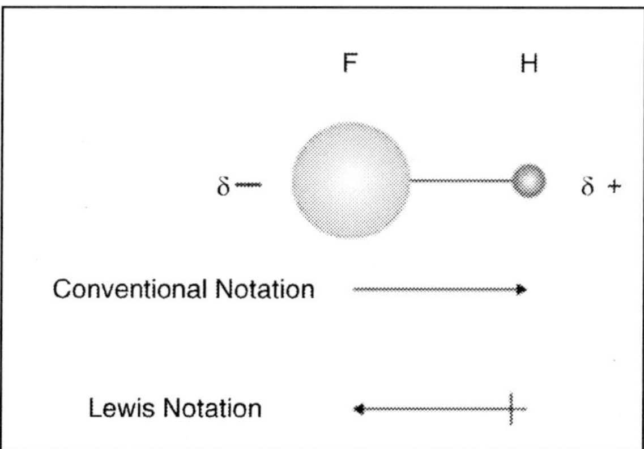

Under certain circumstances nonpolar neutral molecules may possess a transient dipole moment. An external electric field may distort the otherwise symmetrical distribution of charge within the molecule to induce a dipole moment. The magnitude of the induced dipole moment will depend upon both the polarisability of the molecule α and the strength of the external electric field

$$\mu = \alpha E$$

Polarisability is an anisotropic, tensor property that depends upon the direction within the molecule. In general, molecules are more polarisable along bonds than perpendicular to bonds. Furthermore, molecules with π bonds are more polarisable than those with only σ bonds.

Random fluctuations in the instantaneous positions of the electrons within a molecule may also result in a transient dipole moment in a nonpolar molecule.

Quadrupoles and Higher Order Moments

Symmetrical arrays of supplementary than two charges result in higher order multipole moments such as quadrupoles and octopoles. Molecules that possess a quadrupole moment, such as carbon dioxide and acetylene, are electrically neutral overall and do not possess a dipole moment. Similarly, molecules with octopole moments, such as methane are neutral and do not possess either a dipole or quadrupole moment. In general, the electrostatic potential energy resulting from the interaction of two multipole moments of order n and m varies with separation r as

$$V \propto \frac{1}{r^{m+n-1}}$$

A point charge may formally be considered as a monopole of order 1, so that the potential energy of interaction between two such changes must vary with separation as

$$V \propto \frac{1}{r^{m+n-1}} = \frac{1}{r^{1+1-1}} = \frac{1}{r}$$

Similarly, the potential energy arising from the interaction of a charge and a dipole must vary with separation as

$$V \propto \frac{1}{r^{m+n-1}} = \frac{1}{r^{1+2-1}} = \frac{1}{r^2}$$

For a given separation therefore, charge-charge interactions tend to be stronger than charge-dipole interactions and so on.

SINGLE MOLECULE DETECTION AND MANIPULATION

The purpose of this program announcement is to encourage basic research on the detection and manipulation of single molecules. Recent advances in optical imaging and biomechanical techniques have demonstrated that it is possible to make observations on the dynamic behaviour of single molecules, to determine mechanisms of action at the level of an individual molecule, and to explore heterogeneity among different molecules within a population. These studies have the potential to provide fundamentally new information about biological processes and are critical for a better understanding of cellular function.

Current high-resolution techniques, such as x-ray crystallography and NMR, have provided a vast array of structural detail for biological molecules, yet the output of these techniques is limited by its static molecular view and ensemble averaging. Single molecule techniques provide an alternative set of approaches that will lead to a more direct view of the action of individual molecules without the need to infer process or function from static structures. Real-time measurements on the spatial and temporal fluctuations of single molecules in living cells, which are not possible using other techniques, are a major goal of this initiative. Despite the promise of single molecule techniques, there are a number of technical challenges that must be met to optimize these studies.

Development of the collateral chemistry and instrumentation required to carry out single molecule studies is essential for progress. New tools and strategies, as well as refinement of current techniques, are also needed. Single molecule techniques are likely to lead to significant advances in understanding molecular movement, dynamics, and function.

Recent advances in the detection and manipulation of single molecules offer great promise for enhancing our understanding of the behaviour of individual biological macromolecules in the living cell. Scanning probe techniques allow imaging of single molecules on surfaces, and specialized optical techniques enable their characterization in complex environments. Single molecule biomechanical studies have been used to manipulate individual molecules and to measure the force generated by molecular motors or covalent bonds. The development of new probe technologies, such as quantum dots and high-resolution laser fluorescence microscopy, allow real-time observations of molecular interactions and trafficking within living cells. These tools enable individual members of a population to be examined, identified, and quantitatively compared within cellular sub-populations and substructures.

Single molecule studies have the potential to provide spatial and temporal information that is impossible to obtain using other, more static techniques. X-ray crystallography, nuclear magnetic resonance, and electron microscopy have provided a wealth of information on molecular structure, yet none of these techniques can be used to make measurements on the in vivo dynamic movements of single molecules in intracellular space or to observe the behaviour of single molecules over extended periods of time. Using single molecule techniques, it should be possible to study time trajectories and reaction pathways of individual members in a cellular assembly without averaging across populations. Cellular processes, such as exocytosis, flux through channels, or the assembly of transcription complexes, could be visualized. Individual differences in structure or function generated by allelic polymorphisms should be detectable at the level of the single molecule. Monitoring the coordinated expression of a gene or group of genes in specific tissues, or at certain developmental stages, is within reach using these technologies. Thus, single molecule techniques are recognized as an important new set of tools that can be applied to high resolution studies in many areas of biology.

Scientific Objectives

The goals of single molecule research are to observe the dynamic behaviour of individual molecules, to explore heterogeneity among molecules, and to determine mechanisms of action. Single molecule studies are uniquely designed to yield information about molecular motion, behaviour and fluctuations over time and space. An important aspect of the research will be to measure features of individual molecules that are masked by ensemble measurements. Real-time observation of single molecules in live cells, relative to in vitro studies, is an important goal.

Scope

This PA emphasizes the need to encourage the participation, in addition to biologists and biophysicists, of chemists, engineers and physicists in single molecule research. Because of the level of experience and skill required, support may include career track, senior postdoctoral scientists with expertise in chemistry, physics, and instrument development. State-of-the-art instruments that are optimized for high-resolution studies on single molecules often require several years to build and are not commercially available. Instrument development is essential for growth in this field and should therefore be recognized as a legitimate research activity on a grant application. As such, this type of research does not have to be "hypothesis-driven" to be considered worthy of support. The funding Insitutes may provide a substantial contribution for the acquisition or development of instruments, when the instrument is justified as part of the supported research effort. In all cases, the cost of the instrument and associated operating support must be consistent with the scope of the research project(s) with which it is associated.

DNA-BASED NANOMECHANICAL DEVICE

DNA has been used since 1999 to produce a variety of nanomechanical devices. Some of these devices are controlled by structural transitions triggered by small molecules or pH. Although interesting, devices controlled by global changes in the environment lack the programmability associated with sequence-dependent devices, controlled by the addition of individual DNA strands. For example, a device based on the B-Z transition of DNA will permit only two states, the B-state and the Z-state, regardless of how many different species are present, although some nuance in Z-forming proclivity might increase this number somewhat. Sequence-dependent devices offer the ability to address a collection of them individually, so that, say, N two-state devices can lead to 2^N structural states. For example, a translation machine containing two different sequence-dependent two-state devices has been reported, leading to four different products.

The first sequence-dependent DNA device was a tweezers-like machine reported by Yurke and colleagues. The key contribution of that group was the introduction of a "toehold" region on a state-setting strand so that it could be removed, thereby enabling an alteration of the structure. A robust nanomechanical device is one that behaves like a macroscopic device: It has well defined endpoints and does not undergo component-changing isomerizations, such as dissociation or dimerization. The tweezers device lacked this robustness in that it could dimerize during the transition between states; in addition, its open state was not geometrically very well defined. Simmel and Yurke later developed a device with two well structured end-states, in addition to a floppy intermediate.

The PX-JX_2 device is a robust two-state rotary nanomechanical DNA device controlled by hybridization topology. As a function of two different pairs of set strands, one end swivels relative to the other by a half turn; the intermediate state is poorly structured because it has a lot of single-stranded character. The PX-JX_2 device lies at the heart of the translation device. Recently, a cassette was developed to insert the PX-JX_2 device into a two-dimensional DNA array, thereby enabling a series of these devices to be associated with each other in a larger context; this system creates situations where the number of devices, N, can grow, and the number of possible states of the system, 2^N, can grow accordingly.

One more way to increase the total states of a system is to increase the number of states available to each device. We report the development of a three-state device by adding a translational contraction/expansion motion to the previous PX-JX_2 device, thereby creating what we call a PX-JX_2-BX device. This alteration would increase significantly the number of states in a multicomponent system, 3^N vs. 2^N. The three states, PX, JX_2, and BX. They are both controlled by a pair of set strands, drawn in green for the PX state and in yellow for the JX_2 state. The green set strands continue a PX structure that is present in the outer regions of the device frame. The yellow set strands also promote the PX structure in the central part of the device, but they lead to two strand juxtapositions near the upper part of the set strand region.

This difference leads to a half-turn difference between the tops of their helical domains (indicated as A and B) and the bottoms (C and D). Both strands are removed by binding their complete complements, including the toeholds, which are octanucleotides drawn extending horizontally from the device.

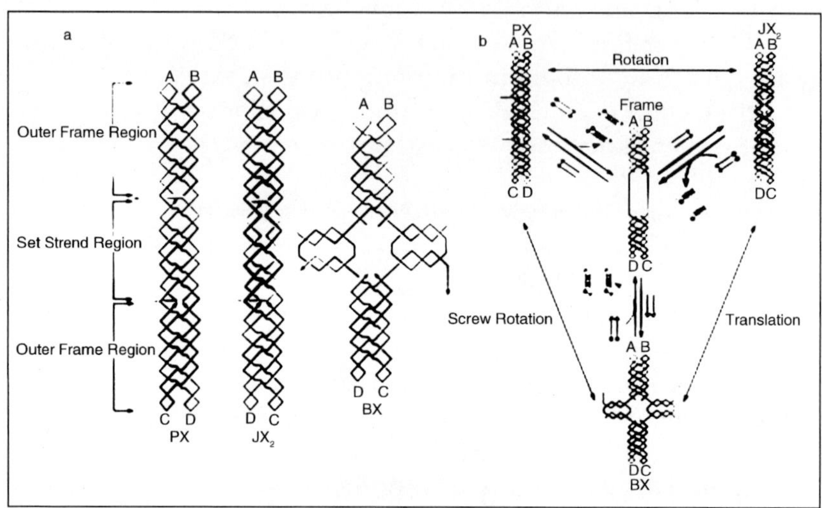

Figure: Schematic drawings of the three-state device. (*a*) The PX, JX_2, and BX motifs. The Frame consists of two strands, one drawn in red and one in black, where arrowheads indicate the 32 ends of strands. The outer Frame regions consist of PX-DNA, two double helices wrapped around each other. The green set strands in the PX motif (left) continue this pattern. The toeholds, used to remove the set strands, are drawn as horizontal lines, one on the 52 end and one on the 32 end. The set strands in the JX_2 motif are drawn in yellow, but the same conventions apply. Note the lack of two crossovers in the set strand region. The set strands in the BX motif are drawn in purple, and the toeholds are vertical. The colour-coding of the strands and labels in Figure *a* indicates that the top ends, A and B, are the same in all of the molecules but that the bottom ends, C and D, are rotated 180° in JX_2 and BX molecules. BX is contracted vertically, relative to PX and JX_2. (*b*) Principles of device operation. The three states are shown at the corners of this diagram, and the nature of the transitions (rotation, translation, or screw rotation) is indicated by labels next to the double-headed arrows. All transitions go through the unstructured Frame, shown in the centre of the diagram. The addition of set strands (drawn as coloured arrows) transforms the Frame to the PX state (green set strands), the JX_2 state (yellow set strands), or the BX state (purple set strands). The set strands can be removed from any motif by the addition of the unset or fuel strands, drawn as arrows with black dots (representing 52 biotin groups), thereby returning the device to the Frame state; the duplex molecules so generated can be removed by the addition of streptavidin-coated magnetic beads.

This portion of the device differs from previously reported PX-JX_2 devices because the set-strand region is much longer (six half-turns of double-helical DNA, rather than three half-turns) and because there is some PX character to the set strand region of the JX_2 state. We contain added the new state BX, which entails contracting the single-stranded portions of the set strand region of the device frame via the addition of the purple strands. The designed structures of the extruded part of the BX state involving the set strands are much like the central portion of a DAO DNA double crossover molecule. The set strand region of the device is much longer than in the original PX-JX_2 device because we needed to have enough single-stranded DNA to produce a detectable contraction, in addition to the need to have a sufficiently long region to stabilize the set strands in each of two domains on each branch.

Another feature of this system is that there are three different transitional motions between the endpoints. The two-state PX-JX_2 device simply rotates a half turn from the PX state to the JX_2 state and then rotates through the inverse twofold rotation back to the PX state. The new device also performs this motion between those two endpoints, but in addition, it contracts and expands between the JX_2 and the BX state. Furthermore, transitions between the PX state and the BX state correspond to a twofold screw rotation. It should be noted that these motions describe the relationships between the starting and ending states of the transitions. The intermediate, shared by all transitions, is known to be floppy and structurally ill defined.

Formation and Operation of the Device

To demonstrate the process of a robust molecular mechanical device, it is necessary to both show the uniform behaviour of the bulk material and visualize the structural transformations of selected molecules. The formation and interconversion of the Frame structure PX, JX_2, and BX DNA by nondenaturing gel electrophoresis. The absence of species other than the PX, BX, or JX_2 molecules at the concentrations used attests to the robustness of the device in bulk. Lane F (at right) contains the unstructured intermediate termed "Frame" (1 μM), lane P contains the device (1 μM) assembled with PX set strands, lane J contains the device (1 μM) assembled with JX_2 set strands, and lane B contains the device (1 μM) assembled with BX set strands. Gel mobility differs because the PX and JX_2 devices are likely to have a more compact time-averaged structure than the BX device; smaller differences between the mobilities of PX and JX_2 were noted previously. Lanes B and J (left of P) contain the products of removing the PX set strands from the material in lane P and replacing them with set strands corresponding to the BX and JX_2 conformations, respectively. Likewise, lanes B and P (left of J) contain the products of removing the JX_2 set strands from the material in lane J and replacing them with those corresponding to the BX and PX conformations. Lanes J and P (left of B) contain the products of removing the BX set strands from the material in lane B and replacing them with those corresponding to the JX_2 and PX conformation. Note the absence of extraneous products in all lanes containing transformation products, indicating the robustness of the transformations.

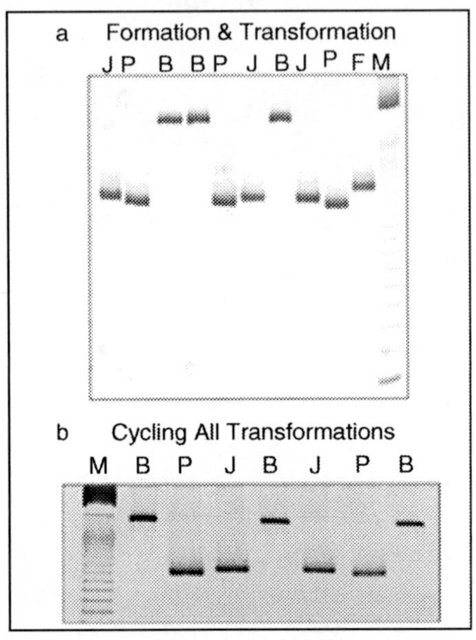

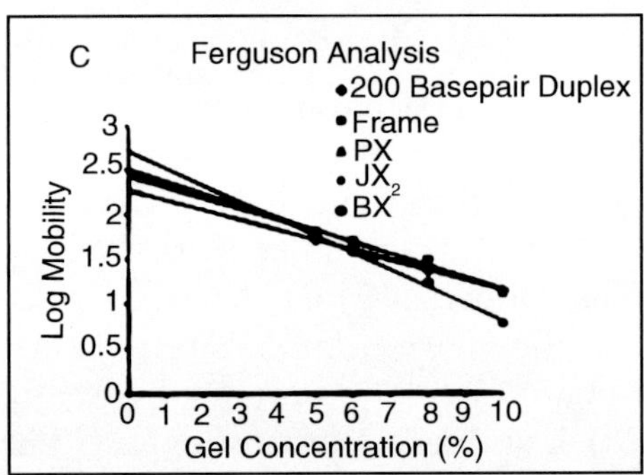

Figure : Gel evidence for the operation of the device. (*a*) Formation and interconversion of Frame structure and PX, JX_2, and BX motifs demonstrated by nondenaturing gel electrophoresis. This is a 10% nondenaturing polyacrylamide gel, run at 22°C and stained with stains-all dye. The lanes are described from the right: lane M, 10-bp ladder marker; lane F, the unstructured intermediate, Frame; lane P, the device assembled with PX set strands; lane J, the device assembled with JX_2 set strands; lane B, the device assembled with BX set strands. Lanes J and B (left of P) contain the products of removing the PX set strands from the material in lane P and replacing them with set strands corresponding to BX and JX_2, respectively. Lanes B and P (left of J) contain the products of removing the JX_2 set strands from the material in lane J and replacing them with those corresponding to BX and PX. Lanes J and P (left of B) contain the products of removing the BX set strands from the material in lane B and replacing them with those corresponding to JX_2 and PX. (*b*) Cycling the device. The gel shows the cycling of the device through six possible transitions starting from any of the three states (here, we have started with BX). Lane M shows the 10-bp ladder marker. Continuing right, lane B is the initial BX conformation, lane P is PX transformed from BX in lane B, lane J is JX_2 from PX in lane P, and lane B is BX from the JX_2. Proceeding right, lane J is JX_2 again from BX in lane B, lane P is PX from JX_2 in lane J, and finally lane B is BX from PX in lane P. Thus, BX can be formed after six steps of operation through all possible transitions. (*c*) Ferguson analysis of the motifs used here. The plots of the PX, JX_2, BX, and Frame molecules are compared with a double-helical molecule of similar length.

A nondenaturing gel that shows the cycling of the device through six possible transitions starting from any of the three states. Lane M shows a 10-bp ladder marker; proceeding right, lane B is the initial BX conformation, and lane P is PX transformed from the material in the BX state in lane B. It is flanked at right in lane J by JX_2 obtained from the material in lane P. To its right in lane B is BX from the JX_2. On its right, this material is transformed back to JX_2 (lane J), then to PX (lane P), and finally to BX again (lane B at right). Thus, BX can be formed after six steps of operation through all possible transitions between three states. The addition of unset strands, followed by set strands, was repeated five times for the cycle. These data establish the robustness of the device as well as the ability to transform it from any state to any other state.

Ferguson Analysis

We determined the Ferguson plots of these species by comparing their mobilities as a function of polyacrylamide concentration. The mobility, *M*, of a molecule as a function of total gel concentration, *T*, may be described by the well known relationship (12) $\log(M) = \log(M_0) - K_R T$, where M_o is the free mobility, and K_R is the retardation coefficient. Rodbard and Chrambach have shown that K_R is an approximately linear function of the exposed surface area (friction constant) of the electrophoresing species. The Ferguson plot for all of the robust states of the device, the floppy frame, and a 200-nt-pair duplex, similar in size to the device.

The largest slope characterizes the BX device, probably because of its "X" shape. The PX and JX_2 molecules are slightly different, as can be seen in the plot, but in both cases they contain large occluded surfaces, which decrease their friction constants relative to the BX structure. Similar behaviour is seen with four-arm junctions, which also occlude part of their surfaces. These results suggest that the BX structure does not occlude its surface significantly and is thus closer to the 4 × 4 structure than to an eight-helix equivalent of the stacked four-arm junction.

Atomic Force Microscopy (AFM) Illustrating the Operation of the Device

Distorted gel mobilities do not guarantee that the construct undergoes the designated structural transformation. We demonstrate this aspect of the device with the system. We have used half-hexagon trapezoidal markers made of three triangles via edge sharing, as done previously. The trapezoids are connected into one-dimensional oligomeric arrays by linkages that include PX-JX_2-BX devices. The drawing shows that if the devices are all in the PX state, the trapezoids have a parallel arrangement, and when the devices are all in the JX_2 state, the trapezoids form a zigzag structure. If the device is in the BX state, the markers will have the same zigzag orientation as in JX_2; however, the distances between individual markers will be smaller, owing to the translational motion, which contracts the device. Although estimated to be small (<3 nm), in one-dimensional arrays the contraction adds up and we see a marked change from end to end. All AFM images have dimensions of 200 nm × 200 nm.

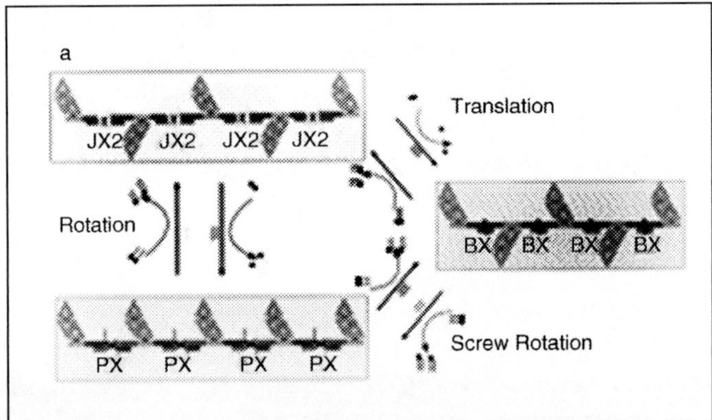

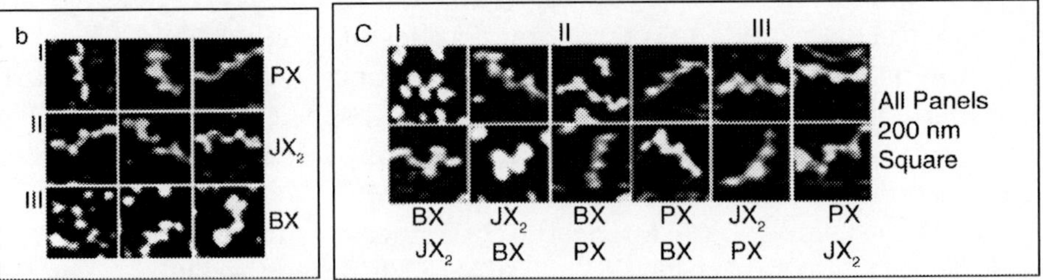

Figure : Visualization of the transition of the devices by AFM. (*a*) A highly simplified representation of the system used. It consists of a one-dimensional array of half-hexagon trapezoids (light blue outer strands, orange inner strands) joined by the device. Each trapezoid consists of three edge-sharing DNA triangles. The colour-coding used for the set strands for each oligomer of the device, and (in a lighter shade) as background with green set strands in the PX state, yellow set strands in the JX_2 state, and purple set strands for the BX state, with constant strands in red and black. The strand topology is both more complex and larger. There are 40 nucleotides between the first device crossover point and the nearest triangle crossover point, a number that was determined empirically to give the most nearly planar structure. In the lower molecule, all of the half-hexagons point in the same direction, whereas they point in opposite directions in the upper molecule. Biotinylated fuel strands (with black circles) are shown removing set strands in all parts of the cycle. (*b*) Control AFM images. The trapezoid oligomer is shown in the PX state in *I*, in the JX_2 state in *II*, and in the BX state in *III*. (*c*) Cycling of the device between two states. *c* illustrates the operation of the device by displaying representative molecules sampled from solutions as the system is cycled. *Upper Left* is the initial state, *Upper Right* and *Lower Left* are a transformed state, and *Lower Right* is transformed back to the same state as *Upper Left*. *I* shows the system originating in the BX state, converting to the JX_2 state, and then back to the BX state. *II* starts from the BX state, is converted to the PX state, and is then converted back to the BX state. *III* starts with the JX_2 state, forms PX, and returns to JX_2.

The contains AFM images of the three states of the oligomers. These images contain control molecules, not devices, which are constrained to be in the PX, BX, or JX_2 motifs. The PX state contains a series of trapezoids extended parallel to each other, much like the extended fingers of a hand. By contrast, the JX_2 state is characterized by a zigzag arrangement of trapezoids, and the BX state is a contracted version of the JX_2 state. The operation of the device by displaying representative molecules sampled from solutions as the system is cycled.

The intermediate state produces a single band on a gel, but it is not well structured when examined by AFM. Three sets of two-step conversions of the device state. *Upper Left* in each is the original state, *Upper Right* and *Lower Left* are aliquots from the transformed state, and *Lower Right* shows the return to the original state. In the first set, the system originates in the BX state and is then converted to the JX_2 state and back to the BX state. The second set also begins in the BX state, which is converted to the PX state and back to the BX state. The third set begins with the JX_2 state, which is converted to PX and then converted back to JX_2.

The PX device arrays are clearly in a parallel arrangement, and the distance between the units is ≈60 nm; the JX_2 device arrays are clearly in the zigzag arrangement, and the half-repeat distance on one side is ≈34 nm; and BX device arrays are in a contracted zigzag orientation with the half-repeat distance on one side at ≈26 nm. Thus, the system operates as designed, both in bulk and in individual cases. The large difference between the PX and JX_2

distances, expected to be similar, makes us suspicious of interpreting the BX-JX_2 difference (≈8 nm) quantitatively.

Fluorescence Resonance Energy Transfer (FRET) Measurements

Two different labelling schemes that have been used to yield FRET measurements. The diagrams for the two schemes are similar to those, and the conventions there apply here as well. A green ellipse indicates the site where a fluorescein dye has been placed, and a red ellipse indicates the site of a Cy3 unit. Scheme I shows the dye-pairs are far apart in PX and JX_2 (theoretical distance: 10.5 nm) but closer in BX (theoretical distance: 4.8 nm). In scheme II, dye-pairs are placed the opposite way where they are closer in PX and JX_2 (theoretical distance: 4.8 nm) but far apart in BX (theoretical distance: 11.0 nm). The donor energy transfer for device molecules in the three different states and for the two different schemes. In scheme I, the energy transfer trend is PX ≈ JX_2 << BX. In scheme II, the trend is PX ≈ JX_2 >> BX. These measurements into distances between the dye-pairs. In scheme I, the distances follow the expected trend PX (9.0 ± 0.7 nm) ≈ JX_2 (10.0 ± 0.8 nm) > BX (7.2 ± 0.5 nm). In scheme II, the trend is PX (6.2 ± 0.1 nm) ≈ JX_2 (6.2 ± 0.4 nm) < BX (7.6 ± 0.3 nm). These distance estimates are in agreement with the models, and the range of the errors does not lead to ambiguity. The similarities between the PX and JX_2 distances lead us to accept the BX-JX_2 differences of 2.1–2.5 nm as being more reliable than the larger differences seen by AFM.

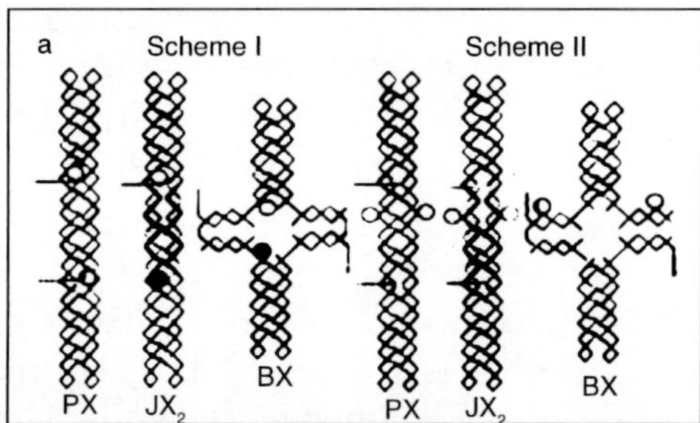

Figure: FRET data for the motifs used here. (*a*) Labelling schemes used for the experiments. Two sets of molecules are shown, corresponding to two different labelling schemes. The molecules are drawn according to the conventions, but they now include green ellipses to represent the locations of fluorescein donor dyes and red ellipses to represent the locations of Cy3 acceptor dyes. (*b*) Donor energy transfer for the frame plus the device molecules in three different states and for two different labelling schemes. The colour-coding is indicated. The ordinate shows the extent of donor energy transfer. (*c*) Distances between the dye-pairs plotted for device molecules in three different states and for two different labelling schemes. Colour-coding is the same as in *b*. The ordinate shows the distance (*D*) of the dye-pairs. Standard deviations are indicated.

Construction of the Device

We have built and demonstrated a robust DNA nanomechanical device with three well structured endpoints, as well as a weakly structured common intermediate. Gel evidence for the end states of ensembles of the device is clear, so that neither multimerization nor breakdown of the motif is seen after the transition. Data from both FRET and AFM clearly support the operation of the device as engineered. The route to a robust three-state device from its predecessor two-state PX-JX_2 device was not straightforward. Extending the control region from three to six half-turns to permit the formation of the BX state was successful, but extending it to seven half-turns was unsuccessful. Seven half-turns was the product of adding a full PX turn to the control region. However, the presence of three major groove half-turns and four minor groove half-turns led to smeared or split bands in the PX and JX_2 states, whereas four major groove half-turns and three minor groove half-turns led to split bands in the BX state. Model building confirmed that it is necessary to have an even number of half turns in the control region of the three-state device. Nevertheless, the frame of the six-half-turn molecule was found to be unstable unless we extended its outer portions from three to five half-turns.

Features of the Device

Robust devices built before had two well defined endpoints, so a collection of N independent devices were capable of forming 2^N independent states, perhaps representative of binary logic or other sets of states. Under similar circumstances, this device is capable of forming 3^N states, leading, perhaps, to trinary logic but certainly to greater diversity of physical states. Currently, this device, like most of its predecessors, is a shape-shifter when free in solution. Were this device to be incorporated into a cassette capable of insertion into a two-dimensional DNA array, tiles of different sizes, as well as different sticky ends, could be directed to form arrays.

The uses of a nanorobotic device can derive either from the number of states in its endpoints or from the number and types of different structural transitions between those endpoints. This device has three different physical transition types between endpoint states under the control of the set strands, regardless of transition mechanism: the twofold rotation between PX and JX_2, the translation between JX_2 and BX, and the twofold screw rotation between PX and BX. In general, there will be $N!/[(N-2)!\ 2!]$ different transition types between endpoints associated with an N-state device; of course, each transition can go in two different directions. It is not hard to imagine extending the device reported here to more than three states and correspondingly more structural transitions. For example, the two ends of the frame could be bent toward one another by an appropriate pair of set strands. Although one can obtain multiple states by combining many different simple, say two-state, devices, combinations of identical devices will not display the variety of transitional movements of which multiple-state devices are capable.

A PRIMER ON RATCHETS AND BROWNIAN MOTORS

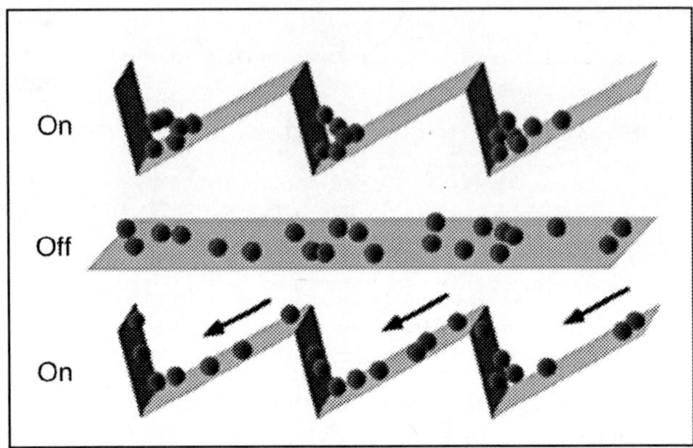

Example: The flashing ratchet. Brownian particles are trapped in a periodic, asymmetric potential that can be turned on and off. The random diffusion when the potential is off is converted into net motion to the left when the ratchet is switched on.

Most research in the Linke lab is based on the ratchet concept: The combination of non-equilibrium and asymmetry generally leads to transport.

The basic physical idea is simple: a system that is not in thermal equilibrium tends towards equilibrium. If this system lives in an asymmetric world, then moving towards equilibrium will usually also involve a movement in space. To keep the system moving, we need to perpetually keep it away from thermal equilibrium, which costs energy - this is the energy that drives the motion.

Ratchets are interesting for a number of reasons:

- The physics of precisely how transport is achieved can be very subtle and interesting, including quantum phenomena.
- The direction of transport often depends on fine details of the system, for instance the temperature or the size of the particles. In principle, ratchets can therefore be used to separate particles, for instance by their size.
- When thermal motion is important for the function of a ratchet, the system is called a thermal ratchet or a Brownian motor. Such systems use non-equilibrium energy to rectify Brownian motion: a subtle, interesting phenomenon that cannot be observed in thermal equilibrium.
- Brownian motors are excellent model systems to understand how nanoscale machines may operate in the presence of substantial thermal motion. For this reason, the Brownian motor concept is often used to model biological, molecular motors.

MOLECULAR MOTOR

Molecular motors are biological molecular machines that are the essential agents of movement in living organisms. In general terms, a motor may be defined as a device that consumes energy in one form and converts it into motion or mechanical work; for example, many protein-based molecular motors harness the chemical free energy released by the hydrolysis of ATP in order to perform mechanical work. In terms of energetic efficiency, this type of motor can be superior to currently available man-made motors. One important difference between molecular motors and macroscopic motors is that molecular motors operate in the thermal bath, an environment in which the fluctuations due to thermal noise are significant.

Frequency-Specific Optical Nano-Actuators?

In order to manage nanomachines via external signals, it will be necessary to use multiple actuators that can be activated independently. Molecular structure is largely determined by where the electrons are. If the electrons can be made to move, the molecule may change shape. For some molecules, hitting them with light can make the electrons move. Thus, light-actuated molecular actuators have already been developed. It would be nice if, instead of just "light," there was a whole palette of actuators, each activated by a different colour. It would be nice if, instead of having to develop each one individually, a general approach could create a whole family of these actuators.

Nanotech solar cell research has explored ways that light energy can be transferred from one molecule to another - connect one molecule that absorbs the photon to another that creates electricity or hydrogen. So what if, instead of building an actuator molecule that both absorbs photons and moves, build one that doesn't absorb any photons - but can accept photon energy from a nanoparticle it's connected to? It seems like this scheme would allow a variety of actuators to be built, sensitive to different frequencies of light, simply by swapping out the nanoparticles. Frequency-specific nanoparticles already exist, and the frequency can be tuned simply by adjusting the size of the nanoparticle when it's fabricated.

MOLECULAR NANOELECTRONICS

Present molecular electronics technology, based on patterning of thin films of organic semiconductors, promises to transform the electronics industry by printing electronics on thin and flexible plastic substrates using a process scaleable for large area, high volume and low cost. This "plastic electronics" involves lengthscales of microns, but ultimate miniaturisation involves individual molecules hundreds of times smaller than the smallest features conceivably possible by semiconductor technology. More speculatively, if a single molecule, or molecular assembly, can be made sufficiently complex to integrate more than one device function, then entirely new nanoelectronic functionalities might open up. But big hurdles exist for the replacement of current semiconductor technology by single molecule devices. For example, (i) there is as yet no solution to scaling up to multi-molecular devices, (ii) incorporation of such devices into silicon processing technology has yet to be developed, and (iii) understanding of electrical contacts to molecules is in its infancy. Our primary aim is to improve fundamental understanding of transport mechanisms and of the molecule-electrode contact. Following from this, we hope to develop reliable contacts and to incorporate silicon-based processing. Longer-term goals include interfacing to supramolecular assemblies based on weak non-covalent bonding.

Single Molecules

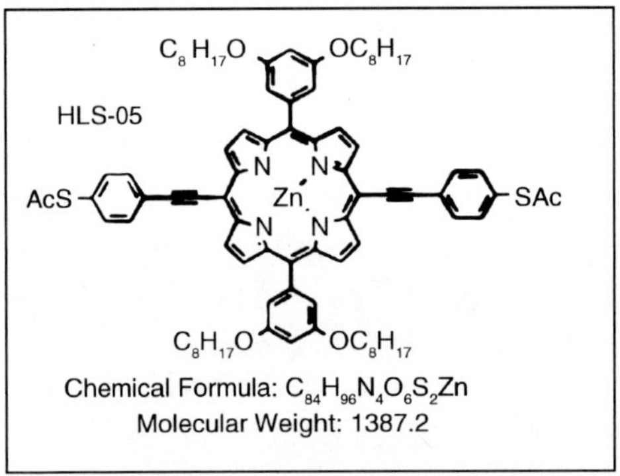

Porphyrin Project

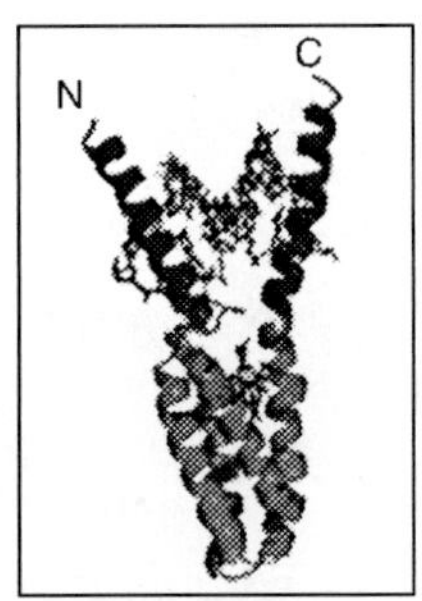

Cytochrome Project

We are presently examining conduction processes in two important molecules (i) a range of porphyrin-based single molecules and (ii) cytochrome b562 (a haem-binding electron-transfer protein). Molecules of both classes are crucial in biological systems, and they form excellent test-beds to further understanding of molecular conduction processes.

Polymers

The major focus of this project is to study semiconducting organic polymers and their properties at metal-polymer interfaces. This uses electrostatic force microscopy (EFM) to examine the detailed behaviour of polymer contacts with various metals. Semiconductor polymer electronics is rather different to conventional inorganic semiconductors such as Si, and these differences have many benefits. Polymer transistors can be constructed by casting polymer films directly from solution, making mass-production very easy and even flexible subtrates can be used. The polymers have useful optical properties, making them an excellent candidate for polymer LEDs and Photovoltaics (Solar Cells).

The polymers studied are the conjugated polymers polythiophene and polyfluorene or part of each (co-polymer). A conjugated structure consists of alternating covalent single and double bonds which results in a path for delocalised electrons along the structure. Delocalised electrons means electrical conduction, so these polymers are conductive. The polymers gain electrons/ holes by doping and these electrons/holes are then free to move through the structure.

When constructing devices such as LEDs and Photovoltaics, one needs to apply electrodes to the polymer films in order to inject or extract charges from the polymer film - this in turn controls the energy levels inside the polymer. Having applied the electrodes, there will be certain physical mechanisms at work at the metal-polymer interface which govern how easily one can inject or extract charge from the polymer.

Having studied the interface under a variety of conditions (electric field, temperature, sample preparation, substrate, atmosphere, electrode material) it is possible to drastically improve the performance of any polymer devices..

OPTOELECTRONIC DEVICES BASED OF SMALL MOLECULES

A positive strategy for synthesizing conjugated small molecules for use in optoelectronic devices is to begin with a chromophore that can absorb most of the visible region of the solar spectrum. Diketopyrrolopyrrole (DPP) is such a chromophore and is also a desirable molecular building block because of its chemical stability, and ease of synthesis and modification. The general molecular architectures of DPP-containing small molecules synthesized in our lab. These compounds are classified into two subgroups: mono-DPPs which have only one DPP unit and bis-DPPs which have two DPP units. Due to the electron deficiency of the lactam structure in DPP, electron donating functional groups such as thienyl and phenyl derivatives are selected to incorporate with DPP as linkage and endcapping groups to further tune properties such as the optical band gap, highest occupied molecular orbital (HOMO), and lowest unoccupied molecular orbital (LUMO) levels. Alkyl chains are introduced on the conjugated backbone to induce solubility in common organic solvents, and to modify the degree of crystallinity and specific packing motif.

Additionally, central groups in bis-DPP compounds provide a further structural handle capable of modifying material properties. This combination of different side chains and central, linkage, and endcapping groups allows the resultant materials optoelectronic properties to be tuned with a high degree of sensitivity. This tunability enables engineering of the material properties necessary for optimal performance in optoelectronic devices. A number of DPP-containing derivatives have been synthesized by our group and several other groups. We highlight several specific examples of DPP-containing conjugated small molecules for use in organic solar cells (compounds 1–3) and FETs (compound 4).

The DPP-containing small molecules displayed are synthesized by the divergent approach illustrate. For the mono-DPPs (compounds 1–3), this approach consists of several reaction steps such as DPP-linkage formation, N-alkylation, linkage bromination, and endcap group coupling. When the linkage group is a thienyl group, the DPP-thiophene can be prepared by a single reaction of a thienyl nitrile with a succinic diester. The resultant DPP-thiophene precursor is not soluble in organic solvents due to its internal hydrogen bonds.

N-alkylation of the DPP-thiophene precursor with an alkyl bromide makes the precursor soluble enough for the subsequent modifications. The following bromination with N-bromosuccinimide (NBS) provides the thienyl groups with the reactive sites necessary to complete the final endcap group coupling. This final step can be achieved via a palladiumcatalyzed Suzuki or Stille coupling. When the linkage is a phenyl group, the above bromination step can be omitted by using bromobenzonitrile as the starting material. For the bis-DPPs, the divergent approach usually consists of iterative Suzuki or Stille couplings. Taking compound 4 as an example, an intermediate is synthesized by reacting an active benzothiadiazole (BT) precusor (benzothiadiazole-bis(boronic acid piancol ester)) with a mono-brominated DPP-thiophene via Suzuki coupling. The obtained intermediate is then brominated by NBS and subsequently reacted with an endcapping group via Suzuki coupling, generating compound 4. It is important to note the purity of the final material is critical for optimal device performance. It has been our experience that the presence of even a small amount of impurity can negatively affect the performance of a device. Therefore, both molecular design and precise purification are essential for synthesis of conjugated small molecules for high-performing optoelectronic devices.

Field Effect Transistors

Conjugated small molecules be able to also be utilized as the charge transporting layer in FETs. Their p-type or n-type charge transport capabilities can be tuned by modifying their chemical structures and energy levels. Most DPP containing compounds we have investigated have p-type transport characteristics. However, by incorporating electron deficient groups, n-type or ambipolar transport can be achieved. The chemical structure of a bis-DPP compound (compound 4), which has a BT unit as the central unit flanked with two DPP-T_3C_6 building blocks. Due to the electron deficiency of the BT unit, compound 4 has lower energy levels and is capable of ambipolar transport characteristics. A bottom-gate, topelectrode device structure was used to measure the ambipolar transport of compound 4. The typical transfer characteristics at different annealing temperatures and the corresponding carrier mobilities. Balanced carrier mobilities up to 10^{-3} $cm^2V^{-1}s^{-1}$ were demonstrated using gold electrodes. The electron and hole mobilities can be further improved up to 10^{-2} $cm^2V^{-1}s^{-1}$ by using a low work function electrode such as barium.

TOWARD INTEGRATION OF *IN VIVO* MOLECULAR COMPUTING DEVICES

What is Molecular Computing?

Molecular computing is a generic word for any computational scheme which uses individual atoms or molecules as a means of solving computational problems. Molecular computing is most frequently associated with DNA computing, because that has made the most progress, but it can also refer to quantum computing or molecular logic gates. All forms of molecular computing are currently in their infancy, but in the long run are likely to replace traditional silicon computers, which suffer barriers to higher levels of performance. A single kilogram of carbon contains 5×10^{25} atoms. Imagine if we could use only 100 atoms to store a single bit or perform a computational operation. Using massive parallelism, a molecular computing weighing just a kilogram could process more than 10^{27} operations per second, more than a billion times faster than today's best supercomputer, which operates at about 10^{17} operations per second. With so much greater computational power, we could achieve feats of calculation and simulation unimaginable to us today.

Different proposals for molecular computers vary in the principles of their operation. In DNA computing, DNA serves as the software whereas enzymes serve as the hardware. Custom-synthesized DNA strands are combined with enzymes in a test tube, and depending on the length of the resulting output strand, a solution can be derived. DNA computation is extremely powerful in its potential, but suffers from major drawbacks. DNA computation is non-universal, meaning that there are problems it cannot, even in principle, solve. It can only return yes-or-no answers to computational problems. In 2002, researchers in Israel created a DNA computer which could perform 330 trillion operations per second, more than 100,000 times faster than the speed of the fastest PC at the time.

Another proposal for molecular computing is quantum computing. Quantum computing takes advantage of quantum effects to perform computation, and the details are complicated. Quantum computing depends upon supercooled atoms locked in entangled states with one another. A major challenge is that as the number of computational elements (qubits) increases, it becomes progressively more difficult to insulate the quantum computer from matter on the outside, causing it to decohere, eliminating quantum effects and restoring the computer to a classical state. This ruins the calculation. Quantum computing may yet be developed into practical applications, but many physicists and computer scientists remain skeptical.

An even more advanced molecular computer would involve nanoscale logic gates or nanoelectronic components conducting processing in a more conventional, universal, and controlled manner. Unfortunately, we currently lack the manufacturing capability necessary to fabricate such a computer. Nanoscale robotics capable of placing each atom in the desired configuration would be necessary to realise this type of molecular computer. Preliminary efforts to develop this type of robotics are underway, but a major breakthrough could take decades.

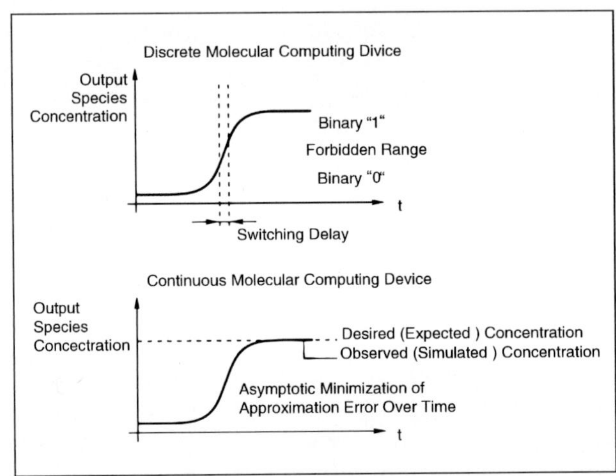

In contrast to the *in vitro* computing devices, *in vivo* MCDs make use of the naturally occurring translational regulation mechanism of the host organism. Since *in vivo* systems have to go through an additional step of protein translation, they are suggested to be implicitly slower than DNA based *in vitro* MCDs. However, the use of error correction mechanisms naturally implemented in the evolutionarily optimized transcription regulation machinery of living cells makes the overall computation more robust and hence justifies the trade-off with speed. A recent theoretical study conducted by Cory and Perkins has laid the focus on the use of a transcriptional regulatory mechanism to solve basic arithmetic operations. The study shows how different parametrizations of a simple chemical kinetic model of transcription regulation can give rise to these different operations. The accuracy of such theoretical arithmetic calculations based on the transcription regulatory mechanism is dependent on the kinetic parameters as well as the transcription factor concentrations.

From today's perspective, two implementation approaches for MCDs using gene regulation mechanisms have emerged: discrete and continuous devices. They reflect the encoding of processed information. In discrete devices, the domain of possible species concentration is divided into several layers separated by intermediate forbidden ranges. Each layer represents a digit, upper part. Within this encoding scheme, binary MCDs have been explored.

The discretized interpretation of species concentrations might offer some preferences concerning MCD implementation. Since a range of concentration levels encodes the same digit, the discrete approach benefits from a certain fault tolerance.

This leads to robust systems according to electronic gates employing electric current instead of species concentration. Slight variations of signal quality (caused by intrinsic stochasticity of chemical reactions in practice) can be compensated this way. Furthermore, an opportunity for signal refresh becomes feasible. By this procedure, the signal levels (*e.g.*, species concentrations) are adjusted within each layer in order to maximize their distance to the forbidden range. Signal refresh can be seen as an essential prerequisite for construction of complex circuits applying feedback loops or consecutive processing units. Typically, those circuits are equipped with a clock whose oscillating binary signals strictly determine the time period for Boolean switching. The clock assures that a computational step is finished before the next one starts (synchronization). Standard devices for Boolean computations were successfully implemented +*in vivo* using gene regulation mechanisms of *E. coli*. While Gardner *et al.* introduced a genetic toggle switch (bit storage unit acting as a flip-flop), the pioneering work of Guet *et al.* resulted in a NAND gate composed of inhibitors. Each Boolean function can be expressed by interconnected NAND gates. A gene regulatory network (repressilator) generating oscillating clock signals was provided by Elowitz and Leibler. Recent studies investigate capabilities of coupling logic gates, toggle switches, and clock generators toward genetic circuits, a computer architecture based on gene regulation. This promising biochemical realization of von-Neumann's concept will probably lead to reliable as well as programmable discrete MCDs, but it lacks some properties to outperform electronic computers. For instance, an *in vivo* binary switching consumes several hours, and the clock rate of the repressilator is rather slow (approximately 2 h for one clock cycle). Moreover, the potential of massively data parallel computation known from *in vitro* molecular computation has not been effectively utilized *in vivo* up to now.

Comparison of discrete and continuous molecular computing devices regarding interpretation of species concentration courses as output data of a computation.

Continuous MCDs as discuss in Cory and Perkins (Cory and Perkins, 2008) might be a clue to overcome insufficiencies of discrete devices. In electronic engineering, analogous computers exploit the continuous principle of operation for computing purposes. Historically, they became established for some niche applications such as functional differentiation or integration. Magnasco and Deckart and Sauro showed that the dynamical behaviour of chemical reaction networks based on a kinetic model can perform computations in a continuous manner. Here, initial species concentrations represent the input data, and dedicated species concentrations reached within the steady state form the output. Particularly, the computation of arithmetic functions seems to be effectively feasible this way. The non-mass-conserving reaction $A+B \xrightarrow{K_1} A+B+C$ along with the decay $C \xrightarrow{K_2} \varnothing$ is a simple example for a reaction network able to multiply two nonnegative numbers: The initial concentration values of species A and B {denoted as $[A](0)$ and $[B](0)$} directly define both multipliers. In mass-action kinetics, the concentration course of output species C over time is specified by the ordinary differential equation $d[C]/dt = k_1.[A].[B].k_2.[C]$ while catalyst concentrations $[A]$ and $[B]$ remain constant. In the steady state, the concentration $[C]$ does not change anymore, such that $d[C]/dt = 0$. Thus, $[C] = [A].[B]$ holds for parametrization of kinetic reaction rates with $k_1 = k_2 = 1$. Assuming a more complex reaction network incorporating several motifs, predetermined parts of the network can be deactivated by setting corresponding reaction rates $k_i = 0$ and activated ($k_i > 0$), respectively. This way, a reaction network becomes capable of performing multiple arithmetic operations. Cory and Perkins seized this idea to propose analogous computation of arithmetic functions by a continuous MCD. The selection of activated reactions controls the executed operation whose result (output species concentration) evolves asymptotically up to the steady state. Therefore, an arithmetic computation appears as a temporal minimization of the approximation error of the output species. To this end, continuous MCDs benefit from the compactness of the underlying reaction networks in combination with the advantages of assisting self-repair mechanisms available *in vivo*. The contribution by Cory and Perkins marks an important step in understanding how arithmetic computations can be done effectively and reliably by a continuous genetic MCD from a modelling and simulation point of view. A thorough understanding of the influence of kinetics in terms of synthetic biology can be used to design a gene regulatory network that can faithfully capture the desired network behaviour.

Tackling Combinatorial Search

In a recently conducted new study, Haynes and co-workers used site-specific DNA recombination to solve a burnt-pancake problem (BPP) *in vivo*. The advantage that an *in vivo* system offers in such a case, where an exponential expansion in search space is known to occur, is that DNA replication and bacterial cell growth are inexpensive, feasible, and scalable. Haynes *et al.* draw parallels between the actual BPP and their *in vivo* computational units by suggesting the pancakes be represented by flippable DNA segments. A flip move is mediated by Hin DNA recombinase that regulates gene expression by switching the orientation of the promoter and the coding region while the two palindromic 26bp *hix* sequences flanking the invertible DNA segment act as recognition sites for cleavage and strand exchange. Cells with sorted DNA segments represent the final state. As a proof of concept, Haynes *et al.* designed a two-pancake BPP comprised of a *lac* promoter (pLac) and a tetracycline resistance coding region, each flanked by *hixC* sites. The reaction proceeds by cotransforming cells with a BPP plasmid containing a *hixC*-flanked RBS-tetA(C)rev coding region, a *hixC*-flanked pLac promoter [permutation (–2,1)], and a HinLVA expression plasmid. As described, the system reaches equilibrium state after more than 11 h, with most cells in the starting pancake arrangement while about one-third of the cells were detected to have undergone simultaneous inversion of both DNA segments. However, it was also found that due to general leakiness of the pLac promoter, HinLVA-mediated inversion did not seem to require induction of the pLac promoter on the HinLVA plasmid. Furthermore, cells with a strong pLac repressor (*lacI*^{Q}) were found to be tetracycline resistant even without pLac activation. Cells were still found to be tetracycline resistant even when the pLac promoter was removed from the plasmid construct, leading to the conclusion that pLac is not necessary for expression in the pBR322-derived cloning vector. Tests with a pLac sensitive vector showed that the pLac promoter demonstrates both forward and backward transcription activity, hence rendering phenotypic characterization of the final state ambiguous. In addition, the experimental setup does not allow for quick reversal of states. However, by revealing the unexpected behaviour of the pLac promoter, these results illustrate the interdependence of synthetic biology and *in vivo* computing.

These consequences were confirmed by the accompanying mathematical simulations in which flipping was modelled as a Markov chain such that each of the possible permutations was considered as a state. The probability of a plasmid to be in the sorted order after *k* flips was determined by dividing the number of paths from the initial to the desired final state by the total number of paths of length *k* from the start to any state. The results from the mathematical simulations showed that 25% of the total plasmids attained the equilibrium between an intermediate state and the final state after five flips. The overall slow speed (time scale of hours to days rather than minutes) and the lack of robustness could be the bottleneck of such an approach that involves a combinatorial search space of the possible states and involves extensive site-specific DNA recombination. For these reasons, it seems that like most other state of the art MCDs, a more complicated network comprised of the basic subunits represented in the study would be difficult to achieve. However, the study provides experimental proof of employing DNA for computing *in vivo*. Furthermore, this approach gives a new perspective to MCDs, where each DNA segment itself can be considered as a state comprising a finite state automaton (FSA), and transition in states could be conceived as a flip in a DNA segment. The final state, defined by a phenotypic expression, is achieved when the automaton has flipped and oriented all the DNA segments in the desired order. So far biomolecular computational units have been demonstrated *in vivo* and *in vitro*. Some of the key issues that still need to be handled are: noise propagation, interfacing output and input, and integrating basic subunits.

Sustenance of a large population of cells, removal of intermediate and final products aiding a fast reversal of states are also problems that need to be looked into. Furthermore, the use of MCDs in drug targeting, where they can be programmed to sense the disease indicators and output the drug based on the computation module, makes it imperative to implement these biomolecular computational units in mammalian cells.

Recently, Rinaudo *et al.* successfully implemented an RNAi based approach to Boolean logic evaluation based on molecular computing, wherein they transfected human embryonic kidney cells with evaluator circuits. An evaluator circuit as described by them consisted of mRNA with fused siRNA targets at the noncoding sites. Based on the presence or absence of complimentary siRNA and the linkage of the mRNAs, various Boolean operations can be performed. Such and other attempts at *in vivo* computing in mammalian cells could benefit from developments in the understanding of genetic circuits and *vice-versa.*

DNA COMPUTING

DNA computing is a form of computing which uses DNA, biochemistry and molecular biology, instead of the traditional silicon-based computer technologies. DNA computing, or, more generally, biomolecular computing, is a fast developing interdisciplinary area. Research and development in this area concerns theory, experiments and applications of DNA computing.

Capabilities

DNA computing is fundamentally similar to parallel computing in that it takes advantage of the many different molecules of DNA to try many different possibilities at once. For certain specialized problems, DNA computers are faster and smaller than any other computer built so far. Furthermore, particular mathematical computations have been demonstrated to work on a DNA computer. As an example, Aran Nayebi has provided a general implementation of Strassen's matrix multiplication algorithm on a DNA computer, although there are problems with scaling. In addition, Caltech researchers have created a circuit made from 130 unique DNA strands, which is able to calculate the square root of numbers up to 15. DNA computing does not provide any new capabilities from the standpoint of computability theory, the study of which problems are computationally solvable using different models of computation. For example, if the space required for the solution of a problem grows exponentially with the size of the problem (EXPSPACE problems) on von Neumann machines, it still grows exponentially with the size of the problem on DNA machines. For very large EXPSPACE problems, the amount of DNA required is too large to be practical.

Techniques

There are multiple techniques for building a computing device based on DNA, each with its own advantages and disadvantages. Most of these build the basic logic gates (AND, OR, NOT) associated with digital logic from a DNA basis. Some of the different bases include DNAzymes, deoxyoligonucleotides, enzymes, DNA tiling, and polymerase chain reaction.

DNAzymes

Catalytic DNA (deoxyribozyme or DNAzyme) catalyze a reaction when interacting with the appropriate input, such as a matching oligonucleotide. These DNAzymes are used to build logic gates analogous to digital logic in silicon; however, DNAzymes are limited to 1-, 2-, and 3-input gates with no current implementation for evaluating statements in series. The DNAzyme logic gate changes its structure when it binds to a matching oligonucleotide and the fluorogenic substrate it is bonded to is cleaved free. While other materials can be used, most models use a fluorescence-based substrate because it is very easy to detect, even at the single molecule limit. The amount of fluorescence can then be measured to tell whether or not a reaction took place. The DNAzyme that changes is then "used," and cannot initiate any more reactions. Because of this, these reactions take place in a device such as a continuous stirred-tank reactor, where old product is removed and new molecules added. Two commonly used DNAzymes are named E6 and 8-17. These are popular because they allow cleaving of a substrate in any arbitrary location. Stojanovic and MacDonald have used the E6 DNAzymes to build the MAYA I and MAYA II machines, respectively; Stojanovic has also demonstrated logic gates using the 8-17 DNAzyme. While these DNAzymes have been demonstrated to be useful for constructing logic gates, they are limited by the need for a metal cofactor to function, such as Zn^{2+} or Mn^{2+}, and thus are not useful in vivo. A design called a *stem loop*, consisting of a single strand of DNA which has a loop at an end, are a dynamic structure that opens and closes when a piece of DNA bonds to the loop part. This effect has been exploited to create several logic gates. These logic gates have been used to create the computers MAYA I and MAYA II which can play tic-tac-toe to some extent.

Enzymes

Enzyme based DNA computers are usually of the form of a simple Turing machine; there is analogous hardware, in the form of an enzyme, and software, in the form of DNA. Benenson, Shapiro and colleagues have demonstrated a DNA computer using the FokI enzyme and expanded on machine; there is analogous hardware, in the form of an enzyme, and software, in the form of DNA. Benenson, Shapiro and colleagues have demonstrated a DNA computer using the FokI enzyme and expanded on their work by going on to show automata that diagnose and react to prostate cancer: under expression of the genes PPAP2B and GSTP1 and an over expression of PIM1 and HPN. Their automata evaluated the expression of each gene, one gene at a time, and on positive diagnosis then released a single strand DNA molecule (ssDNA) that is an antisense for MDM2. MDM2 is a repressor of protein 53, which itself is a tumour suppressor. On negative diagnosis it was decided to release a suppressor of the positive diagnosis drug instead of doing nothing. A limitation of this implementation is that two separate automata are required, one to administer each drug. The entire process of evaluation until drug release took around an hour to complete. This technique also requires transition molecules as well as the FokI enzyme to be present. The requirement for the FokI enzyme limits application *in vivo*, at least for use in "cells of higher organisms". It should also be pointed out that the 'software' molecules can be reused in this case.

Algorithmic self-assembly

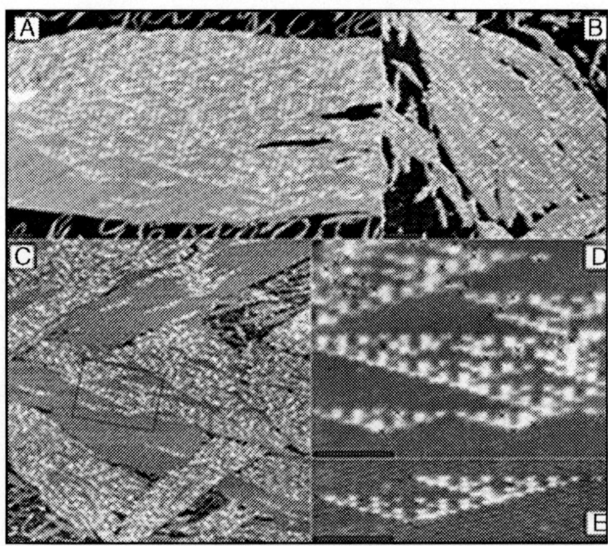

Figure: DNA arrays that display a representation of the Sierpinski gasket on their surfaces. DNA nanotechnology has been applied to the related field of DNA computing. DNA tiles can be designed to contain multiple sticky ends with sequences chosen so that they act as Wang tiles. A DX array has been demonstrated whose assembly encodes an XOR operation; this allows the DNA array to implement a cellular automaton which generates a fractal called the Sierpinski gasket. This shows that computation can be incorporated into the assembly of DNA arrays, increasing its scope beyond simple periodic arrays.

Toehold Exchange

DNA computers have also been constructed using the concept of toehold exchange. In this system, an input DNA strand binds to a sticky end, or toehold, on another DNA molecule, which allows it to displace another strand segment from the molecule. This allows the creation of modular logic components such as AND, OR, and NOT gates and signal amplifiers, which can be linked into arbitrarily large computers. This class of DNA computers does not require enzymes or any chemical capability of the DNA.

MEMRISTIVE NANODEVICE BASED ON PROTEIN

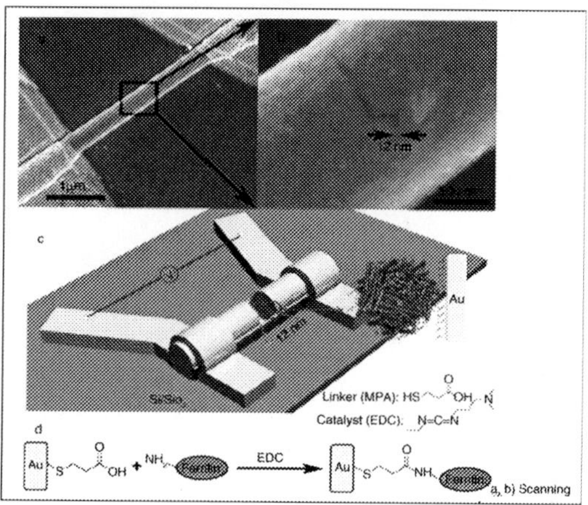

Figure : (a, b) Scanning electron microscope (SEM) images of a device prepared with an OWL-fabricated nanowire with a ~12 nm gap. (c) A diagram of ferritin molecules spanning the ~12 nm gap. (d) The reaction scheme for the chemical immobilization of ferritin onto the sides of the gap. (Reprinted with permission from Wiley-VCH Verlag) "Previous work on memristors were based on man-made inorganic/organic materials, so we asked the question whether it is possible to demonstrate memristors based on natural materials," Xiaodong Chen, an assistant professor in the School of Materials Science & Engineering at Nanyang University, tells Nanowerk. "Many activities in life exhibit memory behaviour and substantial research has focused on biomolecules serving as computing elements, hence, natural biomaterials may have potential to be exploited as electronic memristors." The researchers based their device fabrication on the chemical immobilization of ferritin molecules – a ubiquitous intracellular protein that stores iron – within on-wire lithography (OWL) generated nanogaps. "The programmable resistive switches were due to the electrochemical processes in the active centre of ferritin" explains Chen. "In addition, we demonstrated that such ferritin-based nanodevices with reversible resistance can be used for nonvolatile memory based on write-read-erase cycle tests." This work provides a direct proof that natural biomaterials, especially redox proteins, could be used to fabricate solid state devices with transport junctions, which have potential applications in functional nanocircuits. As Chen notes, the researchers chose ferritin as a model system in this experiment for a variety of reasons: it is the main intracellular iron storage protein with a nearly spherical shell and an active mineral core of hydrous ferric oxide; ferritin exhibits high stability and can withstand a pH range of 2.0-12.0 as well as temperatures up to 80 °C, thus it is a desirable, natural biomaterial for fabricating electronic devices; the conductivity of ferritin has been measured by conductive AFM, highlighting the potential application of this protein in electronics; and the mechanism of iron uptake and release *in vivo* shows biological 'memory' storage, which provides a potential target to develop biomolecule-based memristors. "Based on this work we will further explore the use of bio-integrated nanodevices in terms of fundamental challenges and potential applications," says Chen. "We also want to know how multifunctionality could be integrated into these nanodevices. However, we should also note that natural biomaterials such as DNA or proteins are not as stable as commonly used inorganic materials. Bio-inspired approaches may help us to resolve this problem."

Memristors : The fourth fundamental two-terminal circuit element following the resistor, the capacitor, and the inductor – have attracted intensive attention owing to their potential applications for instance in nanoelectronic memories, computer logic, or neuromorphic computer architectures. R. Stanley Williams and his group at Hewlett Packard were the first to demonstrate a solid-state memristive device based on a titanium dioxide thin film, which was inserted between two platinum electrodes. Scientists have been able to show that various materials such as metal oxides, chalcogenides, amorphous silicon, carbon, and polymer-nanoparticle composite materials exhibit memristive phenomena. One unanswered question so far has been whether natural biomaterials like proteins can be used for the fabrication of solid-state devices with transport junctions. Researchers in Singapore have now demonstrated that proteins indeed can be used to fabricate bipolar memristive nanodevices. This work provides direct proof that natural biomaterials, especially redox proteins, could be used to fabricate solid-state devices with transport junctions and can be the core component in the development of bioelectronic devices.

REFERENCES

- Kapoor, Manish : *BioMEMS and Biomedical Nanotechnology*, Oxford Book Company, Delhi, 2012.
- Parag Diwan and Ashish Bharadwaj: *Nanoelectronics*, Pentagon, Delhi, 2006.

Bibliography

Awasthi, K.P.: *Advanced Magnetic Nanostructures,* Cyber Tech Publications, Delhi, 2008.

Bhatia, A.S.: *Nano Electronics,* Deep and Deep, Delhi, 2011.

Bhatt, R.C.: *Nanostructure and Nanoculture,* Swastik Pub, Delhi, 2009.

Boucher P.M.: *"Nanotechnology : Legal Aspects."* CRC Press Boca Raton, 2008.

Branda Paz: *A Handbook on Nanoelectronics,* Dominant, Delhi, 2008.

Brecket, A.G.: *A Handbook on Nanotechnology,* Dominant Pub, Delhi, 2008.

Cerofolini, G.F.: *Nanoscale Devices,* Springer-Verlag, Berlin, 2009.

Datta, S.: *Electronic Transport in Mesoscopic Systems,*Cambridge University Press, Cambridge, 1995.

Drexler E.: *"Engines of Creation : The Coming Era of Nanotechnology."* Anchor Books New York, 1986.

Drexler, K. Eric: *Nanosystems: Molecular Machinery, Manufacturing, and Computation,* New York: Wiley, 1992.

Gregory Kovacs: *Micromachined Transducers Sourcebook,* McGraw-Hill, London, 1998.

Héctor J. De Los Santos: *Introduction to Microelectromechanical (MEM) Microwave Systems,* Artech House, Delhi, 1999.

Julian W. Gardner, and Vijay K. Varadan: *Microsensors, MEMS and Smart Devices,* Wiley, NY, 2001.

Kapoor, Manish: *BioMEMS and Biomedical Nanotechnology,* Oxford Book Company, Delhi, 2012.

Kaushik, N.L.: *Nanotechnology and Micromachines,* Pearl Books, Delhi, 2007.

Kumar, H.D. : *Material Science: Nanotechnology and Applications,* I.K. International, Delhi, 2011

Kumar, N. Phani: *Principles of Nanotechnology : Materials, Tools and Process at Nanoscale,* Scitech Pub, Delhi, 2010.

Kumar, U.: *A Handbook of Nanotechnology,* Agrobios, Delhi, 2012.

Ljubisa Ristic: *Sensor Technology and Devices,* Artech House, Delhi, 1994.

Maluf, Nadim: *An Introduction to Microelectromechanical Systems Engineering,* Artech House, Delhi, 1999.

Marc Madou: *Fundamentals of Microfabrication,* CRC Press, Delhi, 1997.

Mathur, K.P.: *Nanoscience and Technology*, Rajat, Delhi, 2007.

Mohamed Gad-el-Hak: *The MEMS Handbook*, CRC Press, Delhi, 2001.

Parag Diwan and Ashish Bharadwaj: *Nanoelectronics*, Pentagon, Delhi, 2006.

Parthasarathy, B.K.: *Nanoscience and Nanotechnology*, Isha Books, Delhi, 2007.

Pope M and Sweenberg : *Electronic Processes in Organic Crystals and Polymers*, Oxford: Oxford University Press, 1999.

Randy Frank: *Understanding Smart Sensors*, Artech House, Delhi, 2000.

Rao, M. Balakrishna: *Nanotechnology in Electronics*, Campus Books International, Delhi, 2007.

Reddy, K. Krishna: *Nanotechnology in Electronics*, Campus Books International, Delhi, 2007.

Remco J. Wiegerink, Miko Elwenspoek: *Mechanical Microsensors (Microtechnology and MEMS)*, Springer Verlag, Delhi, 2001.

Sergey Edward Lyshevski: *Nano- and Microelectromechanical Systems*, CRC Press, Delhi, 2000.

Sharma, P.K.: *NanoMEMS Physics : Applications and Principles*, Vista International Pub, Delhi, 2011.

Silinsh E A and Capek: *Organic Molecular Crystals*, New York: AIP Press, 1994.

Suri, Shalini: *Nano Technology : Basic Science to Emerging Technology*, APH, Delhi, 2006.

Tai-Ran Hsu: *MEMS and Microsystems: Design and Manufacture*, McGraw-Hill, London, 2001.

Thakur, Vijay Kumar and Amar Singh Singha: *Nanotechnology in Polymers*, Studium Press, Delhi, 2012.

Vasudev Bhatnagar: *Nanostructure Science and Technology*, Campus Books, Delhi, 2009.

Index

P

Q

R

S

T